T0292590

Springer Finance

Springer Finance

Springer Finance is a programme of books addressing students, academics and practitioners working on increasingly technical approaches to the analysis of financial markets. It aims to cover a variety of topics, not only mathematical finance but foreign exchanges, term structure, risk management, portfolio theory, equity derivatives, and financial economics.

For further volumes:
www.springer.com/series/3674

Antonio Mele · Yoshiki Obayashi

The Price of Fixed Income Market Volatility

 Springer

Antonio Mele
Swiss Finance Institute
University of Lugano
Lugano, Switzerland

Yoshiki Obayashi
Applied Academics LLC
New York, NY, USA

ISSN 1616-0533
ISSN 2195-0687 (electronic)
Springer Finance
ISBN 978-3-319-26522-3
ISBN 978-3-319-26523-0 (eBook)
DOI 10.1007/978-3-319-26523-0

Library of Congress Control Number: 2015958378

Mathematics Subject Classification (2010): 91G20, 91G30, 91G40, 91B25

Springer Cham Heidelberg New York Dordrecht London

Printed on acid-free paper

Springer is part of Springer Science+Business Media (www.springer.com)

Preface

The volatility of major asset classes is a key driver of portfolio performance affecting institutional and individual investors alike. Portfolio volatility may be managed by diversification through asset allocation and security selection decisions as well as by derivatives supplying direct exposure to volatility. Of the two main traditional assets—stocks and bonds—volatility derivatives and methodologies underlying their designs and pricing have been well-developed for equity markets, while fixed income markets have lagged in this respect despite its principal role in capital markets. This book fills, or at least aims to significantly narrow, this gap.

While the exposition of this book is of a theoretical nature, its ultimate objective is to serve as a foundation upon which a market for standardized fixed income volatility trading may be built. In fact, some of the interest rate volatility index designs proposed in this book have already been brought to life in the US and in Japan—by far the two largest government bond markets by notional outstanding. The Chicago Board Options Exchange (CBOE), home to the omnipresent CBOE Volatility Index® (VIX®), launched the CBOE Swap Rate Volatility Index^SM (SRVIX^SM) in 2012, and subsequently partnered with the CME Group to launch the CBOE/CBOT 10-Year US Treasury Note Volatility Index^SM (TYVIX^SM) in 2013. Across the Pacific, S&P Dow Jones Indices and Japan Exchange Group partnered to launch the S&P/JPX JGB VIX in 2015; a Japanese Government Bond analogue of TYVIX. CBOE listed TYVIX futures in 2014 as its first-ever listed derivative for standardized fixed income volatility trading, and counterpart to its popular VIX for broad equity market volatility trading.

This book presents a unified fixed income volatility evaluation framework and in-depth accounts of its application to four major fixed income asset sub-classes: interest rate swaps, government bonds, time deposits, and credit. It develops model-free, forward-looking volatility indexes for each of these asset types, which involves dealing with disparate market conventions and numerous complexities that are absent when pricing equity volatility. Some of these complexities had long been recognized by practitioners as hurdles for creating VIX-like indexes for interest rate volatility, but left bereft of mathematically rigorous solutions for a number of years, which presumably contributed to the stunted development of standardized measurement and trading of fixed income volatility until recently.

Our work draws its origins from a series of research notes and implementation details that we developed over the last six years. The present monograph organizes this work in a self-contained fashion by providing the reader with a comprehensive piece with interconnected parts. Our work does not focus on purely mathematical innovations; rather, it relies on existing methods and develops new tools aimed at facilitating contract evaluation in the fixed income space. It is our hope that this work will lead to significant and positive contributions in the world of financial engineering, and will help investors measure and better manage risks arising from fixed income volatility.

This book will appeal to both applied researchers and theorists. Researchers in academia and at financial institutions are the main audience of this book, while advanced students in finance, economics, and mathematics should also find the material useful for further study in the area of asset pricing. Applied researchers will gain access to the mathematically rigorous background required for undertaking empirical research in relatively new topics such as: time series behavior of forward-looking interest rate volatility indexes; interest rate volatility risk-premiums; linkages between volatility and market liquidity; and the impact of macroeconomic developments and monetary policy on fixed income volatility. Theorists will find contributions to an exciting area in asset pricing regarding interest rate volatility evaluation as well as the evaluation of new financial products referenced to forward-looking gauges of interest rate volatility, such as TYVIX futures.

Last but not least, we are very grateful to seven anonymous reviewers for their insightful comments on our work and valuable suggestions, and to Dr. Catriona Byrne at Springer for coordinating the countless moving parts involved in bringing this project to fruition. We also thank Kodjo Apedjinou, Giovanni Barone-Adesi, Ruslan Bibkov, Peter Carr, Praveen Korapaty, Catherine Shalen, and David Wright for comments and suggestions throughout this project, Shihao Yang for excellent research assistance, and audiences at the IAQF (International Association of Quantitative Finance) Thalesian seminar in New York University, Jane Street Capital, Kellogg School of Management, Morgan Stanley Quantitative Risk Modeling Group in New York, the NYU-Stern Volatility Institute, the 7th Annual Risk Management Conference at National University of Singapore, the 4th Annual Global Derivatives in Chicago, the 19th Annual RISK USA Conference in New York, the 1st and the 3rd CBOE Risk Management Conference in Europe (Dublin), the 8th Annual Meeting of the Swiss Finance Institute in Geneva, and a Swiss Finance Institute Knowledge Transfer Workshop in Lugano for remaining comments. However, we retain full responsibility for any remaining omissions or mistakes.

Lugano, Switzerland   Antonio Mele
New York, USA   Yoshiki Obayashi
September 2, 2015

Contents

Information About the Authors

Antonio Mele holds a Senior Chair at the Swiss Finance Institute and is a Professor of Finance at the University of Lugano, after a decade spent as a tenured faculty at the London School of Economics. He is also a Research Fellow for the Financial Economics program at the Centre for Economic Policy Research (CEPR) in London. He holds a PhD in Economics from the University of Paris. His work spans a variety of fields in financial economics, pertaining to capital market volatility, interest rates and credit markets, macro-finance, capital markets and business cycles, and information in securities markets, and has appeared in journals such as the *Journal of Financial Economics*, the *Review of Economic Studies*, the *Review of Financial Studies*, and the *Journal of Monetary Economics*.

His work outside academia includes developing fixed income volatility indexes for Chicago Board Options Exchange and S&P Dow Jones Indices. He is currently a member of the *Securities and Markets Stakeholder Group* of the European Securities Markets Authority (ESMA), the supra-national supervisor of European financial markets. At ESMA, he is also a member of the Group of Economic Advisers.

Yoshiki Obayashi is a managing director at Applied Academics LLC in New York. The company specializes in developing and commercializing ideas emanating from a growing think tank of academic researchers selected on the basis of their work's relevance to practice in the finance industry. His most recent projects range from running systematic trading strategies for funds to developing fixed income volatility indexes for Chicago Board Options Exchange and S&P Dow Jones Indices.

Yoshiki Obayashi previously managed US and Asian credit portfolios for a proprietary fixed-income trading group at an investment bank. He holds a PhD in Finance and Economics from Columbia Business School.

Chapter 1
Introduction

1.1 Background

On the 6th and 7th of August 2008, the Bank of England's Monetary Policy Committee convened to decide on the course of its monetary policy. The economic environment during the few weeks preceding Lehman Brothers' collapse made this a difficult task. The inflationary pressures mounting in the UK pointed towards higher interest rates, yet the fragility of the global financial system caused by the subprime mortgage crisis, and the UK's substantial exposure to it, pointed towards a policy of low interest rates. The Committee considered three polarized scenarios: an immediate increase in the Bank rate, a cut, and a continuation of the status quo. The majority of the Committee voted in favor of keeping the Bank rate unchanged, with one member voting for an increase and another for a decrease (Bank of England 2008). Many other outcomes were arguably conceivable during the summer of 2008.

How can one hedge against, or express views on, uncertainties surrounding interest rate movements? Interest rate volatility has traditionally been traded through option-based strategies such as a straddle, which is comprised of long positions in both a payer and a receiver swaption, which may pay off if swap rates experience heightened volatility during the life of the trade. However, a common frustration is that such strategies do not necessarily lead to profits consistent with directional volatility views. As with equity option straddles, swaption straddles suffer from "price dependency," which is the sensitivity of returns to not only the absolute movement of the underlying, i.e. volatility, but also to its direction. Together with other market forces, this problem likely contributed to the emergence of variance swap contracts in equity markets that better align directional views with payoffs, as well as dedicated volatility derivatives such as those tied to the Chicago Board Options Exchange's ("CBOE") Volatility Index ("VIX").

Variance swap contracts are defined as forward agreements whereby one party receives payments tied to the realized variance of the price of a security over a given horizon. They overcome challenges that arise when hedging volatility using options (a straddle, say) by insulating the volatility component of a "dirty" P&L that is muddled by path dependency. The CBOE VIX is linked to the fair value of a variance

A. Mele, Y. Obayashi, *The Price of Fixed Income Market Volatility*, Springer Finance,
DOI 10.1007/978-3-319-26523-0_1

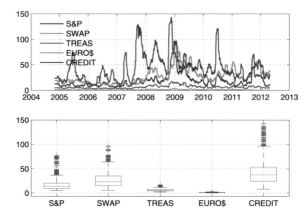

Fig. 1.1 *Top panel*: Estimates of 20 day realized volatility of (i) SPY daily returns, (ii) daily logarithmic changes in the 10-year Swap Rate, (iii) daily returns on the on-the-run 10-year Treasury bond prices, (iv) Eurodollar Future returns, (v) daily changes in the on-the-run CDX Investment Grade Index. Realized volatility is quoted in annualized percentage, calculated as $100\sqrt{12\sum_{t=1}^{21}\ln^2\frac{S_{t-i+1}}{S_{t-i}}}$, where S_t denotes (i) SPY, (ii) 10-year swap rate, (iii) 10-year Treasury bond price, (iv) Eurodollar future three-month LIBOR for a 1-year maturity, (v) on-the-run CDX Investment Grade Index. *Bottom panel*: Boxplot with lower quartile, median, and upper quartile values of the volatilities displayed in the *top panel*. The sample includes daily data from 27 September 2004 to March 27, 2012, for a total of 1881 observations

swap referencing the S&P 500® Index, and is "model-free" in the sense that it does not rely on any parametric assumptions regarding price movements (except standard dynamics in financial economics), and only assumes the absence of arbitrage in frictionless markets. An appealing property of this fair value-based measurement is that it is a *forward-looking* gauge of equity market volatility, corrected by risk.

Model-free indexes for interest rate volatility are a more recent development despite the fact that transaction volumes in fixed income markets far surpass those in equity markets (see, e.g., Mele and Obayashi 2015). CBOE launched the Interest Rate Swap Volatility Index ("SRVIX") in 2012 based on our work in Chap. 3 of this book, which in turn draws from our original technical white paper (Mele and Obayashi 2012). CBOE subsequently teamed up with CME Group to launch the CBOE/CBOT 10-Year US Treasury Note Volatility Index ("TYVIX") in 2013 based on a methodology drawn from Chap. 4 of this book, which in turn draws from our working paper (Mele and Obayashi 2013a). Across the Pacific, S&P Dow Jones Indices and Japan Exchange Group teamed up to launch the Japanese Government Bond analogue of TYVIX (the "S&P/JPX JGB VIX"). Then, in November 2014, CBOE Future Exchange ("CFE") listed futures contracts on TYVIX, the first exchange-traded contracts based on these new standardized fixed income volatility gauges.

Historically, interest rate and equity volatilities have not been tightly linked. Figures 1.1 and 1.2 illustrate that realized volatilities in four fixed income asset classes—swap, government bond, time deposit, and credit—display significantly

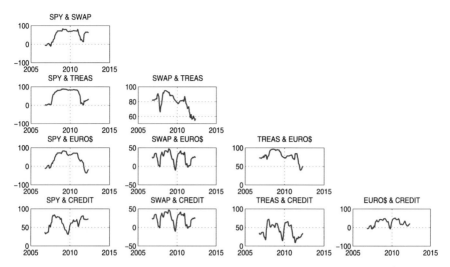

Fig. 1.2 Moving average estimates of the correlations across all volatilities depicted in Fig. 1.1. Each correlation estimate is calculated over the previous two years of data

distinct behaviors. While the volatility in each of these markets reacts to different events, Fig. 1.1 also reveals that the size of fixed income volatility is substantial, overall, with the highest average levels and peaks occurring in swap and credit markets, and sometimes surpassing those in the equity market, labeled as "S&P." The volatility in Treasury markets is also comparable to that in equity markets, although lower than it. The volatility in time deposit markets is the lowest, which is to be expected, since Eurodollar futures reference short-term rates. Remarkably, Fig. 1.2 demonstrates that correlations among these asset classes undergo large swings and even switch signs through time.

While this book does not cover details related to empirical aspects of these markets, these high-level observations suggest that hedging against fixed income volatility requires dedicated instruments for each asset class. Additional motivation regarding this important topic is given in Sect. 1.2.

The aim of this book is to fill a gap in the literature, where virtually no mathematically rigorous attempts have been made to develop methodology for fixed-income volatility pricing in a *model-free* fashion. We aim to develop a unified methodology to design and price variance swaps for fixed income markets in a model-independent framework, which is a key property shared by the VIX. "Model-free" means that under standard assumptions on price dynamics and no-arbitrage, the fair value of these variance swaps can be expressed as a function of the price of already traded assets such as (i) at-the-money interest rate options, (ii) the cross-section of out-of-the-money interest rate options, and (iii) zero-coupon bonds. This approach is appealing as it is immune to the pitfalls of model-misspecification for price dynamics and market risk premiums, and serves as the basis and motivation for the creation of VIX-like, forward-looking indexes of fixed income volatility.

Fig. 1.3 CBOE SRVIX of
interest rate swap volatility
(referenced to 1Y-10Y swaps)
vis-à-vis CBOE VIX index of
equity volatility

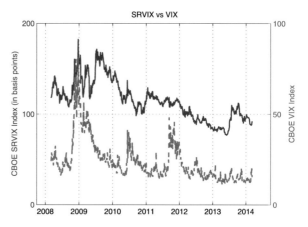

Mele and Obayashi (2012) first considered the pricing of variance swaps in
interest rate swap markets in a model-free fashion, and in subsequent work ad-
dressed government bond volatility (2013a), time-deposit volatility (2013b), and
credit volatility (2013c). This book provides a unifying treatment and additional
methodological details regarding each asset class. The reader seeking a more succinct account of our work is referred to Mele and Obayashi (2014).

This introductory chapter aims to provide perspective on how the complexity of
fixed income security evaluation translates into a comparably complex framework
for pricing fixed income volatility. In Sect. 1.3, we highlight numerous instances in
which a model-free expression for an index of interest rate volatility could be re-
garded as an approximation to the price of an interest rate variance swap contract.
As will be made clear, such variance swap contracts must be designed carefully
for each fixed income asset class to afford this interpretation, and we shall provide
intuition regarding the challenges encountered even when model-free pricing is possible. Section 1.4 provides a roadmap of the book. Before diving into the technical
details, the next section discusses some empirical evidence regarding the behavior
of the newly launched indexes of interest rate volatility based on our work.

1.2 From Realized to Expected Fixed Income Volatility

Realized volatilities in equity and fixed income markets behave differently, and it
is reasonable to expect the same to hold for forward-looking volatility measures
across different asset classes. Figures 1.3 and 1.4 depict the behavior of the SRVIX
and TYVIX indexes in relation to the VIX. The SRVIX references interest rate
swaptions with a 1-year maturity and 10-year underlying tenor, and the TYVIX
references Treasury future options with a 1 month maturity on the 10-year T-Note
futures.

Not surprisingly, "forward-looking," or "expected," volatility in fixed income
markets differs from that in equity markets. While fixed income and equity volatilities seem to respond similarly to pronounced global events, there are several

Fig. 1.4 CBOE/CBOT
TYVIX index of Treasury
volatility (referenced to one
month volatility of 10Y
Treasury Note Future price)
vis-à-vis CBOE VIX index of
equity volatility

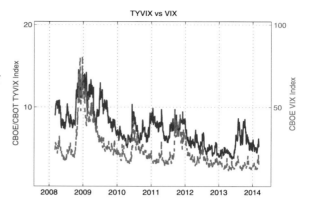

episodes in which the timing and nature of their responses differ significantly. For
example, the correlation between daily changes in the TYVIX and VIX is less than
30 %, while that between TYVIX and SRIVX is greater than 50 %. We have identified at least five historical instances of significant divergence between fixed income
and equity market volatilities.

First, right at the start of the crisis in the summer of 2007, the SRVIX begins to
trend upwards for over one year. During the same period, the VIX fluctuates without
any apparent trend, only to dramatically spike in 2008.

Second, during the spring of 2009, encouraging economic news in the US is accompanied by a major bond sell-off and an equally major stock market rally, which
leads to a surge in both SRVIX and TYVIX but a grind down in VIX. This period
demonstrates that a bond market sell-off arising from good macroeconomic news,
improved investor sentiment, and a stock market rally is not necessarily inconsistent
with a "rate fear" episode.

Third, during the sovereign debt crisis of 2011–2012, fixed income volatility
remains relatively range-bound as a result of monetary policy, while equity volatility
experiences a more pronounced and sustained increase.

Fourth, both SRVIX and TYVIX enter another heightened regime during the QE
"taper tantrum" of May–September 2013. During this period, the TYVIX doubles
in a matter of weeks while the VIX drifts sideways in a seemingly disconnected
manner.

Fifth, Yellen's "regime" seems to have stabilized fixed income volatility while
having no apparent effects on the dynamics of the VIX.

It is instructive to document how expected volatility behaves in both fixed income
and equity markets during periods of distress. Tables 1.1 and 1.2 confirm that fixed
income volatility increases over rate increases, coinciding, for example, with a fall
in the Dow Jones Investment Grade Corporate Bond Index or an increase in 10-year
Treasury yields. During these episodes, the VIX actually decreases. The natural
interpretation of these historical patterns is a "reverse flight-to-quality" effect, in
which investors flock back to equities as in the 2009 episode described above.

Indeed, both Tables 1.1 and 1.2 suggest that during days in which interest rates increase, the S&P500 index also increases. In particular, Table 1.2 reveals that on days

Table 1.1 Behavior of SRVIX, VIX and S&P500 over days of interest in debt and equity markets

	No Obs.	SRVIX average Δ change (std error)	VIX average Δ change (std error)	S&P500 average change (std error)
DJ IG Corp Bond down by:				
< 0	621	0.37 bps (0.09 bps)	−0.54 bps (0.09 bps)	0.47 % (0.06)
< −0.5 pt	138	1.38 bps (0.27 bps)	−1.06 bps (0.25 bps)	1.01 % (0.18)
< −1.0 pt	21	1.63 bps (0.80 bps)	−1.72 bps (1.08 bps)	1.61 % (0.76)
S&P 500 down by:				
< 0	618	0.16 bps (0.10 bps)	1.39 bps (0.09 bps)	
< −2 %	114	0.80 bps (0.28 bps)	4.24 bps (0.31 bps)	
< −5 %	14	2.69 bps (1.02 bps)	9.11 bps (1.26 bps)	

Table 1.2 Behavior of TYVIX, VIX and S&P500 over days with increasing yields

10Y yield up by:	No Obs.	TYVIX average Δ change (std error)	VIX average Δ change (std error)	S&P500 average change (std error)
> 0	685	5.06 bps (1.70 bps)	−63.49 bps (7.23 bps)	0.51 % (0.05)
> 5 bps	297	14.17 bps (2.88 bps)	−90.07 bps (13.32 bps)	0.79 % (0.10)
> 10 bps	112	18.41 bps (6.09 bps)	−113.48 bps (28.77 bps)	1.06 % (0.23)
> 12 bps	73	29.01 bps (6.17 bps)	−101.95 bps (6.09 bps)	0.98 % (0.27)
> 15 bps	25	34.48 bps (12.40 bps)	−114.72 bps (32.70 bps)	0.68 % (0.65)

with moderate increases in the 10-year yield, the S&P increases, possibly driven by this reverse flight-to-quality effect; on days in which the 10-year yield increases more drastically, a possible fear effect seems to curb equity market rallies.

Finally, Table 1.3 documents that both TYVIX and VIX surge when yields experience large falls, possibly driven by a common fear that a fall in long-term yields might reflect bad times to come, which policy makers may have to handle in future rate cuts. This explanation is consistent with the fact that the return on S&P is on average significantly negative on days in which the 10-year yield is down.

Table 1.3 Behavior of TYVIX, VIX and S&P500 over days with falling yields

10Y yield up by:	No Obs.	TYVIX average Δ change (std error)	VIX average Δ change (std error)	S&P500 average change (std error)
< 0	723	−4.61 bps (1.45 bps)	63.39 bps (8.75 bps)	−0.48 % (0.06)
> −2 bps	605	−4.52 bps (1.66 bps)	73.67 bps (10.21 bps)	−0.58 % (0.07)
> −3 bps	468	−3.58 bps (2.04 bps)	91.93 bps (12.79 bps)	−0.76 % (0.08)
> −5 bps	312	−0.19 bps (2.85 bps)	119.17 bps (17.38 bps)	−0.92 % (0.11)
> −10 bps	100	13.71 bps (7.39 bps)	212.69 bps (42.41 bps)	−1.50 % (0.29)
> −12 bps	55	23.63 bps (11.78 bps)	212.90 bps (64.60 bps)	−1.81 % (0.40)
> −15 bps	28	44.03 bps (19.05 bps)	300.64 bps (113.99 bps)	−2.44 % (0.69)

While forward-looking indexes of volatility certainly serve as valuable sources of information, they do not constitute directly tradable products. Indeed, the VIX index itself is not tradable, but there are now numerous securities tied to the VIX, such as futures, options, and exchange traded products (see, e.g., Rhoads 2011). However, the empirical evidence provided above suggests that VIX products are unlikely to serve as proxies for exposure to fixed income volatility. More generally, the "uniqueness" of fixed income volatility is widely acknowledged in the academic literature: it has been well known since at least Collin-Dusfresne and Goldstein (2002) that fixed income markets are incomplete, in that their volatility does not appear to be priced only based on existing fixed income assets. Investable products tied to "standardized" fixed income volatility gauges would therefore fill an important gap.

Standardized products on fixed income volatility are still in their infancy. At the time of writing, TYVIX futures are the only exchange-listed products available for standardized trading of fixed income volatility. Figure 1.5 depicts the time series behavior of both the VIX and the TYVIX (top panel) and their "basis" (bottom panel). The basis in this context is the difference between future prices and spot index levels; as a representative case, we have shown the basis corresponding to one-month futures. The basis for VIX is based on market data whereas the basis for TYVIX is based on theoretical pricing from a three-factor model developed by Mele et al. (2015b).

The evidence in Fig. 1.5 suggests that volatility futures behave quite differently in the fixed income and equity spaces. For equities, the volatility basis is predominantly positive, meaning "contango." The main trading implication of contango is that a rolling long futures strategy "bleeds" from the roll-down; put another way, going long one-month VIX futures and holding the position to expiration leads to

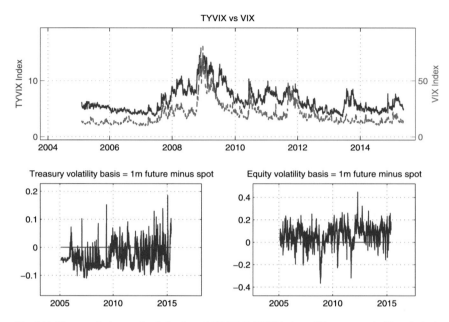

Fig. 1.5 *Top panel*: time-series behavior of CBOE TYVIX index of Treasury volatility vis-à-vis CBOE VIX index of equity volatility. *Bottom panel*: the "basis" for these two indexes, defined as the difference between 1-month futures minus index levels. TYVIX basis is hypothetical and model-based

negative returns on average. One explanation for the persistent contango is that VIX futures provide protection against tail events. Indeed, it is well known that long VIX positions perform like insurance when markets are in turmoil. However, the price for this insurance is that VIX futures are indeed expensive, in that they converge to spot values that are on average lower.

On the other hand, Fig. 1.5 suggests that the volatility basis in the fixed income space is often negative, meaning "backwardation." A rolling long strategy in TYVIX futures may then enjoy a positive P&L drift. Mele et al. (2015b) demonstrate this feature of the strategy, which, at the same time, may also serve as a tail hedge. Naturally, these results have to be interpreted with caution since they refer to hypothetical market conditions in which TYVIX futures did not exist. It is possible that backwardation in the Treasury volatility curve may become muted or reversed as a result of trading in TYVIX futures and/or the emergence of risk premiums from an increase in the demand for Treasury volatility.

1.3 The Right *Numéraire* and Volatility Pricing

The model-free linkage between traded derivatives and the fair value of fixed income market volatility is far from obvious. The complexity of interest rate transactions is such that their volatility is more difficult to price compared to the equity case. Intuitively, we price assets by discounting their future payoffs, but when it comes to fixed income securities, discounting rates are obviously random and, in turn, interest rate volatility depends on this randomness.

A further critical issue pertaining to fixed income markets is the presence of additional sources of randomness, on top of those relating to interest rate developments. For example, in swap markets, uncertainty links to two sources of randomness: a first, direct, source relating to changes in swap rates; and a second, indirect, source pertaining to changes in points of the yield curve that affect the value of swap contracts. Similar difficulties arise in other markets such as credit. The main difficulty for us is to disentangle these sources of risk and price the pure interest rate volatility component in a model-free fashion.

This book introduces security designs for trading uncertainty related to developments in interest rate volatility while addressing the above issues. Our goal is the pricing of these products in a model-free fashion, i.e. only relying on the price of traded assets such as (i) European-style interest rate derivatives, serving as counterparts to the European-style equity options underlying equity volatility, and (ii) zero coupon bonds. For example, we derive model-free expressions for expected volatility in the interest rate swap case by relying on both swaptions and zero coupon bonds, the latter of which are needed to deal with uncertainties arising from the interest rate payments of the underlying swap. In these cases, our contract designs lead to new indexes that reflect market expectations of fixed income volatility, adjusted for the relevant notion of market risk, and relying on minimal assumptions such as absence of arbitrage and standard price dynamics. The reader is referred to the next chapter for references relating to the literature on equity markets, which we have left out of this introduction to focus on the heuristic details of our work.

1.3.1 Market Risk and Model-Free Pricing

What is the relevant notion of market risk-adjustment in our context? Financial theory predicts that in frictionless markets, absence of arbitrage implies that there is a unit of account such that the prices of all securities specified in terms of this unit are martingales under a certain probability—the "market probability." The relevant notion of risk-adjustment in our context is a unit of account, termed *numéraire* to honour Léon Walras' (1874) famous observation that in general equilibrium, any market is always in equilibrium given equilibrium in the other markets so that we can take the price of any commodity as given, and study the equilibrium price of all commodities in terms of the chosen numéraire.

Securities pricing in different markets is facilitated by the right specification of a numéraire that applies to each market. In the equity case, a standard notion of numéraire is the money market account—assuming interest rates are constant—such that discounted asset prices are martingales under the so-called risk-neutral probability. For fixed income asset classes, we consider alternative notions of numéraires that are more suitable to the unique risks inherent in each class. The next chapter provides a unified treatment of these issues, which we shall apply to each market of interest in the remaining chapters of the book.

Our main finding is the following. In frictionless and arbitrage-free markets, a variance swap on any "forward risk" can be evaluated in a model-free fashion once the final payoff is re-scaled by an appropriate numéraire tailored to the particular risk. The meaning of a forward risk is that of a clearing price for a forward contract. One example is a forward swap rate and, accordingly, the variance swap needs to be rescaled by an annuity factor so that the price of swap volatility is an expectation under the so-called annuity-based, or swap, probability. As another example, the fair value of a variance swap on government bond forward volatility is the expectation under the so-called forward probability.

To summarize, our contract variance design suggests the following solution allows us to price variance swaps in a model-free fashion: (i) rescale the payoff of a variance swap by the market numéraire at the time of expiration of the variance swap; and (ii) express the price of variance as the expectation of future variance under the market probability. That is, the probability under which the forward risk is a martingale also works as a "variance swap pricer" in a model-free fashion.

Note that we are not merely stating that in the absence of arbitrage, security prices rescaled by some unit of account are martingales under a certain probability. We are stating something stronger, relating to security design, which is that *model-free* pricing of variance swaps is possible once we rescale the payoff of these contracts with the appropriate numéraire.

These are examples of "variance swap pricers." In Chap. 2, we identify a host of possible variance swap "tilters" that make the fair value of the resulting contracts model-free, with the "numéraire" tilter being a specific case. Given the economic appeal of the market numéraires, and the familiarity with it by both academics and practitioners, we will build upon this notion to derive model-free security designs and volatility indexes for fixed income markets in Chaps. 2 through 5.

There are additional qualifications to be made around practical considerations for model-free pricing and the many pitfalls arising in fixed income markets. For example, pricing variance swaps on government bonds proves to be deceptively challenging due to the high dimensionality of these markets. Suppose we wish to price the 1-month forward volatility of a 3-month future on the 10-year Treasury note, and that available for trading are American-style options expiring in 1 month. Can these options be used to price the desired volatility in a strictly model-free fashion? The practical answer depends on the magnitude of two issues. One relates to the early exercise premium embedded in American option prices. A second relates to the mismatch in maturity between the options and the underlying futures, i.e. 1-month options are used to span risks generated by 3-month futures returns. This second source is specific to the nature of fixed income markets. We now provide heuristic details of such issues.

1.3.2   Getting the Right Volatility with the Right Model

The volatility of a given fixed income security can be priced as an expectation of the security future realized volatility under an appropriate probability. To illustrate the pitfalls using the wrong probability, we provide examples illustrating how heuristic derivations of these prices based on the standard methodology developed in the equity space can actually lead to prices deviating from the true fair value and, hence, to arbitrage opportunities.

1.3.2.1   Rescaling

It is well known (see, e.g., the next chapter) that the price of an equity variance swap is a weighted average of at-the-money (ATM) and out-of-the-money (OTM) options, divided by the factor e^{rT} where r is the short-term rate, assumed to be constant, and T is the maturity of the options making up the index. Now suppose we wish to apply the same methodology using swaptions data with the purpose of calculating a model-free index of swap rate volatility in percentage terms. Accordingly, we aggregate swaptions market data according to the same weighted average and divide by e^{rT} à la VIX. This procedure is wrong. Chapter 3 shows the weighted average of the swaption prices has to be divided by an annuity factor, and not by e^{rT}, which relates to the variance swap "tilting" described in Sect. 1.3.1.

The rescaling is critical for a number of reasons. As another example, in the case of credit variance swaps and indexes covered in Chap. 5, a model-free measure of volatility requires rescaling by a defaultable annuity factor. A wrong rescaling, say dividing by e^{rT} or an annuity factor, may severely affect the dynamics of the index, which depends on an appropriate average default intensity as will be shown. In times of distress, when the default intensity is high, a gauge of credit volatility might underestimate the option-implied volatility unless it relies on the correct rescaling suggested in Chap. 5. Ignoring the rescaling issue would be as fundamentally egregious as calculating mark-to-market P&L of CDS and CDS index contracts using risk-free discount rates instead of default risk-adjusted discount rates. Finally, incorrect rescaling does not provide guidance on how to hedge fixed income variance contracts, a theme covered in some detail in this book.

1.3.2.2   Naïve Model-Free Methodology

The VIX relies on a formula that is well known, but its eminence does not mean that it can be universally used with options in any asset class to generate model-free implied volatility. For example, we show in Chap. 4 that applying this "naïve VIX methodology" to options on cash bonds does not produce model-free implied volatility of the underlying bond returns. We calculate the bias, and show numerically that it can be quite substantial.

It follows that aggregating option prices by a formula such as that used in
CBOE's calculations of VIX for equity market volatility as described in the VIX
white paper (see Chicago Board Options Exchange 2009) is not a procedure we can
implement without checking the internal consistency of the algorithm as applied to
fixed income. One theme of the book is to provide rigorous foundations to a consistent treatment of this issue.

1.3.2.3 Approximations of the Market Probability

There is an issue that we encounter that bears some resemblance to the situation
market practitioners dealt with before the famous *market models* were introduced by
Brace et al. (1997), Jamshidian (1997) and Miltersen et al. (1997). Suppose we want
to price a caplet, which is a European call option on LIBOR. The option strike and
maturity are K and S, and the LIBOR for the period $S - T > 0$ is known at time T.
A basic way to price the caplet at time t relies on the following approximations,

$$\mathbb{E}_t\left(e^{-\int_t^S r_\tau d\tau}\left(L_T(S) - K\right)^+\right) \approx e^{-\int_t^S r_\tau d\tau}\mathbb{E}_t\left(L_T(S) - K\right)^+$$

$$\approx P_t(S)\mathbb{E}_t\left(L_T(S) - K\right)^+, \tag{1.1}$$

where r_τ is the short-term rate at τ, $L_T(S)$ is the LIBOR rate from time T to S,
$\mathbb{E}_t$ is the expectation at t under the risk-neutral probability, and $P_t(S)$ is the price of
a zero-coupon bond expiring at time S,

$$P_t(S) = \mathbb{E}_t\left(e^{-\int_t^S r_\tau d\tau}\right). \tag{1.2}$$

The first approximation in (1.1) relies on the assumption that the short-term rate
is roughly deterministic, so that we can take it out from the expectation. The second approximation then inconsistently reverses the rationale underlying the first approximation, by replacing the exponential of the integrated short-term rate with its
expectation, based on Eq. (1.2).

Naturally, this approach is incorrect. As we know, the right derivation relies on a
change of probability,

$$\mathbb{E}_t\left(e^{-\int_t^S r_\tau d\tau}\left(L_T(S) - K\right)^+\right) = P_t(S)\mathbb{E}_t\left(\frac{1}{P_t(S)}e^{-\int_t^S r_\tau d\tau}\left(L_T(S) - K\right)^+\right)$$

$$= P_t(S)\mathbb{E}_t^{Q_{FS}}\left(L_T(S) - K\right)^+, \tag{1.3}$$

where $\mathbb{E}_t^{Q_{FS}}$ is the expectation under the S-forward probability.

Note that the last terms in Eqs. (1.1) and (1.3) are not the same. Granted, they
can be both implemented through a Black (1976) pricer, which market participants
can feed through Black's implied skew. However, the interpretation of this skew
drastically differs depending on whether we consider the approximation in (1.1)
or the exact formulation in Eq. (1.3). In (1.1), the implied skew is interpreted as

stemming from expectations taken under the risk-neutral probability; in Eq. (1.3), the implied skew is understood under the forward probability.

We face a similar issue while pricing volatility of government bonds and time-deposits. In Chap. 4, we show that the model-free implied variance, defined as the no-arbitrage price of a variance swap, is the expectation of the future variance under the *forward probability*, while at the same time, by incorrectly proceeding with a series of approximations such as those in (1.1), we might end up with a formula that we could interpret as the expectation of the variance taken under the *risk-neutral probability*.

This distinction is not immaterial. For instance, if one were to devise trading strategies around indexes developed in this book and listed products thereon, it is critical for these strategies to be designed so that the variables of interest are modeled under the correct probability space. Consider the following parametric example taken from Chap. 4 regarding the pricing of government bond volatility.

Let v_t be the instantaneous *basis point* volatility of the short-term rate (see Eq. (4.43) in Chap. 4), assumed to be a diffusion process. We shall develop a model that fits the whole initial yield curve, without error, which predicts that the fair value of a percentage variance swap referenced to a zero-coupon bond is (the square of)

$$\text{GB-VI}(v_t; t, T, \mathbb{T}) \equiv \sqrt{\frac{1}{T-t} \int_t^T \bar{\phi}_\tau(t, T, \mathbb{T})d\tau \cdot v_t}, \tag{1.4}$$

where $T - t$ is time-to-maturity of the variance swap, $\mathbb{T} > T$ is the expiration date of a zero coupon bond, and $\bar{\phi}_\tau(t, T, \mathbb{T})$ is a deterministic function, which links to the conditional expectation of a forward bond price under the *forward* probability (see Eq. (4.52)).

Next, suppose we price the same government bond variance swap by taking expectations under the "wrong probability," the risk-neutral. The square of the "fair value" under this misspecified assumption is

$$\widehat{\text{GB-VI}(v_t; t}, T, \mathbb{T}) \equiv \sqrt{\frac{1}{T-t} \int_t^T \phi_\tau(T, \mathbb{T})d\tau \cdot v_t}, \tag{1.5}$$

where $\phi_\tau(T, \mathbb{T})$ is now a function that links to conditional expectations taken under the *risk-neutral* probability (see Eq. (4.54)).

Naturally, pricing a variance swap under the wrong probability (i.e. through Eq. (1.5)) can give rise to arbitrage in a frictionless capital market where, instead, the variance swap would be worth the expression in Eq. (1.4). Consider for example a multiasset strategy that aims to exploit anomalies across different asset classes. The strategy cannot interpret a government bond volatility index as an expectation under the risk-neutral probability, as this would lead to the wrong replication and hedging recipes.

To illustrate, consider a future written on the squared index in Eq. (1.4), and assume that the basis point variance v_t^2 is a martingale under the risk-neutral probability, i.e. the assumption that actually leads to Eq. (1.4). In this hypothetical market, the index future has the same price as the squared index, $\text{GB-VI}^2(v_t; t, T, \mathbb{T})$,

whereas according to a model leading to the index GB-$\widehat{\text{VI}}(v_t; t, T, \mathbb{T})$, there might
be contango or backwardation depending on parameter values. Similar complica-
tions arise while dealing with the risk management of products linked to these indexes. Futures on government bond volatility indexes are risk-neutral expectations
of forward-neutral expectations, and simulating scenarios heavily relies on specify-
ing the correct probability to use in each context of interest. It is another theme of
this book to provide rigorous grounding on these and related topics.

1.3.2.4 Maturity Mismatch

We illustrate another complication that arises when pricing fixed income volatility,
which is due to the "multi-dimensional" nature of rates markets. Suppose that we
wish to price the volatility of a 10-year Treasury note future returns arising over
the next month, that the future expires in 1 year, and that available for trading are
American options expiring in 1 month. Can we price this volatility in a model-free
fashion, i.e. by only exploiting the future options quotes? In Chap. 4, we show that
this is not the case, and that the same issue arises for other markets such as time-
deposits.

   This problem originates from two sources. The first is the early exercise premium
embedded in American option prices, which is a general issue that might also arise
in the equity case, and is discussed in the next section. The second is the maturity
mismatch: one month options are used to span risks generated by one year futures
returns. This second source is specific to the nature of fixed income markets, and
would arise even if European options were traded.

   Suppose that available for trading are European options on forwards with the
10-year Treasury note as an underlying. The options expire at T, whereas the forward expires at some $S \geq T$. Now we have that the "option span" operates under
the T-forward probability, whereas the forward price is a martingale only under the
S-forward, which generates risks in the forward price that are not spanned by the
set of all available options. Unless $T = S$, we would price the forward price volatility with the "wrong" probability. This issue is reminiscent of convexity problems
arising in fixed income security evaluation (see, e.g., Brigo and Mercurio 2006;
Veronesi 2010). In this case, model-free indexes cannot be calculated, and a model-
dependent correction term would be needed to interpret the resulting index as linking to the fair value of a government bond variance swap.

   In particular, let $F_t(S, \mathbb{T})$ be the forward price at t, for delivery at S, of a coupon-
bearing bond expiring at $\mathbb{T}$, with $t \leq S \leq \mathbb{T}$. In a diffusion setting, it satisfies

$$\frac{dF_\tau(S, \mathbb{T})}{F_\tau(S, \mathbb{T})} = v_\tau(S, \mathbb{T}) \cdot dW_{FS}(\tau), \quad \tau \in [t, S],$$

where $v_\tau(S, \mathbb{T})$ is some instantaneous volatility process adapted to $W_{FS}(\tau)$, a Brownian motion under the S-forward probability, Q^{FS}. Note that the forward price is a

martingale under Q^{FS}, whereas options expiring at T are more naturally priced under Q^{FT}. Accordingly, we cannot use these options and evaluate in a model-free fashion a variance swap on, say, the percentage integrated variance,

$$\mathbb{P}(t, T, S, \mathbb{T}) \equiv \mathbb{E}_t^{Q_{FT}} \left(\int_t^T \left\| v_\tau(S, \mathbb{T}) \right\|^2 d\tau \right).$$

Indeed, in Chap. 4, we show (see Eq. (4.134)) that due to this maturity mismatch,

$$\mathbb{P}(t, T, S, \mathbb{T}) = \mathbb{P}_{\mathrm{vix}}(t, T, S, \mathbb{T}) + 2\left[1 - \mathbb{E}_t^{Q_{FT}} \left(e^{\tilde{\ell}(t, T, S, \mathbb{T})} - \tilde{\ell}(t, T, S, \mathbb{T})\right)\right], \quad (1.6)$$

where $\mathbb{P}_{\mathrm{vix}}(t, T, S, \mathbb{T})$ is a weighted average of out-of-the-money and at-the-money options, rescaled by the forward numéraire (and weights equal to the inverse square of the strikes), and $\tilde{\ell}(t, \tau, S, \mathbb{T})$ denotes the integrated drift of the forward price $F_t(S, \mathbb{T})$ from t to T, under the Q^{FT} probability. Naturally, this drift is identically zero once S collapses to T, in which case $\mathbb{P}(t, T, S, \mathbb{T}) = \mathbb{P}_{\mathrm{vix}}(t, T, S, \mathbb{T})$ although in general, the second term in the R.H.S. of Eq. (1.6) is needed to interpret a forward looking gauge of government bond variance as the fair value of a variance swap. Accordingly, the following volatility index,

$$\mathrm{GB\text{-}VI}(t, T, S, \mathbb{T}) = \sqrt{\frac{1}{T - t} \mathbb{P}(t, T, S, \mathbb{T})},$$

is model-dependent as $\mathbb{P}(t, T, S, \mathbb{T})$ is, due to the specific parametric assumption we need to make to determine the dynamics of $\tilde{\ell}(t, T, S, \mathbb{T})$ in Eq. (1.6). In Chap. 4, we shall estimate that the needed correction is small when $S - T$ is small.

1.3.2.5   American Corrections

American options might be available for trading rather than European, and reference futures as the underlying, not forwards. In Chap. 4, we propose a method to deal with the early exercise premium embedded in the price of OTM options by relying on ideas similar to those introduced in the financial econometrics literature.

The main idea of the algorithm we propose can be implemented through three steps: (i) calibration of a "pricing kernel" to the *market* price of American options on futures; (ii) use of the American-implied pricing kernel to derive implications on the (unobservable) price of European options on forwards; (iii) calculation of the volatility indexes through *model*-based data obtained in step (ii). Needless to say, the resulting index calculation is model-dependent and relies on the choice of the pricing kernel hypothesized in the calibration step (i).

In practice, situations arise in which the model-dependent components may be presumed small enough such that the model-free components of the indexes are a reasonable approximation of the true price of volatility. In Chap. 4, we provide further details on how one may estimate these approximation errors. For example, a numerical experiment relying on Vasicek's (1977) model of short term interest rate dynamics under realistic US Treasury market conditions indicates that (i) the early exercise premium embedded into the American options does not inflate the European ones by more than one relative percentage point, and (ii) the impact of the maturity mismatch explained in Sect. 1.3.2.4 is quite small once the maturity mismatch is as small as two months to start with, as in CBOE's current implementation of the government bond volatility indexes of Chap. 4.

1.4 Scope and Plan of the Book

This book puts forth a unified framework for model-free pricing of fixed income volatility, and detail practicable applications for indexing volatility of four major fixed income asset classes: interest rate swaps, government bonds, time deposits, and credit. Each application treats subtle yet important methodological issues arising from the varied market structure of each asset class to make the theory immediately relevant to practice, as in the example of the two interest rate volatility indexes maintained by CBOE and another by S&P Dow Jones Indices and Japan Exchange Group.

Chapter 2 provides a general theory applicable across different fixed income asset classes. Equity options neatly convey information about equity volatility as showcased by the famous VIX Index. Interest rate derivatives, on the other hand, may contain noisy information about interest rate volatility, which Chap. 2 aims to remove for the purpose of pricing volatility.

For example, swaption prices contain information about both interest rate volatility and the value of an annuity. Chapter 3 then applies the framework of Chap. 2 and deals with interest rate swap volatility while insulating the pure interest rate volatility component in a model-free fashion. The annuity factor entering into the payoff of swaptions is worked into the variance swap design to allow for model-free pricing of the contract in both percentage and basis point variance terms, and the CBOE Interest Rate Swap Volatility Index, SRVIX, is based on the latter. The chapter concludes by providing a quick overview of some empirical properties of the SRVIX.

Chapter 4 deals with the volatility of futures on government bonds and time deposits, which share some common characteristics that can be treated together but also have distinct market conventions that distinguish the ultimate index formulations. For one, although futures on both trade on a price basis, the implied volatility of options on government bond futures are quoted in terms of percentage price volatility, whereas that of options on time deposit futures, e.g. Eurodollar, are commonly quoted in terms of equivalent rate volatility. As in the swap case, both percentage and basis point volatility versions of the indexes are proposed, and the former formulation as applied to short-term options on US Treasury Note futures is the basis of CBOE's 10-Year US Treasury Note Volatility Index, CBOE/CBOT TYVIX. Even though a basis point price volatility may not have immediate resonance with practitioners, it comes into play when considering a novel methodology for calculating a duration-based yield volatility index that is nearly model-free and requires far fewer ad hoc assumptions compared to current practices. As explained in this introductory chapter, we also provide instances of model-based adjustments arising due to the many complications interest rate derivatives display.

Chapter 5 deals with credit risk, and shows how to deal with a defaultable annuity factor and use credit derivatives to derive a model-free, forward-looking measure of credit volatility.

Much of our efforts in this book are devoted to the formulation of indexes of fixed income volatility that could be interpreted as, or approximate, the fair value of volatility arising in various segments of fixed income markets. With the exception of a basic example in the government bond case of Chap. 4 (see Sect. 4.2.9.3), we have left out work on the pricing of tradable products on these indexes. Pricing derivatives on fixed income volatility indexes requires efforts that go beyond the scope of this book and forms, instead, the focus of our ongoing research agenda, as in the case of the government bond volatility basis discussed in Sect. 1.2 based on Mele et al. (2015b).

Chapter 2
Variance Contracts: Fixed Income Security Design

2.1 Introduction

Variance swaps are contracts in which the seller pays the amount by which the realized variance of some variable of interest exceeds a threshold predetermined at contract origination. The pricing of variance swaps on equities is well-understood. Decades' worth of financial theory suggests that these contracts can be cast in a "model-free" fashion—the only ingredient that is required for pricing is the price of at-the-money (ATM) and out-of-the-money (OTM) options referencing the stock or stock index of interest.

This model-free methodology still relies on some assumptions, such as absence of arbitrage in frictionless markets. Nevertheless, it is an appealing methodology, which the Chicago Board Options Exchange adopted with a change in the definition of its VIX index in September 2003 to incorporate financial theory developed after the seminal efforts that led to the initial launch of the index in 1993 (see Whaley 1993).

This chapter develops theoretical foundations for variance swaps in the fixed income space. As in the equity case, we seek contract designs that admit model-free pricing. The designs need to be internally consistent in that the contracts' value collapses to a constant, the Black (1976) implied variance, under the hypothetical circumstance of markets with constant uncertainty. The next chapters are self-contained but rely on much of the analysis in this chapter. We shall identify situations where fixed income volatility can be priced in this fashion, and point to cases where it cannot be due to the complex nature of the assets underlying the markets we study.

For example, it is well known that the prices of options on S&P futures can be used to construct an equity-like VIX index. In contrast, Chap. 4 explains that aggregating the prices of options expiring strictly before the expiry of their underlying bond futures (a practically relevant case) results in a model-dependent government bond volatility index; that is, a model-dependent bias arises once a model-free expression is utilized to approximate the true index. A similar bias comes into

© Springer International Publishing Switzerland 2015
A. Mele, Y. Obayashi, *The Price of Fixed Income Market Volatility*, Springer Finance,
DOI 10.1007/978-3-319-26523-0_2

existence when constructing basis point volatility indexes for rates in time-deposit markets (see, again, Chap. 4).

To fully understand the nature of these biases, we start by focusing on ideal situations where these biases do not arise in the first place. We shall establish a connection between this issue and the theory of *market numéraires*. Market numéraires are conceptual tools that allow us to deal with the various complexities arising whilst evaluating interest rate derivatives. It is well known that absence of arbitrage in frictionless markets is equivalent to the property that asset prices, once rescaled by the money market account, are martingales under the risk-neutral probability. While this result is a powerful tool for the analysis of equity derivatives, numéraires other than the money market account are useful when pricing interest rate derivatives, as initially explained by Geman (1989), Jamshidian (1989), and Geman et al. (1995) and, then, further developed by Jamshidian (1997) and Schönbucher (2003, Chap. 7).

The connection we make in this chapter is that the price of fixed income variance swaps is model-free only once the relevant payoffs are rescaled by the numéraire appropriate for each market of interest. We show how to incorporate different market numéraires into the early theory of "spanning contracts" developed over the years in the equity case by Neuberger (1994), Dumas (1995), Demeterfi et al. (1999a, 1999b), Bakshi and Madan (2000), Britten-Jones and Neuberger (2000), and Carr and Madan (2001), among others.

There are additional complications arising in fixed income markets, such as the notion of basis point volatility, which does not arise in equity markets. Basis point volatility is actually not a mere matter of quoting convention, and highlights a fundamental difference between interest rate and equity volatility. The concept of basis point volatility naturally arises because absolute changes describe risk more effectively than relative changes in the context of yields and spreads. A rate increase from 10 bps to 15 bps shares the same percentage change as one from 100 bps to 150 bps, but, all else equal and accounting for convexity, the latter is a nearly tenfold P&L and risk event. In this basic example, it is more useful for rates traders to know whether a position is likely to experience 5 bps moves or 50 bps moves over a given horizon, and a basis point formulation of the problem addresses this by model-free pricing of a variance swap on arithmetic changes in the fixed income instrument of interest instead of logarithmic changes as in the more standard case of equity variance swaps.

Dealing with basis point variance contracts in a context with random interest rates and numéraires leads to issues that have not been considered in the equity literature. We shall deal with these complications through an insight, a linkage between a class of spanning contracts known as "quadratic" to a notion of basis point variability. This insights allows us to price basis point volatility in a model-free and consistent fashion with the notion of numéraire prevailing in each market of interest. The ensuing fixed income volatility indexes in this and the following chapters originate from these dedicated contract designs.

The plan of this chapter is as follows. The next section provides definitions of the risks we wish to price, and the notions of market numéraires needed to achieve this purpose. Section 2.3 introduces dedicated contract designs leading to model-free pricing, and Sect. 2.4 deals with indexes constructed upon these contracts.

Section 2.5 develops further properties of basis point variance swaps, contains indications on how to implement them in the presence of shrinking maturities, and provides estimates of variance risk-premiums based on CBOE's SRVIX index for interest rate swap market volatility. Section 2.6 unveils theoretical properties of basis point and percentage volatility indexes, and compares them in cases in which a limited number of options are available for calculating these indexes. Section 2.7 extends the analysis to markets with discontinuities. Appendix A provides technical details omitted from the main text.

The reader who is not interested in the unified theory may skip the present chapter and directly access the subsequent chapters, in which variance swaps and accompanying model-free volatility indexes are dealt with in a self-contained fashion for various fixed income asset classes.

2.2 Market Numéraires and Volatilities

Consider a forward starting agreement, originated at time t, with a payoff Π_T at time T equal to

$$\Pi_T \equiv N_T \times (X_T - K), \tag{2.1}$$

where both X_T and N_T are measurable with respect to the information set at time T, $\mathbb{F}_T$, and K is chosen at t, so that the value of the contract is zero at inception. Let Q denote the risk-neutral probability, and $\mathbb{E}_t(\cdot)$ the expectation under Q, taken conditionally on $\mathbb{F}_t$, and let r_t denote the short-term rate at t. We assume that N_τ is the price of a tradeable asset for each $\tau \in [t, T]$, and that it is strictly positive.

It is well known (e.g., Mele 2014, Chap. 4) that under regularity conditions, there exist (i) a probability Q^N, and (ii) a martingale process X_τ under Q^N that clears the agreement, i.e. $X_t = K$, so that the value of Π_T is zero at t. Accordingly, we refer to X_τ as the *forward risk process*, and N_T as the value of a *market numéraire* at T, so that any asset price process S_t normalized by N_t is a martingale under Q^N,

$$\frac{S_t}{N_t} = \mathbb{E}_t^{Q^N}\left(\frac{S_T}{N_T}\right),$$

where $\mathbb{E}_t^{Q^N}$ denotes the conditional expectation under Q^N. We call Q^N the *market numéraire probability*. The money market account is the standard notion of numéraire, for which Q^N collapses to the risk-neutral probability.

Examples 2.1 In Chap. 3, N_t is the present value of an annuity of one dollar, Q^N is the *annuity, or swap, probability*, and X_t is the forward swap rate; accordingly, the payoff Π_T in Eq. (2.1) is that of a forward starting swap. In Chap. 4, X_t can be either the forward price of a coupon bearing bond or the forward price of time deposits such as Eurodollars; in both cases, N_t is the price of a zero coupon bond expiring at time T, so that $N_T = 1$, and Q^N is the *forward probability*. Finally, in

Chap. 5, N_t is the present value of a defaultable annuity of one dollar, Q^N is the *survival contingent probability* and, finally, X_t is the loss-adjusted forward default swap index, so that the payoff Π_T in Eq. (2.1) is that of an index default swap.

It is the volatility of X_τ that we are interested in pricing. Unless otherwise stated, we assume that X_τ is a strictly positive diffusion process with stochastic volatility. Section 2.7 contains extensions to jump-diffusions. Let W_τ denote a multidimensional Wiener process under Q^N. Since X_τ is strictly positive, there exists a process σ_τ adapted to W_τ, such that

$$\frac{dX_\tau}{X_\tau} = \sigma_\tau \cdot dW_\tau, \quad \tau \in [t, T]. \tag{2.2}$$

We consider two notions of realized variance. One, based on *arithmetic*, or *basis point* (BP henceforth), changes of X_t in Eq. (2.2), and another based on the *logarithmic*, or *percentage*, changes of X_t. Accordingly, let $V^{\mathrm{bp}}(t, T)$ and $V(t, T)$ denote the realized BP variance and percentage variance in the time interval $[t, T]$,

$$V^{\mathrm{bp}}(t, T) \equiv \int_t^T X_\tau^2 \|\sigma_\tau\|^2 d\tau \quad \text{and} \quad V(t, T) \equiv \int_t^T \|\sigma_\tau\|^2 d\tau. \tag{2.3}$$

While the concept of percentage variance is widely known and used in equity markets, we also consider pricing BP variance to match the fixed income market practice of quoting implied volatilities both in percentage and basis point terms. The aim of the next section is to search for variance swap contract designs, based on $V^{\mathrm{bp}}(t, T)$ and $V(t, T)$, for which the fair value may be expressed in a model-free fashion in a sense to be made precise below. Indexes of expected volatility can then be formulated based on these variance contract designs. We shall return to the definition and properties of $V^{\mathrm{bp}}(t, T)$ in Sect. 2.5.

2.3 Interest Rate Variance Swaps

The risks we study in this book are spanned by interest rate derivatives with payoffs such as those in Eq. (2.1). These payoffs have two components: (i) the forward risk, X_τ at $\tau = T$, which we want to price the volatility of, and (ii) the market numéraire, N_T, which links to the very nature of the derivative involved in the market of interest. We aim to design variance swaps corresponding to the two variances in Eqs. (2.3) so that component (ii) does not affect model-free pricing of these contracts.

2.3.1 Contracts and Model-Free Pricing

We introduce a class of forward contracts for which we replace the standard unit notional with a stochastic notional, as in the following definition:

Definition 2.1 (Forward Contract with Stochastic Multiplier) A forward contract with stochastic multiplier is a contract originated at time t, which promises to pay the following payoff at $T > t$: $\Phi_T \equiv Y_T \times (\Psi(\{X_s\}_{s \in [t,T]}) - K_Y)$, where $\Psi(\cdot)$ is a functional of the entire path of X_t in Eq. (2.2), Y_τ is $\mathbb{F}_\tau$-measurable for $\tau \in [t, T]$, and the strike K_Y is set so that the value of the contract is zero at inception. Y_T is referred to as stochastic multiplier of the security design.

We want to express the fair value, K_Y, as the conditional expectation of the payoff, $\Psi(\cdot)$, taken under an appropriate probability. If interest rates were constant or deterministic, this probability would be the risk-neutral probability once we assume $Y_T \equiv 1$. In the general case, which is relevant to problems arising in fixed income markets, the appropriate probability is obtained once we impose the usual condition that $\mathbb{E}_t(e^{-\int_t^T r_u du} \Phi_T) = 0$, yielding the expression recorded in the next proposition, given without proof. We shall refer to this probability as the *forward multiplier probability*.

Proposition 2.1 (Forward Multiplier Probability) *The fair value of the strike in the forward contract of Definition* 2.1 *is*

$$K_Y = \mathbb{E}_t^{Q^Y}\left(\Psi(\{X_s\}_{s \in [t,T]})\right), \tag{2.4}$$

where $\mathbb{E}_t^{Q^Y}(\cdot)$ denotes the time t conditional expectation under Q^Y, and the Radon–Nikodym derivative of Q^Y with respect to Q is

$$\left.\frac{dQ^Y}{dQ}\right|_{\mathbb{F}_T} = \frac{e^{-\int_t^T r_u du} Y_T}{\mathbb{E}_t(e^{-\int_t^T r_u du} Y_T)}.$$

The probability Q^Y is the forward multiplier probability.

Proposition 2.1 contains a simple but general result, yet our motivation lies in the pricing of interest rate volatility based on $V^{bp}(t, T)$ and $V(t, T)$ in (2.3). Accordingly, we now only consider the two cases, $\Psi(\cdot) = V^{bp}(t, T)$ and $\Psi(\cdot) = V(t, T)$.

Our next step is to identify necessary and sufficient conditions such that K_Y in Eq. (2.4) is model-free and begin with a definition of "model-free" pricing for our context.

Definition 2.2 (Model-Free Pricing) The strike price K_Y in Eq. (2.4) is model-free if we can find a numéraire with value N_τ such that X_τ is a martingale under Q^N as in Eq. (2.2), and a stochastic multiplier Y_T such that K_Y equals the value of a portfolio of European call and options with strike K, say $\text{Call}_t(K)$ and $\text{Put}_t(K)$, where:

$$\frac{\text{Call}_t(K)}{N_t} = \mathbb{E}_t^{Q^N}\left(\frac{\max\{\Pi_T, 0\}}{N_T}\right), \qquad \frac{\text{Put}_t(K)}{N_t} = \mathbb{E}_t^{Q^N}\left(\frac{\max\{-\Pi_T, 0\}}{N_T}\right), \tag{2.5}$$

and Π_T is as in Eq. (2.1).

Absence of arbitrage in frictionless markets implies that there exists a numéraire such that option prices can be expressed as in Eq. (2.5). The additional requirement of the previous definition is that we need to find a stochastic multiplier Y_T such that the value of a variance swap is model-free. We emphasize that our definition of "model-free" pricing does not rely on the replicability of a variance swap. We only require that the value of the variance swap equals the market value of a portfolio of tradeable securities. Section 2.3.3 below contains a more detailed discussion of these issues.

The question arises as to how the two expectations in (2.4) and (2.5) relate to each other. We have:

Proposition 2.2 (Model-Free Contracts)   *The fair value of K_Y in the forward contract of Definition 2.1 is model-free if and only if the Radon–Nikodym derivative of the forward multiplier probability Q^Y with respect to the market numéraire probability Q^N is uncorrelated with $V^{bp}(t, T)$ and $V(t, T)$. For $\Psi(\cdot) = V^{bp}(t, T)$, it is given by:*

$$K_Y = V_t^{bp} \equiv \frac{2}{N_t} \left(\int_0^{X_t} \mathrm{Put}_t(K) dK + \int_{X_t}^{\infty} \mathrm{Call}_t(K) dK \right) \quad (\textit{Basis Point pricing});$$

(2.6)

for $\Psi(\cdot) = V(t, T)$, it is given by:

$$K_Y = V_t \equiv \frac{2}{N_t} \left(\int_0^{X_t} \frac{\mathrm{Put}_t(K)}{K^2} dK + \int_{X_t}^{\infty} \frac{\mathrm{Call}_t(K)}{K^2} dK \right) \quad (\textit{Percentage pricing}).$$

(2.7)

The previous proposition is proven in Appendix A.1. It generalizes Mele and Obayashi (2012), who provide a model-free expression for interest rate variance swaps in the interest rate swap space. The focus of Proposition 2.2 is wider.

First, it provides guidance for model-free pricing of variance swaps regarding other fixed income securities. Notably, it suggests choices for the random multiplier Y_T for each market of interest where different numéraires arise (see, e.g., the previous Examples 2.1).[1]

Second, Proposition 2.2 identifies both necessary and sufficient conditions under which interest rate variance swaps can be priced in a model-free fashion. The most intuitive case arises when the stochastic multiplier of Definition 2.1 coincides with the value of the market numéraire, $Y_T = N_T$, in which case the Radon–Nikodym derivative of Q^Y against Q^N is obviously constant and equal to one, and uncorrelated with the realized variance, $V^{bp}(t, T)$ and $V(t, T)$. Intuitively, tilting an interest rate variance swap through the market numéraire, N_T, leads to a market space where both the strike K_Y and the price of all available options are expectations under Q^N,

---

[1]Proposition 2.2 also covers the standard equity case with constant interest rates. In this case, N_t is the price of a zero-coupon bond expiring at T, i.e. $e^{-\bar{r}(T-t)}$, where $\bar{r}$ denotes the constant interest rate.

with no additional information required to price the contract. The market numéraire is indeed the benchmark in this book. Note, however, that Proposition 2.2 points to a larger set of stochastic multipliers. For example, in Appendix A.2, we show that the following stochastic multiplier is also in the set of those identified by Proposition 2.2, $Y_T = N_T \epsilon_T$, where ϵ_T is any $\mathbb{F}_T$-measurable random variable satisfying $\text{cov}^{Q^N}(V^{\text{bp}}(t, T), \epsilon_T) = 0$.

Finally, we check the internal consistency of the contract design, namely that the variance strikes in Proposition 2.2 collapse to a constant, assuming uncertainty is constant. The notion of uncertainty depends on the assumptions we make regarding the data generating process in Eq. (2.2). Consider the highly idealized case of a constant basis point variance, in which case the risk X_τ could now take on negative values, according to the following Gaussian, or "Bachelier market," model

$$dX_\tau = \sigma_{\text{n}} \cdot dW_\tau, \tag{2.8}$$

for some vector of constants σ_{n}. This assumption is the obvious counterpart to that of a constant percentage volatility underlying the standard Black–Scholes market for equity options. In Appendix A.3, we show that in this case, the variance strike in Eq. (2.6) collapses to

$$K_Y = \frac{2}{N_t}\left(\int_{-\infty}^{X_t} \text{Put}_t(K)dK + \int_{X_t}^{\infty} \text{Call}_t(K)dK\right) = \|\sigma_{\text{n}}\|^2(T - t). \tag{2.9}$$

One can verify that an analogous result holds in the percentage case in Eq. (2.7), using results in Carr and Lee (2009).[2]

Note, finally, that this result relies on the Gaussian assumption in Eq. (2.8). Alternatively, consider a Black–Scholes market, viz

$$\frac{dX_\tau}{X_\tau} = \sigma_{\text{bs}} \cdot dW_\tau, \tag{2.10}$$

for some vector of constants σ_{bs}. In this case, an index of expected annualized BP volatility (not variance) is easily seen to equal,

$$\sqrt{\frac{K_Y}{T - t}} = X_t\sqrt{\frac{e^{\|\sigma_{\text{bs}}\|(T-t)} - 1}{T - t}}. \tag{2.11}$$

Equation (2.11) reveals the intuitive property that expected BP volatility is the product of the forward, X_t, times a pure volatility component. In Sect. 2.4 (see

---

[2]The assumption in Eq. (2.8) that basis point volatility is constant is quite stylized, and is only made for the purpose of neatly illustrating the differences between basis point and percentage volatility. In general, the variance of forward risks in fixed income markets is time-varying. For example, Vasicek (1977) predicts that the basis point volatility of government bond forward prices is time-varying, albeit deterministically (see Chap. 4), $\sigma_{v,\text{n}}^2(\tau)$ say, so that Eq. (2.9) would read, $K_Y = \int_t^T \sigma_{v,\text{n}}^2(\tau)d\tau$.

Proposition 2.3), we generalize the previous formula to the more general case in which X_τ has stochastic volatility.

We now turn to explaining two important features of the variance swap strikes K_Y in Proposition 2.2: (i) their connection to contracts with payoffs linked to the realization of the forward risk at time T, X_T, and hedging issues arising therefrom (Sects. 2.3.2 and 2.3.3); and (ii) the weighting schemes applying to the OTM options, which differ, according to the concept of variance involved—the basis point variance, in Eq. (2.6), and the percentage variance, in Eq. (2.7) (Sect. 2.3.4).

2.3.2 Log Versus Quadratic Contracts

The first part of Proposition 2.2 hinges upon a key insight, namely that the price of a BP variance swap relates to that of a "quadratic contract," one with a payoff equal to $N_T X_T^2$ and fair value $N_t \mathbb{E}_t^{Q^N}(X_T^2)$. To establish this link, note, heuristically, that by Itô's lemma,

$$V^{\mathrm{bp}}(t, T) = X_T^2 - X_t^2 - 2\int_t^T X_\tau dX_\tau, \qquad (2.12)$$

so that, by the martingale property of X_τ under Q^N,

$$\mathbb{E}_t^{Q^N}\left(V^{\mathrm{bp}}(t, T)\right) = \mathbb{E}_t^{Q^N}\left(X_T^2 - X_t^2\right). \qquad (2.13)$$

That is, up to an affine transformation, the BP variance of the forward risk and the quadratic contract on X_T have the same value as claimed.

Therefore, to hedge against BP variance swaps, we need quadratic contracts instead of log-contracts (Neuberger 1994), i.e. those with a payoff equal to $N_T \ln X_T$. To illustrate, set for simplicity $N_\tau \equiv 1$ for all τ, an assumption we relax in each of the next chapters (and in Sect. 2.3.3 below) whilst dealing with the market numéraires of interest. The payoff of a quadratic contract can be approximated by the sum of (i) the payoff of two forwards, and (ii) the payoff of two portfolios comprising OTM options and one ATM option,

$$X_T^2 - X_o^2 \approx 2X_o(X_T - X_o)$$

$$+ 2\left(\sum_{j:K_j<X_o}(K_j - X_T)^+ + \sum_{j:K_j\geq X_o}(X_T - K_j)^+\right)\Delta K \equiv \hat{P}_T^q, \qquad (2.14)$$

where $X_o \equiv X_t$ is the forward and ΔK is the interval between the strikes K_j. The previous approximation turns into an equality when we consider a continuum of options (see Eq. (A.4) in Appendix A.1). In contrast, the payoff of a logarithmic

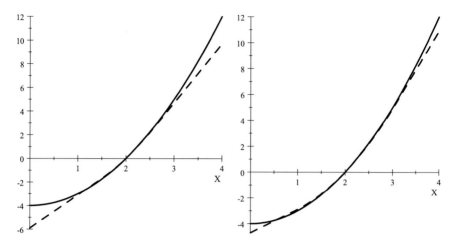

Fig. 2.1 Hedging quadratic contracts with options. In both panels, the *solid line* depicts the terminal value of a quadratic contract, $X^2 - X_o^2$, with $X_o = 2$, and the *dashed line* depicts that of a replicating portfolio, $\hat{P}_T^q$ in Eq. (2.14), comprising: (i) two forwards struck at $X_o = 2$; and (ii) two additional equally weighted portfolios, with $\Delta K = \frac{1}{10}$, each including one ATM option and a number of OTM put and call options. The *dashed line* in the left-hand panel is obtained with a total of 5 puts and 5 calls, and the right-hand panel is with a total of 10 puts and 10 calls

contract can be approximated as

$$\ln \frac{X_T}{X_o} \approx \frac{1}{X_o}(X_T - X_o)$$

$$-\left(\sum_{j:K_j<X_o} \frac{1}{K_j^2}(K_j - X_T)^+ + \sum_{j:K_j\geq X_o} \frac{1}{K_j^2}(X_T - K_j)^+ \right)\Delta K \equiv \hat{P}_T.$$

$$(2.15)$$

Not only do the portfolio weightings in $\hat{P}_T^q$ and $\hat{P}_T$ differ, in that we require (i) $2X_o$ forward contracts in Eq. (2.14) and $\frac{1}{X_o}$ in Eq. (2.15), and (ii) ΔK options in Eq. (2.14) and $\frac{\Delta K}{K_j^2}$ in Eq. (2.15). We also require to go *long* the option portfolio in the case of the quadratic contract, and *short* the option portfolio in the log-contract case.[3]

The portfolio payoff $\hat{P}_T$ has been known to approximate the log-contract payoff quite closely since Demeterfi et al. (1999a, 1999b). Figure 2.1 depicts the quadratic contract payoff and the portfolio payoff $\hat{P}_T^q$, assuming that the forward,

---

[3]Note, however, that for the percentage variance contract, we have, $\mathbb{E}_t^{Q^N}(V(t,T)) = -2\mathbb{E}_t^{Q^N}(\ln \frac{X_T}{X_t})$, so that the option positions have the same sign both when it comes to hedge the basis point and the percentage realized variance, as further clarified in the next chapters. Still, the forward positions have opposite signs as Eq. (2.14) and Eq. (2.15) reveal.

$X_o = 2$, interpreted as an interest rate (say, e.g., a forward swap rate), and that the equally weighted option portfolio has 10 (left panel) and, then, 20 (right panel) out-of-the-money options, with equidistant strikes and distance $\Delta K = \frac{1}{10}$, such that $\min_j K_j = 1.5$ and $\max_j K_j = 2.5$ (left panel) and $\min_j K_j = 1$ and $\max_j K_j = 3$ (right panel). Naturally, the quality of the approximation of the portfolio payoff $\hat{P}_T^q$ to the quadratic contract improves as we increase the number of options in the portfolio. Yet even in the case with fewer options, depicted in the left-hand side of Fig. 2.1, the approximation is still remarkably accurate over a wide range of values of X_T around the forward, $X_o = 2$.

Note that $\hat{P}_T^q$ only aims to hedge the payoff of the quadratic contract. To hedge the BP variance, $V^{\mathrm{bp}}(t, T)$, we also need to hedge the additional term $2 \int_t^T X_\tau dX_\tau$ in Eq. (2.12). The next chapters contain details on these issues for each market and numéraire of interest. We provide preliminary intuition on these details in Sect. 2.3.3 below.

A final issue pertains to the reasons for the constant weights in Eq. (2.6). These weightings follow by spanning arguments that generalize the theory in Bakshi and Madan (2000, Appendix A.3) and Carr and Madan (2001, Eq. (1)) (see, also, Lee 2010) to the case of general numéraires. Section 2.3.4 provides further results and intuition on the origins of these weightings.

2.3.3 Hedging

While our derivations rely on the assumption that the forward risk is a continuous-time process, we can illustrate the main issues arising in our context while relying on a simple discrete-time example. Define the realized annualized variance over n trading days as

$$
\mathrm{var}_{X,n} \equiv \frac{N}{n} \sum_{t=1}^{n} (X_t - X_{t-1})^2
$$

$$
= \frac{N}{n} \left(X_n^2 - X_0^2 \right) - 2 \frac{N}{n} \sum_{t=1}^{n} X_{t-1}(X_t - X_{t-1}), \tag{2.16}
$$

where N denotes the number of trading days over the year. Equation (2.16) is the discrete-time counterpart to Eq. (2.12) and its first term can be expanded just as in Eq. (2.14). In particular, we have:

$$
\mathrm{var}_{X,n} = \frac{N}{n} 2X_0(X_n - X_0) + \frac{N}{n}(X_n - X_0)^2 - 2 \frac{N}{n} \sum_{t=1}^{n} X_{t-1}(X_t - X_{t-1}), \tag{2.17}
$$

where the second term can be replicated through a static position in out-of-the-money options and one at-the-money options as explained. Carr and Corso (2001) consider the variance of price changes rather than returns based on this approach, while only focusing on the second and third term in Eq. (2.17), and markets with constant interest rates.[4] Our work in this book is distinct as we explicitly take into account the first term in Eq. (2.17) and, importantly, consider markets in which interest rates are obviously random. Random interest rates naturally lead to random numéraires, as opposed to the assumption in this section that $N_t \equiv 1$; rescaling by N_t is crucial to model-free evaluation of variance contracts as Eqs. (2.6) and (2.7) reveal.

To summarize, the realized variance, $\text{var}_{X,n}$, cannot be replicated when the market numéraire is a random process. We shall see that within the continuous time setting of the following chapters, we can, instead, replicate the realized variance rescaled by N_T, provided the forward risk X_t is actually traded (or could be replicated), and under additional conditions applying to the specific fixed income securities traded in each market of interest.

To illustrate the reasons $\text{var}_{X,n}$ cannot be replicated in the presence of random numéraires, let us set up the arguments we would use to replicate the last term in Eq. (2.17) in a hypothetical case in which the numéraire is *deterministic* but not necessarily constant. Suppose we are long an amount θ_t of a forward starting agreement at time t and strike K_t such that the value of this agreement at t right after the trade is $\pi_{t+} \equiv \theta_t N_t (X_t - K_t)$ and the value at $t + 1$ before the trade is $\pi_{t+1} \equiv \theta_t N_{t+1}(X_{t+1} - K_t)$. Choosing the strike K_t that clears π_{t+} delivers the forward risk, $K_t = X_t$, such that $\pi_{t+1} = \theta_t N_{t+1}(X_{t+1} - X_t)$. Therefore, the portfolio strategy

$$\theta_{t-1} = \frac{X_{t-1}}{N_t}, \tag{2.18}$$

implies that its overall value at the end of the period is

$$\sum_{t=1}^{n} \pi_t = \sum_{t=1}^{n} X_{t-1}(X_t - X_{t-1}),$$

indicating that the last term in Eq. (2.17) could be replicated.[5] Of course, the crucial assumption underlying the strategy θ_t in Eq. (2.18) is that the market numéraire is deterministic.

If N_t is random and only measurable with respect to the information set at time t, a replication argument based on Eq. (2.18) breaks down. Note that Carr and Corso (2001) deal with markets with constant interest rates and where N_t is just the money market account, such that replication of $\text{var}_{X,n}$ is possible in their setup. Naturally, hedging requires not only dealing with the third term in Eq. (2.17) but also with the first and the second. Still, Proposition 2.2 establishes that in a continuous time

---

[4]Martin (2013) has also recently considered the same setup in Carr and Corso (2001) assuming constant interest rates.

[5]Note that this replication argument hinges upon the forward starting agreement in Eq. (2.1), not the underlying risk X_t, as the latter is not necessarily traded.

setting, the price of interest rate variance swaps is model-free once the realized
variance is re-scaled by the value of an appropriate market numéraire.

In some cases, the realized variance can be replicated by incorporating the
numéraire into the replicating portfolio strategies. Replication is possible through
replication of the forward risk X_t, rather than by trading the forward agreement
through Eq. (2.18). Yet X_t may not be traded in situations of interest. For example,
in Chap. 3, X_t is the forward swap rate, which cannot be perfectly hedged (it is
not traded). However, we show that a model-free expression for the fair value of a
variance swap is available under certain conditions, even when the risk X_t cannot
be replicated.

This point underlies our notion of *model-free contracts* in Definition 2.2, which is
less stringent than one requiring that a variance swap must be perfectly hedged. We
simply require that the fair value of a variance swap equals the value of a portfolio
of tradeable securities rescaled by the value of a market numéraire. Regarding the
previous interest rate swap example, Chap. 3 shows that while the underlying risk
(the forward swap rate) is not traded, variance swaps are priced in a model-free
fashion in terms of Definition 2.2. Finally, note that there are additional instances of
markets in which the underlying risks cannot be perfectly hedged; for example, the
underlying risks can exhibit jumps as in the credit markets studied in Chap. 5.

When hedging is difficult, modeling assumptions are important. Merener (2012)
develops an approach for hedging against interest rate swap volatility by relying on
a set of assumptions regarding the yield curve. He shows how to replicate the third
term in (2.17) (with X_t being the forward swap rate) under the model assumptions.
He utilizes positions in forward-starting interest rate swaps in which gains are rein-
vested in LIBOR accounts at the end of each trading period. His approach does not
focus on model-free indexes of expected interest rate volatility, but is an alternative
to our hedging strategies of interest rate variance swaps on interest rate swaps explained in Chap. 3. We now turn to provide intuition regarding the option weights
in Eq. (2.6).

2.3.4 Constant Gamma Exposure

The weighting in the option portfolio leading to BP expected variance differs from
that underlying percentage variance, which is the standard scheme underlying the
CBOE VIX index for equity volatility. A basic intuition behind the uniform weighting in Eq. (2.6) is that the instantaneous *realized* BP variance simply equals that
of the logarithmic changes of the forward risk rescaled by the squared forward risk,
$\|\sigma_\tau\|^2 \times X_\tau^2$. The implied BP variance shares a similar property: comparing Eq. (2.6)
with Eq. (2.7) reveals that for the BP variance, each option price i carries the same
weight as $\frac{dK_i}{K_i^2}$ (i.e., the contribution of each option price to the implied percentage
variance), rescaled by the squared strike K_i^2, i.e. $\frac{dK_i}{K_i^2} \times K_i^2$.

We can illustrate the proportional weighting in Eq. (2.6) from a different angle.
In their derivation of the fair value of equity variance swaps, Demeterfi et al. (1999a,

1999b) develop an intuitive approach relying on the Black and Scholes (1973) market. They explain that a portfolio of options has a vega that is insensitive to changes in the stock price only when the options are weighted inversely proportional to the squared strike. This property obviously holds in our context with the general market numéraire (see Eq. (2.7)). We develop a similar approach to gain intuition regarding the uniform weightings in Eq. (2.6).

Assume a Gaussian market, i.e. one where the forward risk X_τ in Eq. (2.2) is the solution to Eq. (2.8), and denote the price of an option (be it a put or a call) at time t when the forward risk is X with $\mathcal{O}_t(X, K, T, \sigma_n)$. We create a portfolio with a continuum of these options having the same maturity, and denote the portfolio weights with $\omega(K)$, which are taken to be independent of X. The value of this portfolio is

$$\pi_t(X_t, T, \sigma_n) \equiv \int \omega(K)\mathcal{O}_t(X_t, K, T, \sigma_n)dK. \tag{2.19}$$

We require that the vega of the portfolio, defined as $v_t(X, T, \sigma) \equiv \frac{\partial \pi_t(X, T, \sigma)}{\partial \sigma}$, be insensitive to changes in the forward risk,

$$\frac{\partial v_t(X, T, \sigma)}{\partial X} = 0. \tag{2.20}$$

In Appendix A.3, we show that under regularity conditions, a portfolio of an ATM and all of the OTM options, has vega independent of X if and only if the weightings are independent of K, consistently with Eq. (2.9),

$$\text{Eq. (2.20) holds true} \quad \Longleftrightarrow \quad \omega(K) = \text{const.} \tag{2.21}$$

That is, a portfolio aiming to replicate the BP volatility, σ_n, which is immune to changes in the underlying forward swap rate, is an equally weighted portfolio of out-of-the-money and at-the-money options. In Appendix A.3, we also explain that under the same conditions, the gamma exposure of the options portfolio is constant across different realizations of the forward risk X_t, a result that Bibkov and Misra (2012) illustrate numerically in the case of CBOE's SRVIX index based on the next chapter.

2.4 Implied Volatility Indexes

2.4.1 Model-Free Indexes

Model-free indexes of expected volatility related to the market numéraire N_t follow from Proposition 2.2 in a natural fashion. We define the two indexes,

$$\text{VX}_t^j(T) \equiv \sqrt{(T-t)^{-1}\text{V}_t^j}, \quad j \in \{\text{bp}, \text{p}\}, \tag{2.22}$$

where V_t^{bp} is the strike K_Y for the basis point variance in Eq. (2.6), and V_t^{p} is the strike K_Y for the percentage in Eq. (2.7).

2.4.2 Comparisons to Model-Based Log-Normal and Normal Implied Volatility

2.4.2.1 Skews

The indexes of percentage and basis point volatility in Eq. (2.22) link to the fair value of dedicated interest rate variance swaps in a model-free fashion. As such, they generalize special instances of markets we now describe.

Consider the price of a European call option:

$$\text{Call}_t(K) = \mathbb{E}_t\left(e^{-\int_t^T r_\tau d\tau} N_T (X_T - K)^+\right) = N_t \mathbb{E}_t^{Q^N}(X_T - K)^+, \qquad (2.23)$$

where the second equality follows by the usual change of probability.

We consider two benchmark option pricing models that are derived from specific assumptions on the dynamics of the forward risk.

The first benchmark relies on the Black–Scholes assumption in Eq. (2.10) that the percentage volatility is constant and equal to σ_{bs} (a scalar, say). In this market, the expression for the expectation under the numéraire probability in Eq. (2.23) is given by the Black (1976) formula:

$$\mathbb{E}_t^{Q^N}(X_T - K)^+ = X_t \Phi(d_t) - K\Phi(d_t - \sqrt{T - t}\sigma_{bs}),$$

where

$$d_t \equiv \frac{\ln \frac{X_t}{K} + \frac{1}{2}(T - t)\sigma_{bs}^2}{\sqrt{T - t}\sigma_{bs}},$$

and $\Phi(\cdot)$ denotes the cumulative standard normal distribution. The *log-normal* skew is defined as the mapping $K \longmapsto \text{IV}(K)$ where $\text{IV}(K)$ denotes the value of σ_{bs} such that the option price implied by Black's model coincides with the market price.

The second benchmark relies on the assumption of a Gaussian–Bachelier market with constant basis point volatility σ_n (a scalar, say) (see Eq. (2.8)). In this market, the expectation in the second equality of Eq. (2.23) is

$$\mathbb{E}_t^{Q^N}(X_T - K)^+ = (X_t - K)\Phi(\delta_t) + \frac{\sigma_n \sqrt{T - t}}{\sqrt{2\pi}} e^{-\frac{1}{2}\delta_t^2}, \qquad (2.24)$$

where

$$\delta_t \equiv \frac{X_t - K}{\sigma_n \sqrt{T - t}}.$$

The *normal* skew is defined as the mapping $K \longmapsto \text{IV}^{bp}(K)$, where $\text{IV}^{bp}(K)$ is the value of σ_n such that the option price in this Gaussian market equals the market price.

Clearly, the log-normal skew is flat at σ_{bs} when the forward risk is as in Eq. (2.10) and the normal skew is flat at σ_n should the forward risk be as in Eq. (2.8). In the former case, the index $\text{VX}_t^p(T)$ in Eq. (2.22) equals σ_{bs} and in the latter, the index $\text{VX}_t^{bp}(T)$ in Eq. (2.22) collapses to σ_n, as explained in Sect. 2.3.1.

2.4.2.2 Estimating Expected Volatility from ATM Implied Volatilities

It is well known (see Brenner and Subrahmanyam 1988) that the price of ATM options are approximately linear in volatility. For example, consider the price of an ATM call option in the Black–Scholes market:

$$
\begin{aligned}
\frac{\mathrm{Call}_t^{\mathrm{bs}}(K)|_{K=X_t}}{N_t} &= X_t \left(2\Phi \left(\frac{1}{2}\sqrt{T-t}\,\sigma_{\mathrm{bs}} \right) - 1 \right) \\
&\approx X_t \left(2\left(\Phi(0) + \phi(0)\frac{1}{2}\sqrt{T-t}\,\sigma_{\mathrm{bs}} \right) - 1 \right) \\
&= X_t \frac{1}{\sqrt{2\pi}}\sqrt{T-t}\cdot\sigma_{\mathrm{bs}},
\end{aligned}
\tag{2.25}
$$

with obvious notation. One can easily recover σ_{bs} from the previous formula. Similarly, in Bachelier's market,

$$
\frac{\mathrm{Call}_t^{\mathrm{n}}(K)|_{K=X_t}}{N_t} = \frac{1}{\sqrt{2\pi}}\sqrt{T-t}\cdot\sigma_{\mathrm{n}},
$$

where the L.H.S. of this equation is the expectation of the L.H.S. of Eq. (2.24) evaluated at $K = X_t$.

Carr and Wu (2006, Appendix A) rely on these insights and provide an estimator for the expected variance under the risk-neutral probability, assuming that the instantaneous changes in the standard deviation of a forward risk are independent of those of the forward risk, just as in the seminal paper of Hull and White (1987): in terms of Eq. (2.2), the instantaneous standard deviation, σ_τ (taken to be a scalar), is independent of the entire path of X_τ, up to maturity. Under this assumption, and in terms of the model of this chapter, their results imply that the price of an ATM option, $\mathrm{Call}_t^{\mathrm{sv}}(\cdot)$ is given by

$$
\frac{\mathrm{Call}_t^{\mathrm{sv}}(K)|_{K=X_t}}{N_t} \approx \frac{X_t}{\sqrt{2\pi}}\mathbb{E}_t^{Q^N}\left(\sqrt{V(t,T)}\right),
\tag{2.26}
$$

where $V(t,T)$ denotes the percentage variance, defined as in the second of Eqs. (2.3). Note also that by Eq. (2.25),

$$
\frac{\mathrm{Call}_t^{\$}(K)|_{K=X_t}}{N_t} \approx \frac{X_t}{\sqrt{2\pi}}\sqrt{T-t}\cdot\mathrm{IV}(X_t),
\tag{2.27}
$$

where $\mathrm{Call}_t^{\$}(K)$ denotes the market call price, and $\mathrm{IV}(X_t)$ is the implied volatility for strike $K = X_t$. Therefore, setting $\mathrm{Call}_t(K) = \mathrm{Call}_t^{\mathrm{sv}}(K)$, and then comparing Eq. (2.26) and Eq. (2.27), leaves:

$$
\mathbb{E}_t^{Q^N}\left(\sqrt{\frac{V(t,T)}{T-t}}\right) \approx \mathrm{IV}(X_t).
$$

The authors explain that the approximation is accurate. The assumption under-lying this approximation is that the forward risk volatility is independent of the forward risk path over the life of the option.

2.4.3 Index Decompositions

Let $\sigma(X, K)$ denote Black's implied volatility, defined as usual as $\sigma(\cdot, \cdot)$: $\text{Call}_t(K) = \mathcal{C}_t(X_t, K, \sigma(X_t, K))$, where $\mathcal{C}_t(X, K, \sigma)$ is the price of a call given by the Black (1976) formula when the forward risk is X, the strike is K and the volatility σ_τ in Eq. (2.2) is constant and equal to σ. The next proposition provides basic properties of the indexes in Eq. (2.22), which rely on a standard assumption on implied volatility.

Proposition 2.3 (Index Skew Factors) *Suppose the implied volatility surface has the sticky delta property, or that implied volatilities are homogeneous of degree zero in X and K, i.e. $\sigma(X, K) = \sigma(\lambda X, \lambda K)$, for any constant $\lambda > 0$. Then, (i) there exists a function $\xi(t, T)$ independent of X, such that V_t^{bp} in Eq. (2.6) can be written as $V_t^{bp} = X_t^2 \times \xi(t, T)$, and (ii) V_t^{p} in Eq. (2.7) is independent of X_t.*

Appendix A.4 provides a proof of this proposition. A well-known example of models for which the zero homogeneity property of $\sigma(X, K)$ holds are diffusion processes such that the volatility σ_τ in Eq. (2.2) is a Markov process (see, e.g., Renault 1997), as is the case for the celebrated Heston (1993) model. For example, in the special case of a Black–Scholes market (e.g., Eq. (2.10)), expected volatility is as in Eq. (2.11), consistently with the previous Proposition 2.3. More generally, this proposition provides novel characterizations of the indexes in Eq. (2.22), which are potentially useful in empirical analyses as discussed below. For example,

$$VX_t^{bp}(T) = X_t \times \sqrt{(T-t)^{-1}\xi(t, T)} \equiv X_t \times \hat{\xi}(t, T). \qquad (2.28)$$

Proposition 2.3 tells us that under mild conditions, the function $\hat{\xi}(t, T)$ in Eq. (2.28) is independent of the forward X_t, and can be time-varying, subsuming movements in fundamental uncertainty unrelated to the level of X_t, a "skew factor." Therefore, a BP volatility index moves linearly with the forward, holding uncer-tainty constant. In contrast, a percentage volatility index, such as VIX, should not change with the forward risk, provided uncertainty remains the same. In this sense, the rescaled skew factor, $\hat{\xi}(t, T)$, shares similarities (and orders of magnitude) with a percentage index.

Mele et al. (2015a) perform an empirical analysis of CBOE's SRVIX, the interest rate swap BP volatility index developed in the next chapter. They document that this index is at times led by movements of the forward swap rate, and at other times by uncertainty (see Fig. 2.2). For example, the surge in expected BP rate volatility over

Fig. 2.2 *Top panel*: The
CBOE SRVIX index relating
to the 1Y-10Y forward swap
rate, and the 1Y-10Y forward
swap rate. *Bottom panel*: The
CBOE SRVIX index relating
to the 1Y-10Y forward swap
rate, and the skew factor,
$\hat{\xi}(t, T)$ in Eq. (2.28)

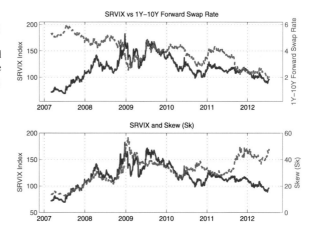

the global financial crisis in 2008 is led by uncertainty, whereas the decline in the
same volatility over 2012 is partially explained by the extraordinarily low interest
rate climate. Mele and Obayashi (2015) provide additional empirical properties of
this index.

2.5 Implementing Basis Point Variance Swaps

The realized variance $V^{bp}(t, T)$ in Eq. (2.3) is a notion that captures the up and
down movements the forward risk (e.g., the forward swap rate) might experience
over a given period. It measures the dispersion of interest rate changes as the sum
of the dispersions occurring over each trading period. It is the relevant notion in the
context of volatility trading and risk management for its potential to track episodes
of sustained and prolonged uncertainty. In Sect. 2.5.1, we provide one additional
definition of realized variance, and in Sect. 2.5.2 we provide estimates of variance
risk-premiums based on these two notions of realized variance.

2.5.1 Incremental Versus Point-to-Point Realized Variance

2.5.1.1 Basis Point

Mele et al. (2015a) consider an additional definition of realized basis point volatility

$$V^{bp}_{p\text{-}t\text{-}p}(t, T) \equiv \sqrt{\frac{(X_T - X_t)^2}{T - t}}, \qquad (2.29)$$

which they label "point-to-point" basis point volatility. While $V^{bp}(t, T)$ in Eq. (2.3)
is "incremental" in nature, point-to-point volatility captures the dispersion of

changes in the forward risk over two distinct points in time. For example, regarding swap markets, point-to-point volatility measures the distance of the future forward rate from the current one, thereby ignoring anything that occurs during the trading period—it may take a small value even after a prolonged period of market turbulence.

While incremental and point-to-point basis point realized variance are obviously not the same, they have the same expectation under the market probability. Consider, for example, basis point variance contracts in swap markets, in which the relevant probability is the annuity. Thus, while the fair values of basis point variance swaps in Sect. 2.3 correctly track the expected variance relevant for trading purposes (i.e. $V^{bp}(t, T)$ in Eq. (2.3)), they can also be interpreted in terms of numéraire-adjusted expected dispersion of the relevant risk, X_T.

The claim that the expectation of the point-to-point variance under the market probability is the same as that of $V^{bp}(t, T)$ follows by the so-called isometry property of Itô integrals (e.g., Øksendal 1998; p. 26), i.e. the second of the next equalities:

$$\mathbb{E}_t^{Q^N}(X_T - X_t)^2 = \mathbb{E}_t^{Q^N}\left(\int_t^T X_\tau \sigma_\tau \cdot dW_\tau\right)^2 = \mathbb{E}_t^{Q^N}\left(\int_t^T X_\tau^2 \|\sigma_\tau\|^2 d\tau\right)$$

$$= \mathbb{E}_t^{Q^N}\left(V^{bp}(t, T)\right), \qquad (2.30)$$

where the first equality follows by Eq. (2.2), and the last is the definition of the incremental basis point variance in Eq. (2.3).

Note, then, that we may define a new variance contract, delivering the following payoff,

$$\Pi_T^* = N_T \times \left((X_T - X_t)^2 - K^*\right),$$

where the fair value, K^*, is:

$$K^* = \frac{1}{N_t}\mathbb{E}_t\left(e^{-\int_t^T r_u du}N_T(X_T - X_t)^2\right) = \mathbb{E}_t^{Q^N}(X_T - X_t)^2 = K_Y,$$

and the last equality follows by Eq. (2.30), with K_Y defined as in Eq. (2.6) of Proposition 2.2. That is, the fair values of point-to-point and basis point variance swaps are the same. Mele et al. (2015a) rely on this property and calculate approximate confidence bands for the forward swap rate forecasts, based on the CBOE SRVIX, reproduced in Fig. 2.3.[6]

2.5.1.2 Percentage

The previous equivalence property—two distinct variance contracts with the same price—does not hold in the case of percentage variance contracts. Consider a contract with payoff referenced to the variance of the cumulative log-return on an asset

---

[6]The forward swap rate is a martingale under the swap market probability but is not necessarily Gaussian. Therefore, the bands in Fig. 2.3 are approximate.

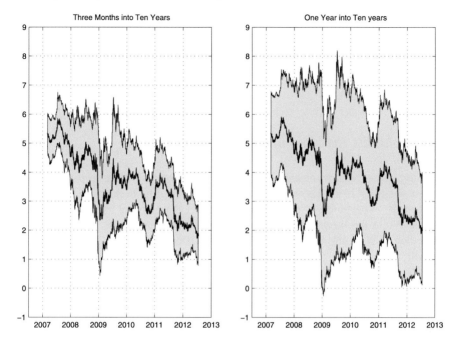

Fig. 2.3 Forward swap rate ∓ 1.96 times the CBOE-SRVIX volatility index for the 3m-10Y forward swap rate (*left panel*) and 1Y-10Y forward swap rate (*right panel*)

price with process X_τ solution to Eq. (2.2), $(\ln \frac{X_T}{X_t})^2$. Such a contract links to the second moment of the cumulative return $\ln \frac{X_T}{X_t}$, rather than the percentage variance. By Itô's lemma, its expectation equals,

$$\mathbb{E}_t^{Q^N}\left[\left(\ln \frac{X_T}{X_t}\right)^2\right] = \mathbb{E}_t^{Q^N}\left[2\int_t^T \ln\left(\frac{X_\tau}{X_t}\right)\frac{dX_\tau}{X_\tau} + \int_t^T \left(1 - \ln\frac{X_\tau}{X_t}\right)\left(\frac{dX_\tau}{X_\tau}\right)^2\right],$$

and can be expressed in a model-free format as

$$\mathbb{E}_t^{Q^N}\left[\left(\ln \frac{X_T}{X_t}\right)^2\right]$$
$$= \frac{2}{N_t}\left(\int_0^{X_t} \frac{1}{K^2}\left(1 - \ln\frac{K}{X_t}\right)\mathrm{Put}_t(K)dK + \int_{X_t}^\infty \frac{1}{K^2}\left(1 - \ln\frac{K}{X_t}\right)\mathrm{Call}_t(K)dK\right),$$
$$(2.31)$$

with the usual notation.[7]

---

[7]The expression in Eq. (2.31) was first derived by Bakshi et al. (2003) in the equity case and constant interest rate $\bar{r}$ (i.e. for $N_\tau = e^{-\bar{r}(T-\tau)}$), in their attempt to determine model-free measures of skewness.

A percentage variance swap based on a cumulative, point-to-point notion is unusual at the time of writing—the standard notion one typically relies on is the incremental realized variance, $\int_t^T (\frac{dX_\tau}{X_\tau})^2$, rather than the point-to-point realized variance, $(\ln \frac{X_T}{X_t})^2$. However, the fair value of *basis point* variance swaps based on the previous two notions coincide.

2.5.2 Volatility Risk Premiums

Volatility risk premiums measure how much investors are willing to pay to hedge against volatility rising above a given threshold. Roughly, they are the difference between the expectation of future volatility under the physical and market probabilities. While the literature on equity volatility risk premiums is large (see, e.g., Bollerslev et al. 2009; Carr and Wu 2008; or Corradi et al. 2013, and references therein), relatively little is known about volatility risk premiums in fixed income markets. Mele et al. (2015a) undertake an empirical study pertaining to swap markets based on the CBOE SRVIX, and here we extend some of their findings by calculating variance risk premiums for both incremental and point-to-point formulations of basis point volatility.

We use realized variance as a proxy for the expectation of future realized variance under the physical probability. Accordingly, we define the incremental and point-to-point variance risk premiums as follows:

$$\pi_{t+S}^{\text{incr}} \equiv \text{SRVIX}_t^2(S) - \text{Vol}_{t+S}^2 \quad \text{and} \quad \pi_{t+S}^{\text{p-t-p}} \equiv \text{SRVIX}_t^2(S) - \text{Vol}_{\text{p-t-p},t+S}^2,$$

where $\text{SRVIX}_t(S)$ denotes the CBOE SRVIX for tenor equal to 10 years, maturity S expressed in fraction of years, and the two realized volatilities, Vol_t and $\text{Vol}_{\text{p-t-p},t}$, are defined below. Note the following important interpretation of π_{t+S}^{incr} and $\pi_{t+S}^{\text{p-t-p}}$: they are P&Ls at time $t + S$ of a short position in a variance swap contract originated at time t.[8]

The realized basis point (incremental) volatility in Eq. (2.3) is estimated as the annualized "quadratic variation" of the daily changes in the forward swap rate,

$$\text{Vol}_t \equiv \sqrt{\frac{251}{21 \cdot n} \sum_{i=1}^{21 \cdot n} \Delta F_{t+1-i}^2},$$

where ΔF_t denotes the change at t of the forward swap rate for a n-month forward starting swap with 10-year tenor.

---

[8]In the equity literature, one usually defines a variance risk premium as the difference between the expectation of future realized variance under the risk-neutral and the physical probabilities (see, e.g., Bollerslev et al. 2009). Our notion of variance risk premium is consistent with the purpose of defining payoffs that have zero value under the market probability, as is the case with π_{t+S}^{incr}. The expectation of π_{t+S}^{incr} under the physical probability *is* the variance risk premium as usually defined in the literature, although we shall keep on referring to π_{t+S}^{incr} as variance risk premium.

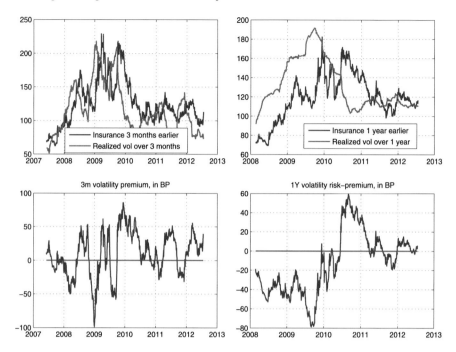

Fig. 2.4 Price of insurance in the interest rate swap volatility space (CBOE SRVIX), realized basis point *incremental* volatility, and swap volatility premiums for three month (*left panels*) and one year (*right panels*) horizons

Estimating asset price volatility is one of the most important and studied topics in financial econometrics (Engle 2004). Andersen et al. (2010) and Aït-Sahalia and Jacod (2014) survey the literature on realized variance and related measurement methods. One complication arising whilst dealing with fixed income market volatility is that the squared changes of variables of interest, such as forward swap rates, may be unobservable due to the lack of constant maturity contracts. To calculate realized volatility, we follow Mele et al. (2015a), and estimate missing forward swap rates through linear interpolation.

The annualized realized point-to-point volatility in Eq. (2.29) is simply

$$\text{Vol}_{\text{p-t-p},t+S} \equiv \sqrt{\frac{(R_{t+S} - F_t(S))^2}{S}},$$

where $F_t(S)$ denotes the forward swap rate at t for a forward starting swap at $t + S$ with ten year tenor, and $R_{t+S} = F_{t+S}(0)$ is the spot swap rate at time $t + S$.

Figures 2.4 and 2.5 plot the price of volatility (SRVIX) and realized volatility, along with *volatility* risk premiums defined as $\hat{\pi}_{t+S}^{\text{incr}} \equiv \text{SRVIX}_t(S) - \text{Vol}_{t+S}$ and $\hat{\pi}_{t+S}^{\text{p-t-p}} \equiv \text{SRVIX}_t(S) - \text{Vol}_{\text{p-t-p},t+S}$. For comparison, Fig. 2.6 depicts equity counterparts, in percentage, utilizing data on the VIX and the three-month horizon

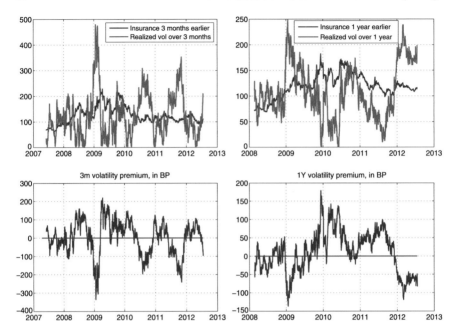

Fig. 2.5 Price of insurance in the interest rate swap volatility space (CBOE SRVIX), realized basis point *point-to-point* volatility, and swap volatility premiums for three month (*left panels*) and one year (*right panels*) horizons

VXV.[9] Finally, Fig. 2.7 provides volatility risk premiums in the Treasury space based on the CBOE/CBOT TYVIX index (see Chap. 4) and realized volatility of the underlying 10-year Treasury Note future price.

Note that the persistence of the volatility risk premium increases with the horizon length in both interest rate and equity cases. Figures 2.4 and 2.6 suggest that at a three month horizon, volatility risk premiums in swap and equity markets display similar persistence properties, although they do not quite react in the same way in times of distress. Mele et al. (2015a) provide further analysis of these issues in swap and equity markets.

2.6 Skew Shifts and the Dynamics of Volatility Indexes

In practice, there are only a finite number options for calculating a volatility index. This section analyzes theoretical properties of the index in the presence of unavoidable truncations. It also develops a few basic numerical experiments that illustrate these properties.

---

[9]Fornari (2010) documents early estimates of volatility risk premiums in the interest rate swap space. His estimates regard percentage volatility, not basis point as in this section, and rely on proxies for model-free implied volatility based on the standard equity methodology instead of the interest rate methodology, which was subsequent to his work.

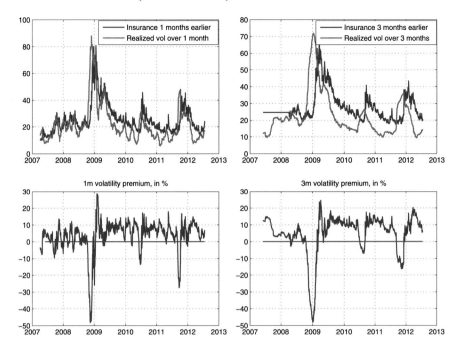

Fig. 2.6 Price of insurance in the equity volatility space (CBOE VIX and VXV), realized percentage volatility, and equity volatility premiums for one month (*left panels*) and three month (*right panels*) horizons

Fig. 2.7 Price of insurance in the government bond volatility space (CBOE/CBOT TYVIX), realized percentage volatility, and volatility premiums for one month horizons. The TYVIX is referenced to one month volatility in the 10-year Treasury Note Future price

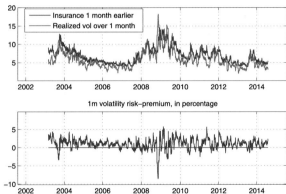

2.6.1 Truncations

Consider the following approximations to the expressions for V^{bp} and V in Eq. (2.6) and Eq. (2.7),

$$V_\ell^{bp} \equiv \frac{2}{N} \left(\int_{X-\ell}^{X} \mathrm{Put}\big(X, K, \sigma(X, K)\big) dK + \int_{X}^{X+\ell} \mathrm{Call}\big(X, K, \sigma(X, K)\big) dK \right),$$
$$(2.32)$$

and

$$V_\ell \equiv \frac{2}{N}\left(\int_{X-\ell}^{X}\frac{\text{Put}(X,K,\sigma(X,K))}{K^2}dK + \int_{X}^{X+\ell}\frac{\text{Call}(X,K,\sigma(X,K))}{K^2}dK\right),$$
(2.33)

where $\ell \in (0,X)$ is a constant, and $\sigma(X,K)$ is Black's implied volatility introduced in Sect. 2.4.

The indexes in Eqs. (2.32) and (2.33) are calculated with a strip of options centered at X over a range equal to 2ℓ. In Appendix A.5, we show that for all $\ell \in (0,X)$:

$$\frac{\partial V_\ell^{\text{bp}}}{\partial X} = \frac{2}{X}V_\ell^{\text{bp}} - \frac{2\ell}{NX}\left(\text{Put}(X,K,\sigma(X,K))\right)\big|_{K=X-\ell}$$
$$+ \text{Call}(X,K,\sigma(X,K))\big|_{K=X+\ell}),$$
(2.34)

and

$$\frac{\partial V_\ell}{\partial X} = -\frac{2\ell}{NX}\left(\frac{\text{Put}(X,K,\sigma(X,K))|_{K=X-\ell}}{(X-\ell)^2} + \frac{\text{Call}(X,K,\sigma(X,K))|_{K=X+\ell}}{(X+\ell)^2}\right).$$
(2.35)

According to Proposition 2.3 in Sect. 2.4, the theoretical percentage index is unresponsive to movements in the forward. Instead, Eq. (2.35) shows that its approximation based on a finite strip of option prices, V_ℓ in Eq. (2.33), moves inversely with X: that is, V_ℓ moves even if the skew remains the same. We now discuss these issues in detail.

2.6.1.1 The Effects on the BP Volatility Index

How does the weighting scheme affect the index behavior in the presence of approximations? It is instructive to analyze the basis point index first. Consider a market in which uncertainty is constant but the forward X increases from X_0 to X_1, say, with $X_1 - X_0 < \ell$. The usable strip of option prices is then re-centered towards the right tail of the available strike distribution, with (i) a new set of call prices entering into the index calculations (those with strikes between $X_0 + \ell$ and $X_1 + \ell$), (ii) a set of call prices leaving the index basis (those with strikes between X_0 and X_1), (iii) a new set of put prices entering into the index (those with strikes between X_0 and X_1), and, finally, (iv) a set of put prices leaving the index basis (those with strikes between $X_0 - \ell$ and $X_1 - \ell$). The index value changes due to the options entering and leaving the index as described *and* due to the change in value of the options remaining in the index. Marginally, when $X_1 - X_0$ is small, we have that

$$\frac{N}{2}\frac{\partial V_\ell^{\text{bp}}}{\partial X} = \overbrace{\text{Put}(X,K,\sigma(X,K))\big|_{K=X}}^{\text{new puts (iii)}} - \overbrace{\text{Put}(X,K,\sigma(X,K))\big|_{K=X-\ell}}^{\text{old puts (iv)}}$$
$$+ \int_{X-\ell}^{X}\partial_X\text{Put}(X,K,\sigma(X,K))dK$$

$$+ \underbrace{\mathrm{Call}\big(X, K, \sigma(X, K)\big)\big|_{K=X+\ell}}_{\text{new calls (i)}} - \underbrace{\mathrm{Call}\big(X, K, \sigma(X, K)\big)\big|_{K=X}}_{\text{old calls (ii)}}$$

$$+ \int_{X}^{X+\ell} \partial_X \mathrm{Call}\big(X, K, \sigma(X, K)\big) dK, \tag{2.36}$$

where ∂_X denotes the total derivative, e.g. $\partial_X \mathrm{Put}(X, K, \sigma(X, K)) = \mathrm{Put}_X(X, K, \sigma(X, K)) + \mathrm{Put}_\sigma(X, K, \sigma(X, K))\sigma_X(X, K)$, and subscripts denote partial derivatives. Naturally, the two sets in (ii) and (iii) collapse to ATM call and put prices; therefore, their combined effect is zero.

In Appendix A.5, we show that if the skew has the sticky delta property, Eq. (2.36) is indeed consistent with Eq. (2.34). Equation (2.34) shows that the sign of $\frac{\partial V_\ell^{bp}}{\partial X}$ depends on two terms. The first is always positive, and is intuitively so because V_ℓ^{bp} is proportional to X^2 just like its theoretical counterpart (see Proposition 2.3 in Sect. 2.4). Consistent with this intuition, in Appendix A.5 we show that the approximating BP variance index, $VX_{\ell,t}^{bp}(T)$, is

$$VX_{\ell,t}^{bp}(T) = X \times \sqrt{(T - t)^{-1}\xi_{\ell,X}}, \tag{2.37}$$

for a function $\xi_{\ell,X}$ of the forward risk (see Eq. (A.12) in Appendix A.5). Instead, the second term in Eq. (2.34) is negative, although it is likely significantly less than the first term in absolute value.

2.6.1.2   The Percentage Index

The percentage index behaves quite differently, as $\frac{\partial V_\ell}{\partial X}$ in Eq. (2.35) is *always* negative. That is, a percentage volatility index (such as the VIX or TYVIX (see Chap. 4)) can change (driven by movements of X) even while uncertainty is constant. Naturally, a drop in the market level may well be accompanied by increased uncertainty, but the analysis of this section identifies an additional, mechanical effect, arising from the index reliance on a moving window of out-of-the-money option prices. We now develop numerical experiments and illustrate these properties based on realistic market data inputs.

2.6.2   Numerical Experiments and Interpretation of Actual Index Behavior

We consider numerical experiments in which we take as given a log-normal skew, representing hypothetical conditions in the swaption market as of December 21, 2011, and summarized by Table 2.1. Table 2.1 also provides the values of the percentage and BP volatility indexes, $VX_{\ell,t}(T)$ and $VX_{\ell,t}^{bp}(T)$, and the value of the BP ATM implied volatility.

Table 2.1 Black's skew and volatility indexes

	atm								
K	0.305	1.305	1.805	2.055	2.305	2.555	2.805	3.305	4.305
Black Vol	79.7	52.9	47.5	45.7	44.4	43.5	42.9	42.2	41.7

$\mathrm{VX}_{\ell,t}(T) = 51.2536, \quad \mathrm{VX}^{\mathrm{bp}}_{\ell,t}(T) = 108.3121, \quad \mathrm{BP\ ATM} = 102.3420$

Fig. 2.8 The Black's skew in Table 2.1 (*circles*) along with the cubic lines fit (*dashed line*)

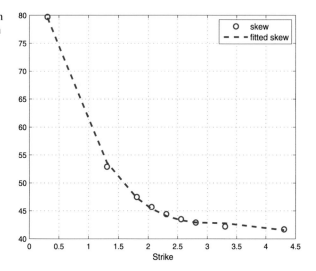

We fit the skew in Table 2.1 with a cubic spline, obtaining the results depicted in Fig. 2.8 where the fitted skew in Fig. 2.8 is denoted by $\hat{\sigma}_K$, and is a continuous function of the strike swap rate K.

In the first experiment, we fix uncertainty and vary the forward rate from 2.10 to 3. For each value of the forward X, we select a set of strikes centered at the forward swap rate X and the same range and coarseness as those in Table 2.1, and calculate a discretized version of Eqs. (2.32) and (2.33) using as an input the fitted skew $\hat{\sigma}_K$.

The right panel of Fig. 2.9 shows that the approximating percentage volatility index is inversely related to the forward, even though its theoretical value is not, as established by Proposition 2.3. The left panel of Fig. 2.9 shows, instead, that the approximating index $\mathrm{VX}^{\mathrm{bp}}_{\ell,t}(T)$ in Eq. (2.37) increases roughly linearly with X, as the theoretical value predicts in Eq. (2.28). The left hand panel also depicts the BP ATM volatility, obtained as,

$$\mathrm{BP\ ATM} \equiv 100 \times (44.4 \cdot X). \tag{2.38}$$

Note that when uncertainty is fixed, the relative magnitude of the BP volatility index vis-à-vis the BP ATM volatility could change. In this example, the BP volatility index is lower than the ATM only when the forward swap rate is relatively high.

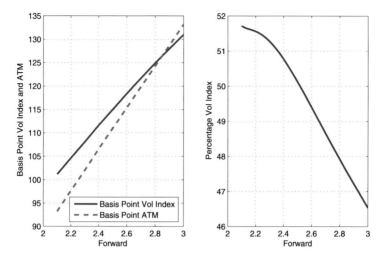

Fig. 2.9 *Left panel*: A Basis Point expected volatility index, calculated as in Eq. (2.32), compared to the Basis Point ATM volatility, depicted as a function of the forward. *Right panel*: A percentage expected volatility index, calculated as in Eq. (2.33). The indexes in the left and right panel are calculated using a strip of swaption prices and the skew in Table 2.1

Fig. 2.10 The forward swap rate for 1 year maturity and 10 year tenor (1Y-10Y) versus the CBOE SRVIX index regarding expected volatility of the 1Y-10Y forward rate

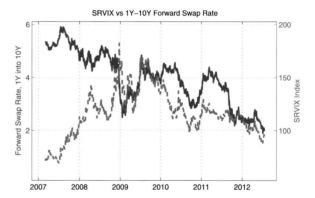

We emphasize that this exercise is one of comparative statics. We would expect that uncertainty also changes while the forward swap rate changes.

Accordingly, we consider an additional experiment in which we vary uncertainty, assuming that the swaption market experiences a parallel and positive shift in the skew of 35 percentage points, which is a realistic figure over periods of stress such as those experienced during the 2007–2009 crisis (see Sect. 3.7 in Chap. 3).

A negative change in the forward swap rate coupled with increased uncertainty can be interpreted as the result of an aggressive monetary policy action aimed at stabilizing market expectations and liquidity conditions. An historical instance of these events occurred in the last months of 2008 that followed Lehman Brothers' collapse, which culminated with a spike in both the CBOE SRVIX index (an upward spike) and the forward swap rate (a downward spike) (see Fig. 2.10, which reproduces the top panel of Fig. 2.2).

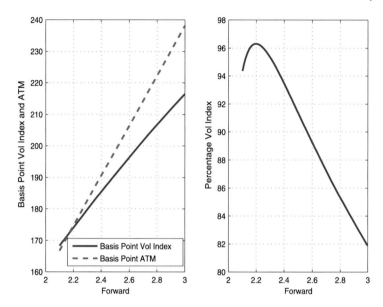

Fig. 2.11 *Left panel*: A Basis Point expected volatility index, calculated as in Eq. (2.32), compared to the Basis Point ATM volatility, depicted as a function of the forward. *Right panel*: A percentage expected volatility index, calculated as in Eq. (2.33). The indexes in the left and right panel are calculated using a strip of swaption prices and the skew in Table 2.1 increased by a positive parallel shift of 35 percentage points

Finally, an increase in the forward swap rate associated with increased uncertainty can be interpreted as the result of decreased risk appetite, a credit crunch, or a combination of the two, such as during some periods in the early part of 2009.

Figure 2.11 illustrates that for each level of the forward, the expected volatility indexes increase after a positive parallel shift in the skew underlying the calculations in Fig. 2.9. Interestingly, the BP volatility index increases less than the ATM for most of the possible range of variation of the forward. An increase in the skew and, hence, an increase in the ATM implied volatility from a value of 44.4 to a value of 79.4, leads to a higher slope of 79.5 in the BP ATM of Eq. (2.38).

In contrast, Figs. 2.9 and 2.11 reveal that the BP expected volatility increases simply through a positive parallel shift for each X. This property is to be expected because, as Fig. 2.9 suggests in this example, $VX_{\ell,t}^{bp}(T)$ in Eq. (2.37) increases roughly linearly with X; that is, the term $\xi_{\ell,X}$ moves primarily through changes in uncertainty. The behavior of the percentage index is quite different as it is always higher than the ATM volatility in the examples depicted in Fig. 2.11 (right panel). Figure 2.12 depicts the spread of the expected volatility indexes versus their ATM counterparts. The experiments in this section predict that during periods of distress, a percentage volatility *spread* (index minus ATM) peaks up as can be seen in the right panel of Fig. 2.12. In contrast, assuming interest rates are relatively unresponsive, a BP interest rate volatility can even change sign.

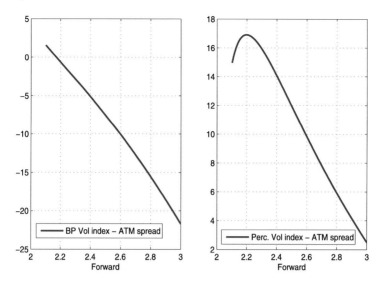

Fig. 2.12 *Left panel*: The spread between the Basis Point volatility index calculated as in Eq. (2.32) and Basis Point ATM volatility, depicted as a function of the forward. *Right panel*: The spread between the percentage volatility index calculated as in Eq. (2.33) and ATM percentage volatility, depicted as a function of the forward

2.7 Jumps

This section examines pricing and indexing of expected volatility in markets where the forward risk is a jump-diffusion process with stochastic volatility:

$$\frac{dX_\tau}{X_\tau} = -\left(\mathbb{E}_\tau^{Q^N}\left(e^{j_\tau}-1\right)\eta_\tau\right)d\tau + \sigma_\tau \cdot dW_\tau + \left(e^{j_\tau}-1\right)dJ_\tau, \quad \tau \in [t, T], \quad (2.39)$$

where σ_τ is a diffusion component, adapted to W_τ, J_τ is a Cox process under Q^N with intensity equal to η_τ, and j_τ is the logarithmic jump size.[10] By applying Itô's lemma for jump-diffusion processes to Eq. (2.39), we have that

$$d \ln X_\tau = (\cdots)d\tau + \sigma_\tau \cdot dW_\tau + j_\tau dJ_\tau, \quad (2.40)$$

for a drift function given in Appendix A.6 (see Eq. (A.18)).

Next, define the realized variance of the *arithmetic changes* of the forward risk over $[t, T]$, as

$$V_J^{\mathrm{bp}}(t, T) \equiv \int_t^T X_\tau^2 \|\sigma_\tau\|^2 d\tau + \int_t^T X_\tau^2 \left(e^{j_\tau}-1\right)^2 dJ_\tau \quad (2.41)$$

---

[10]See, e.g., Jacod and Shiryaev (1987, pp. 142–146), for a succinct discussion of jump-diffusion processes.

and the realized variance of the *logarithmic changes* of the forward risk over $[t, T]$
as

$$V_J(t, T) \equiv \int_t^T \|\sigma_\tau\|^2 d\tau + \int_t^T j_\tau^2 dJ_\tau. \tag{2.42}$$

The definitions in Eqs. (2.41) and (2.42) generalize those in Eqs. (2.3). In Appendix A.6, we show the remarkable property that the fair value K_Y of the *basis point* variance swaps in Proposition 2.2 is resilient to the presence of jumps, in that $K_{J,Y} = K_Y$ where $K_{J,Y}$ denotes the fair value of the contract in a market with jumps and K_Y is as in Eq. (2.6). Accordingly, the basis point index of expected volatility, $\mathrm{VX}_{Jt}^{bp}(T)$, is the same as that we derived in the absence of jumps:

$$\mathrm{VX}_{Jt}^{bp}(T) = \mathrm{VX}_t^{bp}(T), \tag{2.43}$$

where $\mathrm{VX}_t^{bp}(T)$ is as in Eq. (2.22).

In contrast, the fair value of a *percentage* variance swap is

$$K_{J,Y} \equiv K_Y - 2\mathbb{E}_t^{Q^N}\left[\int_t^T \left(e^{j_\tau} - 1 - j_\tau - \frac{1}{2}j_\tau^2 \right) dJ_\tau \right], \tag{2.44}$$

where K_Y is as in Eq. (2.7) of Proposition 2.2. Suppose, for example, that the distribution of jumps is skewed towards negative values. The fair value $K_{J,Y}$ should then be higher than it would be in the absence of jumps.

To illustrate, assume that the distribution of jumps collapses to a single point, $\bar{j} < 0$ say, and that the jump intensity equals some positive constant $\bar{\eta}$, in which case the fair value in Eq. (2.44) collapses to, $K_{J,Y} = K_J + 2(T - t) \cdot \bar{\eta} \mathcal{J}$, where $\mathcal{J} \equiv -(e^{\bar{j}} - 1 - \bar{j} - \frac{1}{2}\bar{j}^2) > 0$. In this example, the percentage volatility index is

$$\mathrm{VX}_{Jt}^p(T) = \sqrt{\mathrm{VX}_t^p(T) + 2\bar{\eta}\mathcal{J}},$$

where $\mathrm{VX}_t^p(T)$ is as in Eq. (2.22).

Remark 2.4 Carr and Wu (2008) derive an expression for a *percentage* variance swap strike incorporating information about jumps, which Eq. (2.44) generalizes to general market numéraires. Mele and Obayashi (2012) derive the "jumps irrelevance result" in Eq. (2.43) in the context of interest rate swap markets. This result is extended to general numéraires in this chapter.

Remark 2.5 Consider the definition of $V_J(t, T)$ in Eq. (2.42),

$$V_J(t, T) \equiv \underbrace{\int_t^T \|\sigma_\tau\|^2 d\tau}_{\equiv \mathcal{V}(t,T)} + \underbrace{\int_t^T j_\tau^2 dJ_\tau}_{\equiv \mathcal{J}(t,T)}.$$

In the statistics literature, one usually refers to $V_J(t, T)$ as the *total variation* of $\ln X_\tau$ over the interval of time $[t, T]$ (an analogous definition can be provided regarding $V_J^{\mathrm{bp}}(t, T)$). Thus, $\mathcal{V}(t, T)$ is the contribution to the total variation of $\ln X_\tau$ due to its continuous component, whereas $\mathcal{J}(t, T)$ is the jump contribution. In this book, we shall still refer to $V_J(t, T)$ as *variance* rather than *variation* to keep the presentation simple. Aït-Sahalia and Jacod (2014) summarize the state of the art regarding filtering methods for both the continuous and jump contributions of a process in a general context and in the presence of market microstructure noise.

Appendix A: Appendix on Security Design and Volatility Indexing

A.1 Proof of Proposition 2.2

We begin with the following preliminary result, from which Eqs. (2.6)–(2.7) follow for $Y_t = N_t$. Note that the arguments in this proof rely on spanning arguments similar to those utilized by Bakshi and Madan (2000) and Carr and Madan (2001) in the equity case, although centered around the notion of a market numéraire. We then provide the proof of the proposition with general stochastic multipliers.

Lemma A.1 *We have*:

$$\mathbb{E}_t^{Q^N}\left(V^{\mathrm{bp}}(t, T)\right) = \frac{2}{N_t}\left(\int_0^{X_t} \mathrm{Put}_t(K)dK + \int_{X_t}^{\infty} \mathrm{Call}_t(K)dK\right), \qquad (A.1)$$

and

$$\mathbb{E}_t^{Q^N}\left(V(t, T)\right) = \frac{2}{N_t}\left(\int_0^{X_t} \frac{\mathrm{Put}_t(K)}{K^2}dK + \int_{X_t}^{\infty} \frac{\mathrm{Call}_t(K)}{K^2}dK\right). \qquad (A.2)$$

Proof We provide the proof of Eq. (A.1), as that of Eq. (A.2) follows as a special case of the arguments leading to Eq. (A.19) and Eq. (A.20) in Appendix A.5 regarding the jump-diffusion case. By Itô's lemma,

$$\mathbb{E}_t^{Q^N}\left(V^{\mathrm{bp}}(t, T)\right) = \mathbb{E}_t^{Q^N}\left(X_T^2 - X_t^2\right). \qquad (A.3)$$

Moreover, by a Taylor expansion with remainder,

$$X_T^2 - X_t^2 = 2X_t(X_T - X_t) + 2\left(\int_0^{X_t}(K - X_T)^+ dK + \int_{X_t}^{\infty}(X_T - K)^+ dK\right). \qquad (A.4)$$

Multiplying both sides of the previous equation by $e^{-\int_t^T r_u du} N_T$, and taking expectation under the risk-neutral probability leaves

$$
\begin{aligned}
\mathbb{E}_t\left(e^{-\int_t^T r_u du} N_T \left(X_T^2 - X_t^2\right)\right) &= 2X_t \mathbb{E}_t\left(e^{-\int_t^T r_u du} N_T (X_T - X_t)\right) \\
&+ 2\left(\int_0^{X_t} \mathbb{E}_t\left(e^{-\int_t^T r_u du} N_T (K - X_T)^+\right) dK\right. \\
&\left. + \int_{X_t}^{\infty} \mathbb{E}_t\left(e^{-\int_t^T r_u du} N_T (X_T - K)^+\right) dK\right) \\
&= 2\left(\int_0^{X_t} \text{Put}_t(K) dK + \int_{X_t}^{\infty} \text{Call}_t(K) dK\right),
\end{aligned}
$$
(A.5)

where the last line follows by a change of probability, from Q to Q^N,

$$
\left.\frac{dQ^N}{dQ}\right|_{\mathbb{F}_T} = \frac{e^{-\int_t^T r_u du} N_T}{N_t},
$$

the martingale property of X_τ under Q^N, and the expressions for $\text{Put}_t(K)$ and $\text{Call}_t(K)$ in Definition 2.2. By the assumption that N_τ is the price of a traded asset, and $N_\tau > 0$, dQ^N integrates to one. Similarly, by a change of probability,

$$
\mathbb{E}_t\left(e^{-\int_t^T r_u du} N_T \left(X_T^2 - X_t^2\right)\right) = N_t \mathbb{E}_t^{Q^N}\left(X_T^2 - X_t^2\right).
$$
(A.6)

Combining Eqs. (A.5) and (A.6) with Eq. (A.3) yields Eq. (A.1).   □

Next, we prove the claims of Proposition 2.2 regarding basis point variance, $V^{\text{bp}}(t, T)$; those for percentage variance $V(t, T)$ follow through a mere change in notation. We only prove the "only if" part, as the "if" part is trivial from the derivation of the proof to follow. Consider the Radon–Nikodym derivative of Q^Y against Q^N,

$$
\zeta_T \equiv \left.\frac{dQ^Y}{dQ^N}\right|_{\mathbb{F}_T},
$$

and suppose on the contrary that there exists a stochastic multiplier Y_T such that

$$
\text{cov}^{Q^N}\left(V^{\text{bp}}(t, T), \zeta_T\right) \neq 0,
$$

and that at the same time,

$$
\mathbb{E}_t^{Q^Y}\left(V^{\text{bp}}(t, T)\right) = \mathbb{E}_t^{Q^N}\left(V^{\text{bp}}(t, T)\right).
$$

In this case, Eqs. (A.3) and (A.5) would imply that the conclusions of Lemma A.1 hold. However, we also have:

$$\mathbb{E}_t^{Q^Y}\left(V^{\mathrm{bp}}(t,T)\right) = \mathbb{E}_t^{Q^N}\left(V^{\mathrm{bp}}(t,T)\right) + \mathrm{cov}^{Q^N}\left(V^{\mathrm{bp}}(t,T),\zeta_T\right).$$

Then, $\mathrm{cov}^{Q^N}(V^{\mathrm{bp}}(t,T),\zeta_T) = 0$, a contradiction. Proposition 2.2 follows by Proposition 2.1 in the main text and Lemma A.1. $\square$

A.2 A Stochastic Multiplier Beyond the Market NumÉraire

Consider the following stochastic multiplier, $Y_T = N_T \epsilon_T$, where ϵ_T is $\mathbb{F}_T$-measurable, and such that $\mathrm{cov}^{Q^N}(V^{\mathrm{bp}}(t,T),\epsilon_T) = 0$. We have,

$$\zeta_T \equiv \frac{dQ^Y}{dQ^N}\bigg|_{\mathbb{F}_T} = c_t \epsilon_T, \quad \text{with } c_t \equiv \frac{\mathbb{E}_t(e^{-\int_t^T r_u du} N_T)}{\mathbb{E}_t(e^{-\int_t^T r_u du} N_T \epsilon_T)}.$$

Heuristically,

$$\zeta_T = \left(\frac{dQ^Y}{dQ} : \frac{dQ^N}{dQ}\right)\bigg|_{\mathbb{F}_T} = \frac{e^{-\int_t^T r_u du} Y_T}{\mathbb{E}_t(e^{-\int_t^T r_u du} Y_T)} : \frac{e^{-\int_t^T r_u du} N_T}{\mathbb{E}_t(e^{-\int_t^T r_u du} N_T)} = \epsilon_T c_t.$$

Next, we claim that

$$\mathrm{cov}^{Q^N}\left(V^{\mathrm{bp}}(t,T),\zeta_T\right) = c_t \cdot \mathrm{cov}^{Q^N}\left(V^{\mathrm{bp}}(t,T),\epsilon_T\right) = 0,$$

as in the class of multipliers identified by Proposition 2.2. Indeed, we have:

$$\mathbb{E}_t^{Q^Y}\left(V^{\mathrm{bp}}(t,T)\right) = c_t \mathbb{E}_t^{Q^N}(\epsilon_T) \mathbb{E}_t^{Q^N}\left(V^{\mathrm{bp}}(t,T)\right) = \mathbb{E}_t^{Q^N}\left(V^{\mathrm{bp}}(t,T)\right),$$

where the first equality follows by the fact that ϵ_T is uncorrelated with $V^{\mathrm{bp}}(t,T)$, and the second follows by the definition of c_t,

$$\mathbb{E}_t^{Q^N}(\epsilon_T) = \frac{1}{N_t}\mathbb{E}_t\left(e^{-\int_t^T r_u du} N_T \epsilon_T\right) = \frac{1}{c_t}.$$

A.3 Vega and Gamma in Gaussian Markets

We first show Eq. (2.9) ("Constant volatility"), then prove that the statement in (2.21) is true ("Constant vega") and, finally, validate our claims regarding the implications on gamma of option portfolios ("Constant gamma exposure"). In what follows, we assume that the portfolio weightings and their first order derivative are integrable with respect to the Gaussian distribution, (i) $E(|\omega(\tilde{y})|) < \infty$ and $E(\int_{-\infty}^{0} |\omega(u)u|\Phi(u)du) < \infty$, and (ii) $E(|\omega'(\tilde{y})|) < \infty$, where $\tilde{y}$ is Gaussian. The first two conditions are necessary for the existence of an abstract volatility index based on at-the-money and out-of-the-money options in a Gaussian market (see Eqs. (A.7) and (A.8) below). The second is a regularity condition needed while dealing with boundedness of the sensitivity of vega with respect to the forward risk, X_t.

We shall rely on the following result. Consider the market in Sects. 2.2 and 2.3. If the forward risk X_t is a solution to Eq. (2.8), the prices of put and call options are given by the "Bachelier formulae" (see Eq. (2.24) in the main text):

$$
\begin{aligned}
\mathcal{O}_t^P(X_t, K, T, \hat{\sigma}_n) &\equiv N_t \cdot \mathcal{Z}_t^P(X_t, K, T, \hat{\sigma}_n), \\
\mathcal{O}_t^C(X_t, K, T, \hat{\sigma}_n) &\equiv N_t \cdot \mathcal{Z}_t^C(X_t, K, T, \hat{\sigma}_n),
\end{aligned}
\tag{A.7}
$$

where $\hat{\sigma}_n \equiv \sqrt{\|\sigma_n\|^2}$,

$$
\mathcal{Z}_t^P(X, K, T, \sigma) = (K - X)\Phi\left(\frac{K - X}{\sigma\sqrt{T - t}}\right) + \sigma\sqrt{T - t}\phi\left(\frac{X - K}{\sigma\sqrt{T - t}}\right),
$$

$$
\mathcal{Z}_t^C(X, K, T, \sigma) = (X - K)\Phi\left(\frac{X - K}{\sigma\sqrt{T - t}}\right) + \sigma\sqrt{T - t}\phi\left(\frac{X - K}{\sigma\sqrt{T - t}}\right),
$$

and ϕ denotes the standard normal density.

CONSTANT VOLATILITY. Plugging Eq. (A.7) into Eq. (2.6) leaves, after substituting the expressions $\text{Put}_t(K) \equiv \mathcal{O}_t^P(X, K, T, \hat{\sigma}_n)$ and $\text{Call}_t(K) \equiv \mathcal{O}_t^C(X, K, T, \hat{\sigma}_n)$, and extending the left limit of the first integral in Eq. (2.6) to $-\infty$,

$$
\begin{aligned}
\frac{1}{2} K_Y &= \sigma^2(T - t) + \int_{-\infty}^{X} (K - X)\Phi\left(\frac{K - X}{\sigma\sqrt{T - t}}\right) dK \\
&\quad + \int_{X}^{\infty} (X - K)\Phi\left(\frac{X - K}{\sigma\sqrt{T - t}}\right) dK \\
&= \sigma^2(T - t)\left[1 + 2\int_{-\infty}^{0} u\Phi(u)du\right] \\
&= \sigma^2(T - t)\left[1 + 2\left(-\frac{1}{2}\int_{-\infty}^{0} u^2\phi(u)du\right)\right] \\
&= \frac{1}{2}\sigma^2(T - t),
\end{aligned}
$$

where the first equality follows by the property of the normal distribution that, $\int_{-\infty}^{\infty} \phi(\frac{X-K}{v})dK = v$ for $v > 0$, and the second holds by a change in variables, the third by an integration by parts, and the fourth by a basic property of the standard normal distribution.

CONSTANT VEGA. We prove that in Gaussian markets, a portfolio with all out-of-the-money and at-the-money European options has constant vega if and only if these options are equally weighted. The value of the portfolio we consider is a special case of Eq. (2.19),

$$\pi_t(X, T, \sigma)$$

$$= N_t \left(\int_{-\infty}^{X} \omega(K) \mathcal{Z}_t^{\mathrm{P}}(X, K, T, \sigma) dK + \int_{X}^{\infty} \omega(K) \mathcal{Z}_t^{\mathrm{C}}(X, K, T, \sigma) dK \right),$$

(A.8)

where $\mathcal{Z}_t^{\mathrm{P}}$ and $\mathcal{Z}_t^{\mathrm{C}}$ are as in Eqs. (A.7).

Note that the vega of a put is the same as the vega of a call, and equals $N_t v_t^{\mathcal{O}}(X, K, T, \sigma)$, where:

$$v_t^{\mathcal{O}}(X, K, T, \sigma) \equiv \frac{\partial \mathcal{Z}_t^{\mathrm{P}}(X, K, T, \sigma)}{\partial \sigma} = \frac{\partial \mathcal{Z}_t^{\mathrm{C}}(X, K, T, \sigma)}{\partial \sigma}$$

$$= \sqrt{T - t} \phi \left(\frac{X - K}{\sigma \sqrt{T - t}} \right),$$

so that the vega of the portfolio is:

$$v_t(X, T, \sigma) \equiv \frac{\partial \pi_t(X, T, \sigma)}{\partial \sigma} = N_t \sqrt{T - t} \int \omega(K) \phi \left(\frac{X - K}{\sigma \sqrt{T - t}} \right) dK. \quad \text{(A.9)}$$

As for the "if" part in (2.21), let $\omega(K) = \mathrm{const.}$, so that by Eq. (A.9), and the fact that the Gaussian density ϕ integrates to one, we have that the vega is independent of X:

$$v_t(X, T, \sigma) = N_t \sqrt{T - t} \cdot \mathrm{const.}$$

As for the "only if" part, let us differentiate $v_t(X, T, \sigma)$ in Eq. (A.9) with respect to X:

$$\frac{\partial v_t(X, T, \sigma)}{\partial X} = -\frac{N_t}{\sigma^2 \sqrt{T - t}} \int \omega(K) \phi \left(\frac{X - K}{\sigma \sqrt{T - t}} \right) (X - K) dK.$$

We claim that the constant weighting is the only function ω independent of X, and such that $\frac{\partial v_t(X, T, \sigma)}{\partial X} = 0$. Suppose not, and note that $\frac{\partial v_t(X, T, \sigma)}{\partial X}$ is zero if and only if,

$$X \int \omega(K) \phi \left(\frac{X - K}{\sigma \sqrt{T - t}} \right) dK = \int K \omega(K) \phi \left(\frac{X - K}{\sigma \sqrt{T - t}} \right) dK.$$

Let us moreover define a random variable $\tilde{y} \sim N(\mu, \sigma^2(T - t))$. In terms of $\tilde{y}$, the previous equality is, by Stein's Lemma,

$$\mu E[\omega(\tilde{y})] = E[\tilde{y}\omega(\tilde{y})] = \mu E[\omega(\tilde{y})] + \mathrm{cov}[\tilde{y}, \omega(\tilde{y})]$$

$$= \mu E[\omega(\tilde{y})] + E[\omega'(\tilde{y})] \sigma^2(T - t),$$

which is a contradiction unless $\omega(\cdot)$ is constant.

Remark A.1 By Proposition 2.2, and previous results in this appendix, the normalized value of the portfolio in Eq. (A.8) *is* σ^2, in the special case $\omega(K) = 2$,

$$\sqrt{\frac{\pi_t(X_t, T, \sigma)}{N_t}} = \sigma\sqrt{T - t}.$$

We illustrate this fact numerically. We set $X_t = 5\,\%$, $T - t = 1$ (one year) and $\sigma = 150$ bps, and approximate the integral in Eq. (A.8),

$$100^2 \times \sqrt{\frac{\hat{\pi}_t(X_t, T, \sigma)}{N_t}}$$

$$\approx 100^2 \times \sqrt{2\left(\sum_{i:K_i < X_t} \mathcal{Z}_t^P(X_t, K_i, T, \sigma) + \sum_{i:K_i \geq X_t} \mathcal{Z}_t^C(X_t, K_i, T, \sigma)\right)\Delta K}$$

$$\approx 150.3907,$$

where $\min_i\{K_i\} = 0$, $\max_i\{K_i\} = 10\,\%$, $\Delta K = 0.0001$.

CONSTANT GAMMA EXPOSURE. Our claim of a constant gamma exposure in a Gaussian market follows because the option price in Eq. (A.7) satisfies,

$$\frac{\partial^2 \mathcal{O}_t^U(X, K, T, \sigma)}{\partial X^2} = N_t \frac{1}{\sigma(T - t)} v_t^{\mathcal{O}}(X, K, T, \sigma),$$

where $\mathcal{O}_t^U(X, K, T, \sigma)$, $U \in \{P, C\}$ denotes an out-of-the-money option price (see Eq. (A.7)). Therefore, we have

$$\frac{\partial^2 \pi_t(X, T, \sigma)}{\partial X^2} = \int \omega(K) \frac{\partial^2 \mathcal{O}_t^U(X, K, T, \sigma)}{\partial X^2} dK$$

$$= \frac{1}{\sigma(T - t)} \int \omega(K) \frac{\partial \mathcal{O}_t^U(X, K, T, \sigma)}{\partial \sigma} dK$$

$$= \frac{1}{\sigma(T - t)} \frac{\partial \pi_t(X, T, \sigma)}{\partial \sigma}. \tag{A.10}$$

The L.H.S. of this equation is independent of X if and only if the R.H.S. is. That is, the gamma exposure of the portfolio is constant if and only if the portfolio vega is independent of X, as claimed in the main text.

A.4 Proof of Proposition 2.3

For simplicity, we suppress the dependence of all the variables and functions on t and T. Assuming the zero homogeneity assumption is satisfied by the implied volatility $\sigma_K \equiv \sigma(X, K)$, we have that the two Black pricers, $\mathcal{P}(X, K, \sigma_K) \equiv$ Put(K) and $\mathcal{C}(X, K, \sigma_K) \equiv$ Call(K), collapse to

$$\mathcal{P}(X, K, \sigma_K) = K\varphi_{\mathrm{p}}^1(u) - X\varphi_{\mathrm{p}}^2(u), \qquad \mathcal{C}(X, K, \sigma_K) = X\varphi_{\mathrm{c}}^1(u) - K\varphi_{\mathrm{c}}^2(u),$$
$$(\mathrm{A}.11)$$

for some functions $\varphi_{\mathrm{p}}^i(x)$ and $\varphi_{\mathrm{c}}^i(x)$ of the moneyness $u \equiv \ln(\frac{K}{X})$. Substituting the two expressions in Eqs. (A.11) into Eq. (2.6) of Proposition 2.2, and making the change of variable $K \mapsto u$, leaves,

$$V^{\mathrm{bp}} = X^2 \cdot \xi,$$

$$\xi \equiv \frac{2}{N}\left(\int_{-\infty}^0 \left(e^u \varphi_{\mathrm{p}}^1(u) - \varphi_{\mathrm{p}}^2(u)\right) e^u du + \int_0^\infty \left(\varphi_{\mathrm{c}}^1(u) - e^u \varphi_{\mathrm{c}}^2(u)\right) e^u du \right),$$

where N is the market numéraire at t, and the function ξ is independent of X, establishing Part (i) of the proposition. Part (ii) is similar. Substituting $\mathcal{P}(X, K, \sigma_K)$ and $\mathcal{C}(X, K, \sigma_K)$ in Eq. (A.11) into Eq. (2.7) of Proposition 2.2, and by changing the variable of integration to $u = \ln \frac{K}{X}$, leaves

$$V = \frac{2}{N}\left(\int_{-\infty}^0 \left(\varphi_{\mathrm{p}}^1(u) - e^{-u}\varphi_{\mathrm{p}}^2(u)\right) du + \int_0^\infty \left(e^{-u}\varphi_{\mathrm{c}}^1(u) - \varphi_{\mathrm{c}}^2(u)\right) du \right),$$

which is independent of X. $\square$

A.5 Approximating Indexes

We derive Eq. (2.34) and Eq. (2.35). We begin with Eq. (2.35). Substituting Eq. (A.11) into Eq. (2.33) yields, for $\ell \in (0, X)$,

$$V_\ell = \frac{2}{N}\left(\int_{\ln(\frac{X-\ell}{X})}^0 \left(\varphi_{\mathrm{p}}^1(u) - e^{-u}\varphi_{\mathrm{p}}^2(u)\right) du + \int_0^{\ln(\frac{X+\ell}{X})} \left(e^{-u}\varphi_{\mathrm{c}}^1(u) - \varphi_{\mathrm{c}}^2(u)\right) du \right),$$

so that,

$$\frac{\partial V_\ell}{\partial X} = \frac{2}{N}\left(-\left(\varphi_{\mathrm{p}}^1(u) - e^{-u}\varphi_{\mathrm{p}}^2(u)\right)\big|_{u=\ln(\frac{X-\ell}{X})} \cdot \frac{\ell}{X(X-\ell)} \right.$$
$$\left. + \left(e^{-u}\varphi_{\mathrm{c}}^1(u) - \varphi_{\mathrm{c}}^2(u)\right)\big|_{u=\ln(\frac{X+\ell}{X})} \cdot \frac{-\ell}{X(X+\ell)} \right).$$

Utilizing the expressions in Eq. (A.11) delivers Eq. (2.35). The proof of Eq. (2.34) proceeds similarly: substitute Eq. (A.11) into Eq. (2.32) to obtain, for $\ell \in (0, X)$,

$$V_\ell^{\mathrm{BP}} = X^2 \times \xi_{\ell, X}, \qquad (\mathrm{A}.12)$$

where,

$$\xi_{\ell,X} \equiv \frac{2}{N}\left(\int_{\ln(\frac{X-\ell}{X})}^{0} \left(e^u \varphi_{\mathrm{p}}^1(u) - \varphi_{\mathrm{p}}^2(u)\right)e^u du + \int_{0}^{\ln(\frac{X+\ell}{X})} \left(\varphi_{\mathrm{c}}^1(u) - e^u \varphi_{\mathrm{c}}^2(u)\right)e^u du\right).$$

We have

$$\frac{\partial \xi_{\ell,X}}{\partial X} = \frac{2}{N}\left(-\left(e^u \varphi_{\mathrm{p}}^1(u) - \varphi_{\mathrm{p}}^2(u)\right)e^u\big|_{u=\ln(\frac{X-\ell}{X})} \cdot \frac{\ell}{X(X-\ell)}\right.$$

$$\left.+ \left(\varphi_{\mathrm{c}}^1(u) - e^u \varphi_{\mathrm{c}}^2(u)\right)\big|_{u=\ln(\frac{X+\ell}{X})} \cdot \frac{-\ell}{X(X+\ell)}\right)$$

$$= -\frac{2\ell}{NX^3}\left(\mathrm{Put}\left(X, K, \sigma(X,K)\right)\big|_{K=X-\ell} + \mathrm{Call}\left(X, K, \sigma(X,K)\right)\big|_{K=X+\ell}\right),$$

where the second equality follows by the expressions in Eq. (A.11). Equation (2.34) follows by straightforward differentiation of Eq. (A.12).

Next we show that Eqs. (2.34) and (2.36) are mutually consistent. We have

$$\frac{\partial}{\partial X}\int_{X-\ell}^{X} \mathrm{Put}\left(X, K, \sigma(X,K)\right)dK$$

$$= \mathrm{Put}\left(X, K, \sigma(X,K)\right)\big|_{K=X} - \mathrm{Put}\left(X, K, \sigma(X,K)\right)\big|_{K=X-\ell}$$

$$+ \int_{X-\ell}^{X} \partial_X \mathrm{Put}\left(X, K, \sigma(X,K)\right)dK$$

$$= \mathrm{Put}\left(X, K, \sigma(X,K)\right)\big|_{K=X} - \mathrm{Put}\left(X, K, \sigma(X,K)\right)\big|_{K=X-\ell}$$

$$+ \frac{1}{X}\int_{X-\ell}^{X} \mathrm{Put}\left(X, K, \sigma(X,K)\right)dK$$

$$- \frac{1}{X}\int_{X-\ell}^{X} K \cdot \partial_K \mathrm{Put}\left(X, K, \sigma(X,K)\right)dK, \tag{A.13}$$

where the second equality follows by the assumption the implied volatilities are homogenous of degree zero in (X, K), so that $\mathrm{Put}(\cdot) = X \cdot \partial_X \mathrm{Put}(\cdot) + K \cdot \partial_K \mathrm{Put}(\cdot)$. An integration by parts of the last term in Eq. (A.13) produces

$$\int_{X-\ell}^{X} K \cdot \partial_K \mathrm{Put}\left(X, K, \sigma(X,K)\right)dK$$

$$= X \cdot \mathrm{Put}\left(X, K, \sigma(X,K)\right)\big|_{K=X} - (X-\ell) \cdot \mathrm{Put}\left(X, K, \sigma(X,K)\right)\big|_{K=X-\ell}$$

$$- \int_{X-\ell}^{X} \mathrm{Put}\left(X, K, \sigma(X,K)\right)dK.$$

Substituting this term into Eq. (A.13) leaves

$$\frac{\partial}{\partial X}\int_{X-\ell}^{X}\text{Put}\big(X,K,\sigma(X,K)\big)dK$$

$$=\frac{2}{X}\int_{X-\ell}^{X}\text{Put}\big(X,K,\sigma(X,K)\big)dK-\frac{\ell}{X}\text{Put}\big(X,K,\sigma(X,K)\big)\big|_{K=X-\ell}.\quad\text{(A.14)}$$

Similarly,

$$\frac{\partial}{\partial X}\int_{X}^{X+\ell}\text{Call}\big(X,K,\sigma(X,K)\big)dK$$

$$=\frac{2}{X}\int_{X}^{X+\ell}\text{Call}\big(X,K,\sigma(X,K)\big)dK-\frac{\ell}{X}\text{Call}\big(X,K,\sigma(X,K)\big)\big|_{K=X+\ell}.$$

$$\text{(A.15)}$$

Equation (2.34) now follows by taking derivatives with respect to X in Eq. (2.32), using Eqs. (A.14)–(A.15), and rearranging terms.

A.6 Jumps

We derive the expression for the fair value of K_Y in Proposition 2.2 under the assumption that the forward risk X_t is a solution to the jump-diffusion process in Eq. (2.39).

BASIS POINT. Apply Itô's lemma for jump-diffusion processes to Eq. (2.39), obtaining,

$$\frac{dX_\tau^2}{X_\tau^2}=-2\big(\mathbb{E}_\tau^{Q^N}\big(e^{j_\tau}-1\big)\eta_\tau\big)d\tau+2\sigma_\tau\cdot dW_\tau+\|\sigma_\tau\|^2 d\tau+\big(e^{2j_\tau}-1\big)dJ_\tau$$

$$=-2\big(\mathbb{E}_\tau^{Q^N}\big(e^{j_\tau}-1\big)\eta_\tau\big)d\tau+2\big(e^{j_\tau}-1\big)dJ_\tau+2\sigma_\tau\cdot dW_\tau+\|\sigma_\tau\|^2 d\tau$$

$$+\big(e^{j_\tau}-1\big)^2 dJ_\tau.$$

By integrating, taking expectations under Q^N, and using the definition of basis point variance, $V_J^{\text{bp}}(t,T)$ in Eq. (2.41), leaves:

$$\mathbb{E}_t^{Q^N}\big(X_T^2-X_t^2\big)$$

$$=\underbrace{-2\mathbb{E}_t^{Q^N}\bigg(\int_t^T X_\tau^2\big(\mathbb{E}_\tau^{Q^N}\big(e^{j_\tau}-1\big)\eta_\tau\big)d\tau\bigg)+2\mathbb{E}_t^{Q^N}\bigg(\int_t^T X_\tau^2\big(e^{j_\tau}-1\big)dJ_\tau\bigg)}_{=0}$$

$$+\underbrace{2\mathbb{E}_t^{Q^N}\bigg(\int_t^T X_\tau^2\sigma_\tau^2\cdot dW_\tau\bigg)}_{=0}+\mathbb{E}_t^{Q^N}\big[V_J^{\text{bp}}(t,T)\big],\quad\text{(A.16)}$$

where the first term is zero as,

$$
\mathbb{E}_t^{Q^N}\left(\int_t^T X_\tau^2(e^{j_\tau}-1)dJ_\tau\right)=\mathbb{E}_t^{Q^N}\left(\int_t^T \mathbb{E}_\tau^{Q^N}\left(X_\tau^2(e^{j_\tau}-1)dJ_\tau\right)\right)
$$

$$
=\mathbb{E}_t^{Q^N}\left(\int_t^T X_\tau^2\left(\mathbb{E}_\tau^{Q^N}\left(e^{j_\tau}-1\right)\eta_\tau\right)d\tau\right).\quad(A.17)
$$

Comparing Eq. (A.16) with Eqs. (A.5) and (A.6) and then (2.6) leads to the conclusions of the main text, and in particular to Eq. (2.43).

PERCENTAGE. First, apply Itô's lemma to Eq. (2.39), obtaining Eq. (2.40), viz

$$
d\ln X_\tau = -\left(\mathbb{E}_\tau^{Q^N}\left(e^{j_\tau}-1\right)\eta_\tau\right)d\tau - \frac{1}{2}\|\sigma_\tau\|^2 d\tau + \sigma_\tau\cdot dW_\tau + j_\tau dJ_\tau
$$

$$
= -\frac{1}{2}\left(\|\sigma_\tau\|^2 d\tau + j_\tau^2 dJ_\tau\right) + \sigma_\tau\cdot dW_\tau - \left(\mathbb{E}_\tau^{Q^N}\left(e^{j_\tau}-1\right)\eta_\tau\right)d\tau
$$

$$
+ j_\tau dJ_\tau + \frac{1}{2}j_\tau^2 dJ_\tau,\quad(A.18)
$$

so that, by the definition of $V_J(t,T)$ in Eq. (2.42) and (by arguments similar to those leading to Eq. (A.17)),

$$
\mathbb{E}_\tau^{Q^N}\left(e^{j_\tau}-1\right)dJ_\tau = \left(\mathbb{E}_\tau^{Q^N}\left(e^{j_\tau}-1\right)\eta_\tau\right)d\tau,
$$

we obtain,

$$
-2\mathbb{E}_t^{Q^N}\left(\ln\frac{X_T}{X_t}\right) - 2\mathbb{E}_t^{Q^N}\left[\int_t^T\left(e^{j_\tau}-1-j_\tau-\frac{1}{2}j_\tau^2\right)dJ_\tau\right] = \mathbb{E}_t^{Q^N}\left(V_J(t,T)\right).
$$
$$(A.19)$$

Next, consider the standard Taylor's expansion with remainder,

$$
\ln\frac{X_T}{X_t} = \frac{1}{X_t}(X_T - X_t) - \left(\int_0^{X_t}\frac{1}{K^2}(K - X_T)^+ dK + \int_{X_t}^\infty\frac{1}{K^2}(X_T - K)^+ dK\right).
$$

Taking the expectation under Q^N yields,

$$
\mathbb{E}_t^{Q^N}\left(\ln\frac{X_T}{X_t}\right) = -\frac{1}{N_t}\left(\int_0^{X_t}\frac{1}{K^2}\mathrm{Put}_t(K)dK + \int_{X_t}^\infty\frac{1}{K^2}\mathrm{Call}_t(K)dK\right),\quad(A.20)
$$

where we have made use of the martingale property of X_τ under Q^N, and the expressions for $\mathrm{Put}_t(K)$ and $\mathrm{Call}_t(K)$ in Definition 2.2. Combining Eq. (A.19) and Eq. (A.20), and using Proposition 2.1, leaves the expression for $K_{J,Y}\equiv \mathbb{E}_t^{Q^N}[V_J(t,T)]$ in Eq. (2.44) of the main text.

Chapter 3
Interest Rate Swaps

3.1 Introduction

One key motivation of this book is that volatility in equity and fixed income markets behave quite differently, yet standardized measurement rooted in rigorous theory and related volatility derivatives for the latter have lagged in development. Figure 3.1 depicts the 10-year spot swap rate over nearly two decades and compares its 20-day realized percentage volatility with that of the S&P 500 Index. While these two volatilities share some common trends and spikes, they display significantly distinct dynamics and have an average correlation of just 51 %. Moreover, Fig. 3.2 shows that the correlation fluctuates wildly over time; for instance, correlation goes from negative before the financial crisis starting in 2007 to highly positive during it—most likely due to concerns about disorderly tail events—and then back to negative by 2011.

This chapter hinges upon the theory developed in Chap. 2, and develops (i) variance swap designs tailored to interest rate swap transactions, and (ii) resulting indexes of interest rate swap volatility, which are currently being calculated by the Chicago Board Options Exchange under the name SRVIX (see Chicago Board Options Exchange 2012).[1]

The plan of this chapter is the following. Section 3.2 provides a concise summary of the risks inherent in relevant interest rate swap transactions, and highlights pitfalls arising from option-based trading strategies aimed to express views on developments in swap rate volatility. These pitfalls motivate the introduction of new interest rate variance contracts.

Section 3.3 develops the design, pricing, and hedging for these contracts for both "basis point" or "Gaussian" volatility (related to absolute changes in the forward swap rate) and "percentage" volatility (i.e. the volatility of the logarithmic changes in the forward swap rate). These contracts are designed to match the market practice

---

[1]The index was launched by CBOE in June 2012 under the ticker name "SRVX." A ticker change to "SRVIX" occurred in June 2015.

© Springer International Publishing Switzerland 2015

A. Mele, Y. Obayashi, *The Price of Fixed Income Market Volatility*, Springer Finance,

DOI 10.1007/978-3-319-26523-0_3

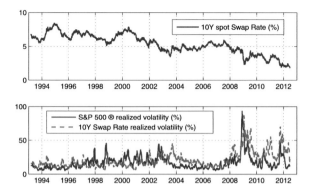

Fig. 3.1 *Top panel*: The 10-year swap rate, annualized, percent. *Bottom panel*: estimates of 20-day realized volatilities of (i) the S&P 500® daily returns (*solid line*), and (ii) the daily logarithmic changes in the 10-year swap rate, annualized, percent, $100\sqrt{12\sum_{t=1}^{21}\ln^2\frac{X_{t-i+1}}{X_{t-i}}}$, where X_t denotes either the S&P 500 Index® or the 10-year swap rate. The sample includes daily data from January 29, 1993 to May 22, 2012, for a total of 4865 observations

Fig. 3.2 Moving average estimates of the correlation between the 20-day realized volatilities of the S&P 500® and the 10-year swap rate logarithmic changes. Each correlation estimate is calculated over the previous one year of data, and the sample includes daily data from January 29, 1993 to May 22, 2012, for a total of 4865 observations

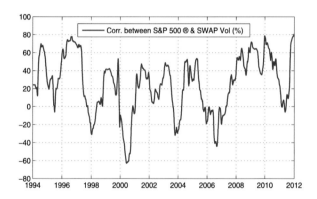

of quoting swaption implied volatility both in terms of percentage and basis point implied volatility. They are priced in a model-free format, and, consistently with the analysis in Chap. 2, hinge upon the "spanning arguments" of Bakshi and Madan (2000) and Carr and Madan (2001), with a few nuances as explained further in this chapter.

The remaining sections of the chapter focus on the implications of these contract designs, and compares them with their equity counterparts. Section 3.4 suggests potential trading strategies for the variance contracts of Sect. 3.3. Section 3.5 develops indexes of interest rate swap volatility stemming from these contracts, and extensions to markets with jumps where our basis point index is still expressed in a model-free format, consistently with results in Chap. 2. Section 3.6 offers an overview of the differences between interest rate swap variance contracts and indexes and those already in place in the equity case. Finally, Sect. 3.7 uses market data to illustrate a numerical implementation of one of the interest rate swap volatility indexes. Appendix B contains details omitted from the main text.

3.2   Risks Regarding Interest Rate Swaps

Chapter 2 emphasizes that interest rate derivatives with the potential to span volatility in fixed income markets also contain information that obfuscates this volatility. For example, swaption prices contain information about both swap rates and the value of an annuity paid over the underlying swap—the latter muddles the information content of swaption prices regarding rate volatility. This section describes this additional source of randomness, the annuity factor, or PVBP (the "price value of the basis point"), and analyzes how this affects the P&L of option-based volatility trading. These issues motivate the variance swaps designed in Sect. 3.3, which are theoretically grounded within the general framework of Chap. 2.

3.2.1   The Annuity Factor

3.2.1.1   The Forward Swap Rate: Definition

A forward-starting interest rate swap is a contract agreed at time t and starting at time T, whereby a party exchanges a fixed against a floating interest rate (the LIBOR, in the sequel) payment over n reset dates $T_0, \ldots, T_{n-1}$ and payment periods $T_1 - T_0, \ldots, T_n - T_{n-1}$, with $T \leq T_0$. The length, $T_n - T_0$, is known as the *tenor* of the interest rate swap. Standard market practice sets $T = T_0$, as in Fig. 3.4 to be discussed below, although this assumption is not critical to the pricing and contract designs in this section.

Let $L(T_{i-1}, T_i)$ denote the LIBOR rate applying to the period from T_{i-1} to T_i. The payoff at T_i to the counterparty *paying* the fixed rate is $\delta_{i-1}(L(T_{i-1}, T_i) - K)$, where K is the fixed interest rate and $\delta_{i-1} = T_i - T_{i-1}$ is the length of the reset interval. Therefore, the value at t of a forward starting interest rate swap is

$$\text{IRS}_t(K; T_1, \ldots, T_n) \equiv \sum_{i=1}^{n} \mathbb{E}_t\left(e^{-\int_t^{T_i} r_\tau d\tau} \delta_{i-1}\left(L(T_{i-1}, T_i) - K\right)\right), \qquad (3.1)$$

where r_τ denotes the instantaneous interest rate as of time τ, and $\mathbb{E}_t$ is the expectation taken under the risk-neutral probability Q, and conditional on the information at time t. The forward swap rate is defined as the fixed rate $K \equiv R_t(T_1, \ldots, T_n)$ such that $\text{IRS}_t(K; T_1, \ldots, T_n) = 0$. Imposing this condition leaves

$$\sum_{i=1}^{n} \delta_{i-1} \mathbb{E}_t\left(e^{-\int_t^{T_i} r_\tau d\tau} L(T_{i-1}, T_i)\right) = \text{PVBP}_t(T_1, \ldots, T_n) \cdot R_t(T_1, \ldots, T_n), \qquad (3.2)$$

where $PVBP_t(T_1, \ldots, T_n)$ is the "price value of the basis point,"

$$PVBP_t(T_1, \ldots, T_n) = \sum_{i=1}^{n} \delta_{i-1} P_t(T_i), \qquad (3.3)$$

and $P_t(T)$ is the time t price of a zero coupon bond expiring at time $T > t$.

The L.H.S. of Eq. (3.2) is the fair value at t of the floating leg of the swap. The R.H.S. is the fair value at t of the fixed leg: $PVBP_t(T_1, \ldots, T_n)$ is the present value of receiving one dollar at each fixed payment date of the swap. We can simplify the L.H.S. as follows. Consider a forward rate agreement (FRA), i.e. an agreement that costs nothing at t, whereby one party shall deliver the LIBOR $L(T_{i-1}, T_i)$ at T_i against a fixed rate $f_t^\delta(T_{i-1}, T_i)$ determined at t. The fixed rate is known as the forward rate, and satisfies

$$\mathbb{E}_t\left(e^{-\int_t^{T_i} r_\tau d\tau} \delta_{i-1} L(T_{i-1}, T_i)\right) = P_t(T_i)\delta_{i-1} f_t^\delta(T_{i-1}, T_i) = P_t(T_{i-1}) - P_t(T_i), \qquad (3.4)$$

where the first equality follows by imposing the condition that the value of the FRA is zero at t, and the second is a famous relation linking the price of zero coupon bonds to forward rates (see, e.g., Mele 2014, Chap. 11). Substituting Eq. (3.4) into Eq. (3.2) leads to the following expression for the forward swap rate:

$$R_t(T_1, \ldots, T_n) = \frac{P_t(T_0) - P_t(T_n)}{PVBP_t(T_1, \ldots, T_n)}. \qquad (3.5)$$

Naturally, $R_\tau(T_1, \ldots, T_n)$ is both a forward and a spot swap rate when $\tau = T$. Finally, note that substituting Eqs. (3.2) and (3.5) into Eq. (3.1) leaves the following expression for the value of forward starting interest rate swap:

$$IRS_t(K; T_1, \ldots, T_n) = PVBP_t(T_1, \ldots, T_n)\left[R_t(T_1, \ldots, T_n) - K\right]. \qquad (3.6)$$

We now rely on this expression to explain the main issues encountered while pricing the volatility of the forward swap rate, $R_t(T_1, \ldots, T_n)$, in a model-free fashion.

3.2.1.2 The Risks in the Interest Rate Swap Space

Our objective is to determine the market expectation of the volatility of the forward swap rate, $R_t(T_1, \ldots, T_n)$, without relying on any strong assumption going beyond the absence of arbitrage—a "model-free" approach, as defined in Chap. 2. To implement a model-free pricing of volatility, we need to rely on options that have $R_t(T_1, \ldots, T_n)$ as the underlying. Applying this approach to fixed income markets requires a treatment of the market numéraire specific to each market. We now illustrate this issue as an introduction.

Consider a *swaption contract* agreed at time t, whereby one counterparty has the right, but not the obligation, to enter a swap contract at $T_0 \in (t, T_1)$ either as a fixed rate payer or as a fixed rate receiver. The agreed fixed rate is equal to some strike K. For example, consider a swaption *payer*. If this swaption is exercised at T, its value at T from the perspective of the fixed rate payer is that of the underlying swap. By Eq. (3.6), it is

$$\text{IRS}_T(K; T_1, \ldots, T_n) \equiv \text{PVBP}_T(T_1, \ldots, T_n)\big[R_T(T_1, \ldots, T_n) - K\big]. \tag{3.7}$$

Equation (3.7) points to an important feature of swaptions. The underlying swap value is unknown until the swaption expiration not only because forward swap rates fluctuate but also because the PVBP of the forward swap's fixed leg fluctuates. Options on equities typically have the stock price as the single source of risk; instead, interest rate swaps are tied to two sources of risk: (i) the forward swap rate, and (ii) the swap's PVBP realized at time T, which measures the present value impact of a one basis point move in the *spot* swap rate at a future date T, $R_T(T_1, \ldots, T_n)$.

In other words, the value of a swaption depends on future values of both the swap rate and the entire yield curve. These facts complicate the definition and pricing of swap market volatility. In Sect. 3.3.7, we shall explain that this complexity would persist even if one were to price interest rate swap volatility through hypothetical derivatives referencing just R_T as if it were a stock option. We now proceed to explain both the effectiveness and limitations of expressing volatility views with on options-based strategies.

3.2.2 Option-Based Volatility Trading

Swaption payers and receivers have traditionally been the primary instruments used for trading volatility in interest rates. In particular, consider the following two trades:

(i) A delta-hedged swaption position, i.e. a long position in a swaption, be it payer or receiver, hedged through Black's (1976) delta.
(ii) A straddle, i.e. a long position in a payer and a receiver with the same strike, maturity, and tenor.

In Appendix B.1 (see Eq. (B.8)), we show the delta-hedged swaption position leads, approximately, to the following daily P&L at the swaption's expiry:

$$\text{P\&L}_T \approx \frac{1}{2}\sum_{t=1}^{T} \Gamma_t^\$ \cdot \big[(\sigma_t^2 - \text{IV}_0^2)\text{PVBP}_T\big] + \sum_{t=1}^{T} \text{Track}_t \cdot \text{Vol}_t(\text{PVBP}) \cdot \frac{\widetilde{\Delta R_t}}{R_t}, \tag{3.8}$$

where $\Gamma_t^\$$ is the Dollar Gamma, i.e. the swaption's Gamma times the square of the forward swap rate, σ_t is the instantaneous volatility of the forward swap rate at day t, IV_0 is the swaption implied percentage volatility at the time the strategy is first implemented, Track_t is the tracking error of the hedging strategy, $\text{Vol}_t(\text{PVBP})$

is the volatility of the PVBP rate of growth at t, and $\frac{\widetilde{\Delta R_t}}{R_t}$ denotes the series of shocks affecting the forward swap rate.

In Appendix B.1 (see Eq. (B.11)), we show that a straddle strategy leads to a P&L similar to that in Eq. (3.8), with the first term being twice that in Eq. (3.8), and with the straddle value replacing the tracking error in the second term:

$$\text{P\&L}_T^{\text{straddle}} \approx \sum_{t=1}^{T} \Gamma_t^{\$} \cdot \left[(\sigma_t^2 - \text{IV}_0^2)\text{PVBP}_T\right] + \sum_{t=1}^{T} \text{Straddle}_t \cdot \text{Vol}_t(\text{PVBP}) \cdot \frac{\widetilde{\Delta R_t}}{R_t},$$

$$(3.9)$$

where Straddle_t is the value of the straddle as of time t.

We emphasize that the P&Ls in Eqs. (3.8) and (3.9) only approximate the exact expressions given in Appendix B.1. For example, the exact version of the P&L in Eq. (3.9) has an additional term because the delta of a straddle could very well drift significantly away from zero, say during episodes of pronounced volatility. Naturally, not including this term in the approximations favorably biases the assessment of option-based strategies for trading interest rate swap volatility.

Equations (3.8) and (3.9) show that the two option-based volatility strategies lead to quite unpredictable profits that fail to isolate changes in interest rate swap volatility. The first terms of Eqs. (3.8) and (3.9) are related to the familiar issue of "price-dependency" (see, e.g., El Karoui et al. 1998, or Bossu et al. 2005). The second terms arise due to the randomness of the PVBP_t and the shocks affecting the forward swap rate, $\frac{\widetilde{\Delta R_t}}{R_t}$. More precisely,

- The first component of the P&L is proportional to the sum of the daily P&L for both strategies. It is given by the "volatility view" $\sigma_t^2 - \text{IV}_0^2$, weighted by the Dollar Gamma, $\Gamma_t^{\$}$, and the PVBP at the swaption expiry. This weighting is similar to, but due to the PVBP at T, distinct from that in the equity case. Indeed, *equity* variance contracts are attractive compared to option-based strategies as they overcome the price dependency in the latter. This price dependency shows up in our context as well. Even when $\sigma_t^2 - \text{IV}_0^2 > 0$ for most of the time, bad realizations of volatility, $\sigma_t^2 - \text{IV}_0^2 < 0$, may occur precisely when the Dollar Gamma $\Gamma_t^{\$}$ is high. In other words, the first terms in Eq. (3.8) and Eq. (3.9) can be negative even if σ_t^2 is higher than IV_0^2 for most of the time.
- The second terms in Eqs. (3.8) and (3.9) show that option-based P&Ls depend on the shocks affecting the forward swap rate, $\frac{\widetilde{\Delta R_t}}{R_t}$. That is, even if future volatility is *always* higher than the current implied volatility, $\sigma_t^2 > \text{IV}_0^2$, the second terms can dwarf the first due to adverse realizations of $\frac{\widetilde{\Delta R_t}}{R_t}$. These terms do not appear in equity volatility option-based P&Ls. They are specific to what swaptions are: instruments that have as underlying a swap contract, whose underlying swap value's PVBP is random and, hence, unknown up to the swaption maturity (see Eq. (3.7)).

These observations lead to *theoretical* reasons for which options might not be appropriate instruments to trade interest rate swap volatility. How reliable are option-

Fig. 3.3 Empirical
performance of directional
volatility trades

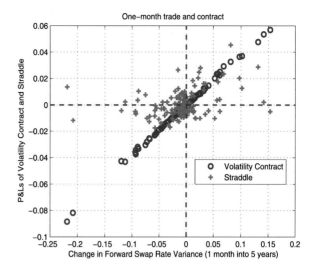

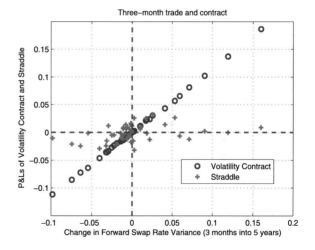

based strategies for that purpose in practice? Figure 3.3 plots the P&L of two directional volatility strategies against the "realized variance risk premium," defined below, and calculated using daily data from January 1998 to December 2009.

The two strategies are long an at-the-money (ATM) straddle and one of the interest rate variance contracts introduced in the next section. Both positions relate to contracts expiring in 1 and 3 months and tenors of 5 years. The realized variance risk premium is defined as the difference between the realized variance of the forward swap rate during the relevant holding period (i.e. 1 or 3 months), and the squared implied ATM volatility of the swaption (at 1 or 3 months) prevailing at the beginning of the holding period. Appendix B.1 provides technical details regarding these calculations.

The top panel of Fig. 3.3 depicts the two P&Ls for 1-month trades (147 trades), and the bottom panel displays the two P&Ls for 3-month trades (49 trades). Do these strategies deliver results consistent with directional views? On average, the P&Ls of both the straddle-based strategy and the interest rate variance contract have the same sign as the realized variance risk premium. However, the straddle P&Ls are more "dispersed," in that they do not always preserve the same sign of the realized variance risk premium, whereas interest rate variance contract do. With straddles, the P&L has the same sign as the variance risk premium approximately 62 % of the time for 1-month trades, and for approximately 65 % of the time for 3-month trades. The correlation between the straddle P&L and the variance risk premium is about 33 % for the 1-month trade, and about 28 % for the 3-month trade. Further studies performed by Jiang (2011) under a number of alternative holding periods suggest that if delta-hedged, these trades lead to a higher correlation of 63 % averaged over all the simulation experiments performed. In contrast, note the performance of the interest rate variance contract in our experiment, which is a pure play on volatility with its P&L lining up in a straight line, and a correlation with the variance risk premium indistinguishable from 100 %.

The reason straddles lead to more dispersed profits in the 3-month trading period is that there is longer exposure to "price dependency," since the likelihood of adverse weightings increases with the duration of the holding period, all else being equal. While these facts are well known in the case of equity volatility, the complexity of interest rate transactions makes these facts even more severe when it comes to trading interest rate swap volatility through options.

3.3   Interest Rate Swap Variance Contracts

We now study variance contracts designed for trading interest rate swap volatility while overcoming issues plaguing the option-based strategies as outlined in the previous section. The contracts are priced in a model-free fashion, and lead to model-free indexes of expected volatility over a reference period $[t, T]$ (see Sect. 3.5). We consider contracts referencing both (i) *basis point*, the standard practice in swap markets, and (ii) *percentage*, or *logarithmic* variance.

Section 3.3.2 provides design for contracts referencing the two notions of variance. Section 3.3.3 contains pricing results, and Sects. 3.3.4 and 3.3.5 develop details regarding marking to market and hedging of these contracts. Section 3.3.6 provides a link to how these variance contracts are priced relative to "constant maturity swaps." Section 3.3.7 explores an alternative contract design based on "rate settlement" (rather than swap settlement). Finally, Sect. 3.3.8 provides a succinct account of the "multiple curve" approach to swaption pricing. This approach could be used to extend some of this chapter's results. We begin in Sect. 3.3.1 with the exact definition of the risks we aim to price, and the spanning derivatives that help price these risks in a model-free fashion.

3.3.1 Risks and Spanning Derivatives

Motivated by the analysis of Chap. 2, we aim to find a "market space"—a numéraire and a probability space—in which the forward swap rate $R_t(T_1, \ldots, T_n)$ is a martingale. In this subsection, we review material that has been well known since Jamshidian (1997) in order to make this chapter self-contained.

Consider the expression for $R_t(T_1, \ldots, T_n)$ in Eq. (3.5). Note that

$$
\begin{aligned}
R_t(T_1, \ldots, T_n) &= \frac{P_t(T_0) - P_t(T_n)}{\mathrm{PVBP}_t(T_1, \ldots, T_n)} \\
&= \frac{1}{\mathrm{PVBP}_t(T_1, \ldots, T_n)} \mathbb{E}_t \left(e^{-\int_t^T r_s ds} \left(P_T(T_0) - P_T(T_N) \right) \right) \\
&= \mathbb{E}_t \left(e^{-\int_t^T r_s ds} \frac{\mathrm{PVBP}_T(T_1, \ldots, T_n)}{\mathrm{PVBP}_t(T_1, \ldots, T_n)} \frac{P_T(T_0) - P_T(T_N)}{\mathrm{PVBP}_T(T_1, \ldots, T_n)} \right),
\end{aligned}
\tag{3.10}
$$

where $\mathbb{E}_t$ denotes the risk-neutral expectation conditional on information at t. The previous expression suggests that $R_t(T_1, \ldots, T_n)$ is a martingale under a new probability, Q_{sw}, defined through the following Radon–Nikodym derivative:

$$
\left. \frac{dQ_{\mathrm{sw}}}{dQ} \right|_{\mathbb{F}_T} = e^{-\int_t^T r_s ds} \frac{\mathrm{PVBP}_T(T_1, \ldots, T_n)}{\mathrm{PVBP}_t(T_1, \ldots, T_n)},
\tag{3.11}
$$

where $\mathbb{F}_T$ is the information set as of time T. It is straightforward to verify that under standard regularity conditions, Q_{sw} integrates to one. This probability is typically referred to as the *annuity probability*, and PVBP is the numéraire in this market.

In terms of Q_{sw}, Eq. (3.10) is

$$
R_t(T_1, \ldots, T_n) = \mathbb{E}_t^{Q_{\mathrm{sw}}} \left(\frac{P_T(T_0) - P_T(T_N)}{\mathrm{PVBP}_T(T_1, \ldots, T_n)} \right) = \mathbb{E}_t^{Q_{\mathrm{sw}}} \left(R_T(T_1, \ldots, T_n) \right),
$$

where $\mathbb{E}_t^{Q_{\mathrm{sw}}}$ denotes the expectation taken under Q_{sw}, and the second equality follows by the definition of the forward swap rate in Eq. (3.5). That is, the forward swap rate is a martingale under the annuity probability.

We can now specify Eq. (2.2) in Chap. 2 to the forward swap rate case, $X_t \equiv R_t(T_1, \ldots, T_n)$. Define

$$
\frac{dR_\tau(T_1, \ldots, T_n)}{R_\tau(T_1, \ldots, T_n)} = \sigma_\tau(T_1, \ldots, T_n) dW_\tau^{\mathrm{sw}}, \quad \tau \in [t, T],
\tag{3.12}
$$

where W_τ^{sw}, a multidimensional Brownian motion defined under Q_{sw}, and $\sigma_\tau(\cdot)$ is some vector process adapted to W_τ^{sw}. We refer to $\sigma_\tau(\cdot)$ in Eq. (3.12) as the instantaneous volatility of the forward swap rate. We let $V_n^{\mathrm{bp}}(t, T)$ and $V_n(t, T)$ denote the BP and percentage forward swap rate variance corresponding to Eqs. (2.3) in

Chap. 2:

$$V_n^{\text{bp}}(t, T) \equiv \int_t^T R_\tau^2(T_1, \ldots, T_n) \big\| \sigma_\tau(T_1, \ldots, T_n) \big\|^2 d\tau \quad \text{and}$$

$$V_n(t, T) \equiv \int_t^T \big\| \sigma_\tau(T_1, \ldots, T_n) \big\|^2 d\tau.$$

(3.13)

We wish to price variance contracts referencing $V_n^{\text{bp}}(t, T)$ and $V_n(t, T)$ using swaptions. The fair value of these swaptions is determined by taking expectations of payoffs under the annuity probability.

Consider, for example, the price of a payer swaption. The payoff is the positive part of $\text{IRS}_T(K; T_1, \ldots, T_n)$ in Eq. (3.7), such that the price of payer swaptions expiring at T is

$$\text{Swpn}_t^P(K, T; T_1, \ldots, T_n) = \text{PVBP}_t(T_1, \ldots, T_n) \mathbb{E}_t^{Q_{\text{sw}}} \big[R_T(T_1, \ldots, T_n) - K \big]^+,$$

(3.14)

and the price of a receiver swaption is determined similarly as

$$\text{Swpn}_t^R(K, T; T_1, \ldots, T_n) = \text{PVBP}_t(T_1, \ldots, T_n) \mathbb{E}_t^{Q_{\text{sw}}} \big[K - R_T(T_1, \ldots, T_n) \big]^+.$$

(3.15)

We now proceed to security designs regarding the variance contracts based on swaption prices.

3.3.2 Contract Designs

We consider three contracts: a forward agreement, which requires an initial payment, and two variance swaps, which are settled at the expiration. These contracts are all equally useful for the purpose of implementing views on swap market volatility (see Sect. 3.4).

As noted previously, uncertainty about interest rate swaps relates to the volatility of the forward swap rate rescaled by the PVBP (see Eq. (3.7)), with the complication that the PVBP is unknown until T. Consider a contract whereby at time t, two counterparties A and B agree that at T, A shall pay B the realized variance of the forward swap rate from t to T, scaled by $\text{PVBP}_T(T_1, \ldots, T_n)$. This contract is a forward agreement. Figure 3.4 illustrates the timing of this forward agreement.

The following definition makes clear how payments regarding the volatility contract are executed.

Definition 3.1 (Interest Rate Variance Forward Agreement) At time t, counterparty A promises to pay counterparty B the product of the forward swap rate variance realized over the interval $[t, T]$ times the swap's PVBP prevailing at time T, i.e., (i) the value $V_n^{\text{bp}}(t, T) \times \text{PVBP}_T(T_1, \ldots, T_n)$ for the basis point contract; and (ii) the

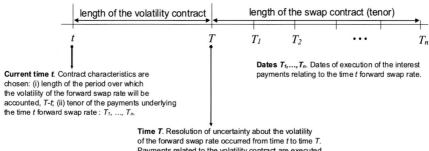

Fig. 3.4 Timing of the interest rate variance contract and the execution of swap payments

value $V_n(t, T) \times \text{PVBP}_T(T_1, \ldots, T_n)$ for the percentage contract. The price that counterparty B must pay counterparty A for this agreement at time t is called the Interest Rate Variance (IRV) Forward rate, and is denoted as (i) $\mathbb{F}_{\text{var},n}^{\text{bp}}(t, T)$, for the basis point contract; and (ii) $\mathbb{F}_{\text{var},n}(t, T)$, for the percentage contract.

The PVBP is the price impact at T of a one basis point move in the swap rate, and is unknown before time T, as discussed in Sect. 3.2.1. Rescaling by the forward PVBP is mathematically unavoidable when the objective is to price volatility in a model-free fashion. Our goal is to simultaneously price interest rate swap volatility, swaps, swaptions and pure discount bonds in such a way that the price of volatility conveys all the information carried by all the traded swaptions and bonds. Equation (3.7) suggests that the uncertainty related to the interest rate swaps underlying the swaptions is tied to both the realized variance, $V_n(t, T)$, and the forward PVBP. To isolate and price volatility, the payoff of the variance contract needs to be rescaled by $\text{PVBP}_T(T_1, \ldots, T_n)$, just as the value of the swap does in Eq. (3.7).

Note that the term $\text{PVBP}_T(T_1, \ldots, T_n)$ also appears in the P&L of the swaption-based volatility strategies, as shown by Eqs. (3.8) and (3.9); the important difference is that the variance contract underlying Definition 3.1 isolates volatility from Dollar Gamma.[2] Finally, from a practical perspective, such a rescaling does not affect the ability of the contract to convey views on future developments of $V_n(t, T)$; recall that the correlation between the variance risk premiums and the P&L of IR-variance contracts is nearly 100 % in the empirical experiment in Fig. 3.3.

In fact, the P&L depicted in Fig. 3.3 relates to a variance contract cast in a "swap format" whereby two parties agree to exchange the following difference with each other at T:

$$\text{Var-Swap}_n(t, T) \equiv V_n(t, T) \times \text{PVBP}_T(T_1, \ldots, T_n) - \mathbb{P}_{\text{var},n}(t, T), \qquad (3.16)$$

---

[2] If the risk-neutral expectation of the short-term rate is insensitive to changes in the short-term rate volatility, the PVBP at time T is increasing in the short-term rate volatility (see Mele 2003). In this case, the IRV forward agreement is a device to lock-in both volatility of the forward swap rate *from t to T, and* short-term swap rate volatility *at T*.

where $\mathbb{P}_{\text{var},n}(t, T)$ is a fixed variance swap rate, determined at time t and set to satisfy a zero value condition at time t. Definition 3.2 below makes this notion precise within both the basis point and percentage variance contexts as summarized by Eqs. (3.13). Accordingly, define the following payoff at time T:

$$\text{Var-Swap}_n^{\text{bp}}(t, T) \equiv V_n^{\text{bp}}(t, T) \times \text{PVBP}_T(T_1, \ldots, T_n) - \mathbb{P}_{\text{var},n}^{\text{bp}}(t, T), \qquad (3.17)$$

where $\mathbb{P}_{\text{var},n}^{\text{bp}}(t, T)$ is defined by a zero value condition, as in the next definition.

Definition 3.2 (IRV Swap Rate) The IRV swap rate is the fixed variance swap rate, $\mathbb{P}_{\text{var},n}^{\text{bp}}(t, T)$ or $\mathbb{P}_{\text{var},n}(t, T)$, which makes (i) the current value of $\text{Var-Swap}_n^{\text{bp}}(t, T)$ in Eq. (3.17) equal to zero (basis point case); or (ii) the current value of $\text{Var-Swap}_n(t, T)$ in Eq. (3.16) equal to zero (percentage case).

Note that the variance contracts of Definition 3.2 are forwards (not swaps). However, we utilize a terminology similar to that already in place for equity variance contracts, and simply refer to the previous contracts as "swaps."

Finally, we consider an alternate definition of the IRV swap rate. This definition is important as it leads to the definition of the index of interest rate swap volatility used in practice by CBOE (see Sects. 3.5 and 3.7). Consider the following two payoffs:

$$\begin{pmatrix} \text{Var-Swap}_n^{\text{bp},*}(t, T) \equiv [V_n(t, T) - \mathbb{P}_{\text{var},n}^{\text{bp},*}(t, T)] \times \text{PVBP}_T(T_1, \ldots, T_n) \text{ and} \\ \text{Var-Swap}_n^*(t, T) \equiv [V_n(t, T) - \mathbb{P}_{\text{var},n}^*(t, T)] \times \text{PVBP}_T(T_1, \ldots, T_n) \end{pmatrix},$$

$$(3.18)$$

where $\mathbb{P}_{\text{var},n}^{\text{bp},*}(t, T)$ and $\mathbb{P}_{\text{var},n}^*(t, T)$, are fixed variance swap rates determined at t, and are set to satisfy a zero value condition at time t:

Definition 3.3 (Standardized IRV Swap Rate) The standardized IRV swap rate is the fixed variance swap rate, $\mathbb{P}_{\text{var},n}^{\text{bp},*}(t, T)$ or $\mathbb{P}_{\text{var},n}(t, T)$, which makes the current value of $\text{Var-Swap}_n^{\text{bp},*}(t, T)$ and $\text{Var-Swap}_n^*(t, T)$ in Eqs. (3.18) equal to zero.

The payoffs in Eqs. (3.18) carry an intuitive meaning. Consider, for example, $\text{Var-Swap}_n^{\text{bp},*}(t, T)$. It is the product of two terms: (i) the difference between the realized BP variance and a fair strike and (ii) the PVBP at time T. This payoff is similar to that of standard equity variance contracts: the holder of the contract would receive $\text{PVBP}_T(T_1, \ldots, T_n)$ dollars for every point by which the realized BP variance exceeds the variance strike price. The added complication of the IRV contract design is that $\text{PVBP}_T(T_1, \ldots, T_n)$ is unknown at time t.

3.3.3 Pricing

In the absence of arbitrage and market frictions, the price of the IRV forward agreement to be paid at time t, $\mathbb{F}_{\text{var},n}^{\text{bp}}(t, T)$ and $\mathbb{F}_{\text{var},n}(t, T)$ in Definition 3.1, are expressed in a model-free format as a combination of the prices of ATM and out-of-the-money (OTM) swaptions.

As for the BP-IRV forward contract, we have that one approximation to $\mathbb{F}^{bp}_{var,n}$, based on a finite number of swaptions, is

$$\mathbb{F}^{bp}_{var,n}(t, T) = 2 \left(\sum_{i: K_i < R_t} \mathrm{Swpn}^R_t(K_i, T; T_n) \Delta K_i + \sum_{i: K_i \geq R_t} \mathrm{Swpn}^P_t(K_i, T; T_n) \Delta K_i \right)$$

(3.19)

where $\mathrm{Swpn}^R_t(K_i, T; T_n)$ (resp., $\mathrm{Swpn}^P_t(K_i, T; T_n)$) is the price of a swaption receiver (resp., payer), struck at K_i, expiring at T and with tenor extending up to time T_n, and $\Delta K_i = \frac{1}{2}(K_{i+1} - K_{i-1})$ for $1 \leq i < M$, $\Delta K_0 = (K_1 - K_0)$, $\Delta K_M = (K_M - K_{M-1})$, where K_0 and K_M are the lowest and the highest available strikes traded in the market, and $M + 1$ is the total number of traded swaptions expiring at time T and with tenor extending up to time T_n.

Appendix B.2 provides more details regarding Eq. (3.19). Given the BP-IRV forward, the BP-IRV swap rates in Definitions 3.2 and 3.3 are

$$\mathbb{P}^{bp}_{var,n}(t, T) = \frac{\mathbb{F}^{bp}_{var,n}(t, T)}{P_t(T)} \quad \text{and} \quad \mathbb{P}^{*bp}_{var,n}(t, T) = \frac{\mathbb{F}^{bp}_{var,n}(t, T)}{\mathrm{PVBP}_t(T_1, \ldots, T_n)}$$

(3.20)

where $P_t(T)$ is the price of a zero-coupon bond maturing at T and time to maturity $T - t$.

The percentage IRV forward contract counterparts to Eqs. (3.19) and (3.20) are

$$\mathbb{F}_{var,n}(t, T) = 2 \left(\sum_{i: K_i < R_t} \frac{\mathrm{Swpn}^R_t(K_i, T; T_n)}{K_i^2} \Delta K_i + \sum_{i: K_i \geq R_t} \frac{\mathrm{Swpn}^P_t(K_i, T; T_n)}{K_i^2} \Delta K_i \right)$$

(3.21)

and

$$\mathbb{P}_{var,n}(t, T) = \frac{\mathbb{F}_{var,n}(t, T)}{P_t(T)} \quad \text{and} \quad \mathbb{P}^*_{var,n}(t, T) = \frac{\mathbb{F}_{var,n}(t, T)}{\mathrm{PVBP}_t(T_1, \ldots, T_n)}.$$

(3.22)

Finally, Appendix B.2 provides an expression for the local volatility surface, which one can use to "interpolate" the missing swaption prices from those that are traded in order to "fill in" Eqs. (3.19) and (3.21). An alternative fitting model could be the popular SABR model (Hagan et al. 2002).

3.3.4   Marking to Market

This section provides expressions for mark-to-market calculations of the previous IRV contracts. These expressions also help assess the value of trading strategies based on these contracts, as explained in Sect. 3.4.

In Appendix B.2, we show that the value, at any time $\tau \in (t, T)$, of the variance contract struck at time t at the IRV swap rate $\mathbb{P}_{\mathrm{var},n}(t, T)$ (see the first of Eqs. (3.22)) is

$$\text{M-Var}_n(t, \tau, T)$$

$$\equiv V_n(t, \tau) \times \text{PVBP}_\tau(T_1, \ldots, T_n) - P_\tau(T)\big[\mathbb{P}_{\mathrm{var},n}(t, T) - \mathbb{P}_{\mathrm{var},n}(\tau, T)\big], \quad (3.23)$$

and the value of the contract struck at the standardized IRV swap rate $\mathbb{P}^*_{\mathrm{var},n}(t, T)$ (see the second of Eqs. (3.22)) is

$$\text{M-Var}^*_n(t, \tau, T) \equiv \text{PVBP}_\tau(T_1, \ldots, T_n)\big[V_n(t, \tau) - \big(\mathbb{P}^*_{\mathrm{var},n}(t, T) - \mathbb{P}^*_{\mathrm{var},n}(\tau, T)\big)\big].$$
$$(3.24)$$

It is immediate to see that the mark-to-market updates for BP contracts are the same as those in Eqs. (3.23) and (3.24), up to a mere change in notation.

3.3.5 Hedging

IRV contracts can be hedged, at least partially, using swaptions. In the exposition below, we take the perspective of a provider of insurance against volatility, i.e., one who sells any of the IRV contracts underlying Definitions 3.1 through 3.3 of Sect. 3.2.

Hedging IRV contracts requires coping with two issues that do not arise in the equity variance case.[3] First, we need to deal with the randomness of the annuity factor, the PVBP_T at time T. Second, we need to hedge basis point variance, which is the dominant convention in swap markets. Last, we shall see that an additional complication is that the forward swap rate is not a tradable asset. That is, it can be thought of as a non-linear function of the price of three tradable assets, as explained in Appendix B.3 (two zero-coupon bonds and an annuity factor, as Eq. (3.5) suggests). However, it cannot be replicated through these three assets through a self-financed strategy.

In this section, we explain how to hedge IRV contracts while assuming that the forward swap rate can be replicated. In Appendix B.3, we give details regarding the portfolio weights needed to replicate the forward swap rate, and explain that this strategy is not self-financed. We begin with the percentage case, which the readers may be more familiar with, although the task we face is still more complex than in the equity case, as we need to address randomness of the annuity factor. We then provide details of the basis point case, which relies on arguments that hinge upon the quadratic contracts reviewed in Chap. 2.

---

[3] See, e.g., Mele (2014, Chap. 10), for details on hedging equity variance swaps.

Table 3.1 Replication of the contract relying on the IRV swap rate of Definition 3.2. ZCB and OTM stand for zero coupon bonds and OTM, respectively, and $R_\tau \equiv R_\tau(T_1, \ldots, T_n)$, $\text{PVBP}_T \equiv \text{PVBP}_T(T_1, \ldots, T_n)$

Portfolio	Value at t	Value at T
(i) long self-financed portfolio of ZCB	0	$(V_n(t, T) + 2\ln\frac{R_T}{R_t}) \times \text{PVBP}_T$
(ii) short swaps and long OTM swaptions	$-\mathbb{F}_{\text{var},n}(t, T)$	$-2\ln\frac{R_T}{R_t} \times \text{PVBP}_T$
(iii) borrow $\mathbb{F}_{\text{var},n}(t, T)$	$+\mathbb{F}_{\text{var},n}(t, T)$	$-\frac{\mathbb{F}_{\text{var},n}(t,T)}{P_t(T)} = -\mathbb{P}_{\text{var},n}(t, T)$
Net cash flows	0	$V_n(t, T) \times \text{PVBP}_T - \mathbb{P}_{\text{var},n}(t, T)$

Table 3.2 Replication of the contract relying on the standardized IRV swap rate of Definition 3.3. ZCB and OTM stand for zero coupon bonds and OTM, respectively, and $R_\tau \equiv R_\tau(T_1, \ldots, T_n)$, $\text{PVBP}_\tau \equiv \text{PVBP}_\tau(T_1, \ldots, T_n)$

Portfolio	Value at t	Value at T
(i) long self-financed portfolio of ZCB	0	$(V_n(t, T) + 2\ln\frac{R_T}{R_t}) \times \text{PVBP}_T$
(ii) short swaps and long OTM swaptions	$-\mathbb{F}_{\text{var},n}(t, T)$	$-2\ln\frac{R_T}{R_t} \times \text{PVBP}_T$
(iii) borrow basket of ZCB for $\mathbb{P}^*_{\text{var},n}(t, T) \times \text{PVBP}_t$	$+\mathbb{F}_{\text{var},n}(t, T)$	$-\mathbb{P}^*_{\text{var},n}(t, T) \times \text{PVBP}_T$
Net cash flows	0	$[V_n(t, T) - \mathbb{P}^*_{\text{var},n}(t, T)] \times \text{PVBP}_T$

3.3.5.1 Percentage

The IRV forward contract in Definition 3.1 can be hedged through the positions in rows (i) and (ii) of Tables 3.1 and 3.2, which we discuss in a moment. Regarding the IRV contracts of Definition 3.2, consider replicating the payoff in Eq. (3.16). As shown in Appendix B.3, we hedge against this payoff through two identical zero-cost portfolios constructed through the following positions:

(i) A dynamic position in a portfolio of zero-coupon bonds replicating the cumulative increments of the forward swap rate over $[t, T]$,

$$\int_t^T \frac{dR_s(T_1, \ldots, T_n)}{R_s(T_1, \ldots, T_n)} = \frac{1}{2}V_n(t, T) + \ln\frac{R_T(T_1, \ldots, T_n)}{R_t(T_1, \ldots, T_n)},$$

and rescaled by the PVBP at time T.

(ii) Static positions in swaps starting at time T and OTM swaptions expiring at T replicating the payoff of the interest rate counterpart to the log-contract for equities (see Chap. 2), rescaled by the PVBP at T. The positions are as follows:

   (ii.1) Short $1/R_t(T_1, \ldots, T_n)$ units of a forward starting (at T) fixed interest payer swap struck at the current forward swap rate, $R_t(T_1, \ldots, T_n)$.

   (ii.2) Long OTM swaptions, each of them carrying a weight inversely proportional to its squared strike, $\frac{1}{K^2}$.

(iii) A static borrowing position aimed to finance the OTM swaption positions in (ii.2).

Details regarding the portfolio in (i) are given in Appendix B.3. Table 3.1 summarizes the costs of each of these trades at t, as well as the payoffs at T, which sum up to be the same as the payoff in Eq. (3.16), as required to hedge the contract. Note that the portfolio in (i) is not self-financed and so it gives rise to hedging costs as explained in Appendix B.3. Table 3.1 does not consider this hedging cost.

Finally, to hedge against the standardized contract of Definition 3.3, we need to replicate the percentage payoff in Eq. (3.18). We create a portfolio, which is the same as that in Table 3.1, except now that row (iii) refers to a borrowing position in a basket of zero-coupon bonds with value equal to $\mathrm{PVBP}_t(T_1, \ldots, T_n)$ and notional of $\mathbb{P}^*_{\mathrm{var},n}(t, T)$, as in Table 3.2. By the expression of $\mathbb{P}^*_{\mathrm{var},n}(t, T)$ in the second of Eqs. (3.22), the borrowing position needed to finance the OTM swaptions in row (ii), is just $\mathbb{F}_{\mathrm{var},n}(t, T) = \mathbb{P}^*_{\mathrm{var},n}(t, T) \times \mathrm{PVBP}_t(T_1, \ldots, T_n)$. Come time T, it will be closed for a value of $-\mathbb{P}^*_{\mathrm{var},n}(t, T) \times \mathrm{PVBP}_T(T_1, \ldots, T_n)$. The net cash flows are precisely those of the contract with the standardized IRV swap rate. Again, Table 3.2 does not account for the hedging costs needed to replicate the forward rate.

If the forward swap rate were replicable through a self-financed portfolio, the previous replication arguments could be alternative means to determine the no-arbitrage value of the IRV forward agreement of Definition 3.1, and the IRV swap rates in Definitions 3.2 and 3.3, $\mathbb{P}_{\mathrm{var},n}(t, T)$ and $\mathbb{P}^*_{\mathrm{var},n}(t, T)$. Suppose, for example, that the market value of the standardized IRV swap rate, $\mathbb{P}^*_{\mathrm{var},n}(t, T)_\$$, is higher than the no-arbitrage value $\mathbb{P}^*_{\mathrm{var},n}(t, T)$ in the second of Eqs. (3.22). Then, one could short the standardized contract underlying Definition 3.3 and simultaneously synthesize it through the portfolio in Table 3.2. This positioning costs zero at time t and yields a sure positive profit at time T equal to $[\mathbb{P}^*_{\mathrm{var},n}(t, T)_\$ - \mathbb{P}^*_{\mathrm{var},n}(t, T)] \times \mathrm{PVBP}_T$. To rule out arbitrage, we need to have $\mathbb{P}^*_{\mathrm{var},n}(t, T)_\$ = \mathbb{P}^*_{\mathrm{var},n}(t, T)$. Naturally, the forward swap rate is not replicable through a self-financed portfolio: replicating the forward incurs a hedging cost. Therefore, these arguments only approximately hold, assuming the hedging cost is small.

3.3.5.2 Basis Point

Replicating BP-denominated contracts requires positioning in a different way. The hedging arguments in Tables 3.1 and 3.2 rely on the replication of *log-contracts* rescaled by the PVBP. To hedge BP contracts, on the other hand, we need to rely on *quadratic contracts*, i.e. those delivering $R_T^2 - R_t^2$, rescaled by the PVBP.

Consider the payoff of the BP-IRV forward contract in Definition 3.2, which can be replicated through the portfolios in rows (i) and (ii) of Tables 3.3 and 3.4, as discussed below. To replicate the contract referencing the BP-IRV swap rate of Definition 3.2, we create portfolios comprised of the following positions:

(i) A dynamic, short position in two identical portfolios of zero-coupon bonds that replicate the cumulative increments of the forward swap rate over $[t, T]$,

Table 3.3 Replication of the payoff relating to the BP-IRV contract of Definition 3.2. ZCB and OTM stand for zero-coupon bonds and OTM, respectively, and $R_\tau \equiv R_\tau(T_1, \ldots, T_n)$, $\text{PVBP}_T \equiv \text{PVBP}_T(T_1, \ldots, T_n)$

Portfolio	Value at t	Value at T
(i) short self-financed portfolio of ZCB	0	$[V_n^{\text{bp}}(t, T) - (R_T^2 - R_t^2)] \times \text{PVBP}_T$
(ii) long swaps and long OTM swaptions	$-\mathbb{F}_{\text{var},n}^{\text{bp}}(t, T)$	$(R_T^2 - R_t^2) \times \text{PVBP}_T$
(iii) borrow $\mathbb{F}_{\text{var},n}^{\text{bp}}(t, T)$	$+\mathbb{F}_{\text{var},n}^{\text{bp}}(t, T)$	$-\frac{\mathbb{F}_{\text{var},n}^{\text{bp}}(t,T)}{P_t(T)} = -\mathbb{P}_{\text{var},n}^{\text{bp}}(t, T)$
Net cash flows	0	$V_n^{\text{bp}}(t, T) \times \text{PVBP}_T - \mathbb{P}_{\text{var},n}^{\text{bp}}(t, T)$

Table 3.4 Replication of the payoff relating to the standardized BP-IRV contract of Definition 3.3. ZCB and OTM stand for zero coupon-bonds and OTM, respectively, and $R_\tau \equiv R_\tau(T_1, \ldots, T_n)$, $\text{PVBP}_\tau \equiv \text{PVBP}_\tau(T_1, \ldots, T_n)$

Portfolio	Value at t	Value at T
(i) short self-financed portfolio of ZCB	0	$[V_n^{\text{bp}}(t, T) - (R_T^2 - R_t^2)] \times \text{PVBP}_T$
(ii) long swaps and long OTM swaptions	$-\mathbb{F}_{\text{var},n}^{\text{bp}}(t, T)$	$(R_T^2 - R_t^2) \times \text{PVBP}_T$
(iii) borrow basket of ZCB for $\mathbb{P}_{\text{var},n}^{*\text{bp}}(t, T) \times \text{PVBP}_t$	$+\mathbb{F}_{\text{var},n}^{\text{bp}}(t, T)$	$-\mathbb{P}_{\text{var},n}^{*\text{bp}}(t, T) \times \text{PVBP}_T$
Net cash flows	0	$[V_n^{\text{bp}}(t, T) - \mathbb{P}_{\text{var},n}^{*\text{bp}}(t, T)] \times \text{PVBP}_T$

weighted by the forward swap rate,

$$\int_t^T R_s(T_1, \ldots, T_n) dR_s(T_1, \ldots, T_n)$$

$$= \frac{1}{2}\big[\big(R_T^2(T_1, \ldots, T_n) - R_t^2(T_1, \ldots, T_n)\big) - V_n^{\text{bp}}(t, T)\big],$$

and rescaled by the PVBP at time T.

(ii) Static positions in swaps starting at T and OTM swaptions expiring at T, which replicates the payoff of a quadratic contract, rescaled by the PVBP at T. The positions are as follows:

   (ii.1) Long $2R_t(T_1, \ldots, T_n)$ units of a forward starting (at T) fixed interest payer swap struck at the current forward swap rate, $R_t(T_1, \ldots, T_n)$.
   (ii.2) Long OTM swaptions, each of them carrying a constant weight.

(iii) A static borrowing position to finance the OTM swaption positions in (ii.2).

Costs and payoffs of these portfolios are summarized in Table 3.3. The result is a replication of the payoff of the contract of Definition 3.2. As for the percentage IRV contracts, the costs in Table 3.3 (and in Table 3.4 below) do not account for the hedging costs needed to replicate the forward swap rate in (i).

Finally, Table 3.4 is obtained by modifying the arguments leading to Table 3.3, in the same way as we did to obtain Table 3.2.

3.3.6 Links to Constant Maturity Swaps

We show that a constant maturity swap (CMS) is approximately the same as a basket of the BP-IRV forwards of Definition 3.1, the important implication being that IRV swaps can thus be used to hedge CMS. The technical reason IRV forwards link to CMS is due to a classical "convexity bias," arising because the value of a CMS is the "wrong expectation" of future payoffs, one under which the forward swap rate is not a martingale. We develop the arguments in Appendix B.4.

In a CMS, one party pays its counterparty the spot swap rate with a fixed tenor over a sequence of N dates, with payment legs set in advance, i.e., $R_{T_j}(T_{j+1}, \ldots, T_{j+n})$ at times $T_j + \kappa$, $j = 0, \ldots, N - 1$, where $\kappa \leq \delta$ is a fixed constant (e.g., $\kappa = \frac{1}{2}$). In Appendix B.4, we show that the value of a CMS is approximately equal to:

$$
\mathrm{CMS}_N(t) \equiv \sum_{j=0}^{N-1} P_t(T_j + k) R_t(T_j, \ldots, T_{j+n})
$$

$$
+ \sum_{j=0}^{N-1} G'\big(R_t(T_j, \ldots, T_{n+j})\big) \mathbb{F}^{\mathrm{bp}}_{\mathrm{var},n}(t, T_j), \qquad (3.25)
$$

where $\mathbb{F}^{\mathrm{bp}}_{\mathrm{var},n}(t, T)$ is the value of the BP-IRV forward contract in Definition 3.1 (see Eq. (3.19)), and $G(\cdot)$ is a function given in Appendix B.4 (see Eq. (B.34)).

That the price of a CMS relates to the entire swaption skew has been known at least since Hagan (2003) and Mercurio and Pallavicini (2006), as further discussed in Appendix B.4. As is also clear from Eq. (3.25), CMS can be hedged through a basket of IRV forwards in BP expiring at the points preceding the CMS payments.

Lastly, note an interesting property of a CMS. Equation (3.25) reveals that the value of BP-IRV forwards can be actually "bootstrapped" from that of a CMS.

3.3.7 Physical Swap Settlement and Variance Contracts

Interest rate swaps settle according to the payoff in Eq. (3.7). We know from Chap. 2 and the present chapter that an IRV contract on interest rate swaps is model-free once the contract is rescaled with the market numéraire in this chapter, the annuity factor, just as in Sect. 3.3.2. A natural question arises as to whether IRV swaps on interest rate swaps could be model-free in the presence of alternative settlements.

We consider a simple situation. Suppose that a settlement is cast in terms of *swap rates*, not the *physical swap* in Eq. (3.7). That is, we consider an interest rate swap in which the payoff in Eq. (3.7) is replaced with

$$
\mathrm{IRS}^*_T(K; T_1, \ldots, T_n) = R_T(T_1, \ldots, T_n) - K. \qquad (3.26)
$$

Option-like derivatives referenced to IRS_T^* do not currently trade. We ask: Could we price swap rate volatility in a model-free format if such options existed? The answer is in the negative. Let us review the usual steps required to price volatility in a model-free format in the interest rate swap space (see Appendix B.2 for more details). It can be shown that

$$R_T^2 - R_t^2 = 2R_t(R_T - R_t) + 2\left(\int_0^{R_t} (K - R_T)^+ dK + \int_{R_t}^\infty (R_T - K)^+ dK \right),$$

where, for simplicity, $R_t \equiv R_t(T_1, \ldots, T_n)$. If options with payoffs referenced to IRS_T^* in Eq. (3.26) were actually traded, we could multiply the previous equation by the discounting factor, $e^{-\int_t^T r_s ds}$, take expectations under the risk-neutral probability Q, and obtain

$$\mathbb{E}_t\left(e^{-\int_t^T r_s ds} R_T^2\right) + P_t(T)R_t^2$$

$$= 2R_t \mathbb{E}_t\left(e^{-\int_t^T r_s ds} R_T\right)$$

$$+ 2\left(\int_0^{R_t} \widehat{\text{Swpn}}_t^{\text{R}}(K_i, T; T_n)dK + \int_{R_t}^\infty \widehat{\text{Swpn}}_t^{\text{P}}(K_i, T; T_n)dK \right), \quad (3.27)$$

where $\widehat{\text{Swpn}}_t^{\text{R}}$ and $\widehat{\text{Swpn}}_t^{\text{P}}$ denote the prices of the hypothetical options with payoff $(-\text{IRS}_T^*)^+$ and $(\text{IRS}_T^*)^+$. Not only are these options hypothetical at the time of writing, they would not help price volatility in a model-free format. For example, consider the price of volatility expressed in basis points, which by results in this chapter can be expressed as,

$$\mathbb{E}_t^{Q_{\text{sw}}}\left(R_T^2\right) - R_t^2$$

$$= \mathbb{E}_t^{Q_{\text{sw}}}\left(V_n^{\text{bp}}(t, T)\right)$$

$$= \frac{1}{\text{PVBP}_t(T_1, \ldots, T_n)} \mathbb{E}_t^{Q_{\text{sw}}}\left(e^{-\int_t^T r_s ds} \text{PVBP}_T(T_1, \ldots, T_n) V_n^{\text{bp}}(t, T)\right). \quad (3.28)$$

There is no way to combine Eqs. (3.27) and (3.28) in such a way to obtain a model-free price of basis point variance.

3.3.8 Multiple Curves

Valuation methodologies in the interest rate swap space have changed after the global financial crisis erupted in 2007. Before the crisis, the standard market convention for collateralized swaps and swaptions relied on the construction of a unique, well-defined zero-coupon curve (typically, the OIS swap curve), the basis for both discounting cash flows (i.e., the floating leg of the swap in Eq. (3.2)) and the determination of forward LIBOR.

This market convention has changed to one in which discounting is done with OIS while forward LIBORs are calculated off of the corresponding forward curve, with the latter typically differing from OIS. The reason underlying such a multiple-curve approach is the market rate segmentation that arose during the crisis. For example, there might be divergences between yield curves corresponding to different interest rate lengths (i.e., the reset dates δ_i); accordingly, one curve is associated with each length δ_i.

Evaluating interest rate variance swaps within a multiple-curve setting would require dealing with numéraires that are modifications of that in this chapter. Mercurio (2009, 2012) explains how to identify numéraires and pricing measures in this new context, and provides new market formulae. For example, swaptions can still be priced with a standard Black pricer[4] under a numéraire constructed with a given discounting curve (e.g., and in the presence of collateral and netting clauses, the OIS), but with the definition of the forward swap rate more general than that in Sect. 3.2.1 (Eq. (3.5)). A clear separation between discount and forward curves and its implications on the pricing of interest rate swap variance swaps go beyond the scope of this chapter and are left for future research.

3.4 Trading Strategies

We consider examples of trading strategies using the IRV contracts of Sects. 3.3. We analyze strategies based on views on: (i) spot IRV swap rate movement (in Sects. 3.4.1 and 3.4.2), and (ii) the value of spot IRV swap rates relative to "forward IRV swap rates," i.e., those implied by the current term-structure of spot IRV swap rates (in Sect. 3.3.4). The strategies we consider apply to both percentage and BP contracts. However, we simplify the presentation and only illustrate strategies regarding percentage contracts.

3.4.1 Spot Trading Through IRV Swaps

First, we consider strategies aimed to express views on IRV contracts that have different maturities and a fixed tenor. For example, Fig. 3.5 illustrates the timing of two contracts: (i) a contract expiring in 3 months, relating to the volatility of $3m$ into $\hat{T}_5 - 3m$ forward swap rates, $m = \frac{1}{12}$; and (ii) a contract expiring in 9 months, relating to the volatility of $9m$ into $T_5 - 9m$ forward swap rates. Time units are expressed in years, and we set $t \equiv 0$. We assume the two contracts have the same tenor, $\hat{T}_5 - 3m = T_5 - 9m$. We refer to these strategies as being "constant tenor trades."

---

[4]See Chap. 2, Sect. 2.4.2.

Fig. 3.5 Constant tenor trade

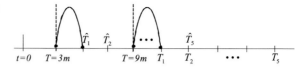

Let $\mathbb{P}_{\mathrm{var},\hat{n}}(t,T)$ and $\mathbb{P}_{\mathrm{var},n}(t,T)$ be the IRV swap rates of these two contracts, and suppose we have the view that at some point $\tau < 3m$,

$$\mathbb{P}_{\mathrm{var},\hat{n}}(\tau,3m) > \mathbb{P}_{\mathrm{var},\hat{n}}(0,3m) \quad \text{and} \quad \mathbb{P}_{\mathrm{var},n}(\tau,9m) < \mathbb{P}_{\mathrm{var},n}(0,9m). \qquad (3.29)$$

A trading strategy for expressing these views can be formulated as follows: (i) go long the $3m$ contract, (ii) short α units of the $9m$ contract, (iii) short the IRV forward contract in Definition 3.1 for the $3m$ maturity, with contract length equal to τ, and (iv) go long α units of the IRV forward contract for the $9m$ maturity, with contract length equal to τ.

By marking-to-market in Eq. (3.23), the value of this strategy at time τ is,

$$\begin{aligned}
\text{M-Var}_{\hat{n}}&(0,\tau,3m) - \alpha \cdot \text{M-Var}_n(0,\tau,9m) \\
&- V_{\hat{n}}(0,\tau) \times \text{PVBP}_\tau(\hat{T}_1,\ldots,\hat{T}_n) + \alpha \cdot V_n(0,\tau) \times \text{PVBP}_\tau(T_1,\ldots,T_n) \\
&= P_\tau(3m)\big[\mathbb{P}_{\mathrm{var},\hat{n}}(\tau,3m) - \mathbb{P}_{\mathrm{var},\hat{n}}(0,3m)\big] \\
&\quad + \alpha \cdot P_\tau(9m)\big[\mathbb{P}_{\mathrm{var},n}(0,9m) - \mathbb{P}_{\mathrm{var},n}(\tau,9m)\big] \\
&> 0. \qquad\qquad\qquad\qquad\qquad\qquad\qquad\qquad\qquad\qquad (3.30)
\end{aligned}$$

Its cost at $t = 0$ is $-\mathbb{F}_{\mathrm{var},\hat{n}}(0,3m) + \alpha \cdot \mathbb{F}_{\mathrm{var},n}(0,9m)$. We choose,

$$\alpha = \frac{\mathbb{F}_{\mathrm{var},\hat{n}}(0,3m)}{\mathbb{F}_{\mathrm{var},n}(0,9m)}, \qquad (3.31)$$

to make the portfolio worthless at $t = 0$. Naturally, the inequality in (3.30) follows only when (3.29) hold true.

The technical reason to enter into IRV forwards is that, by Eq. (3.23), the value at τ of the variance contracts (i) and (ii) entails cumulative realized variance exposures equal to $V_{\hat{n}}(0,\tau) \times \text{PVBP}_\tau(\hat{T}_1,\ldots,\hat{T}_n)$ and $-V_n(0,\tau) \times \text{PVBP}_\tau(T_1,\ldots,T_n)$. These exposures do not offset because they refer to the volatility of two distinct rates: the $3m$ into $\hat{T}_5 - 3m$ forward swap rate and the $9m$ into $T_5 - 9m$ forward swap rate. The device to short α units of the $9m$ contract serves the purpose of offsetting these two exposures at time τ. The expression for α in Eq. (3.31) allows us to create a zero-cost portfolio at $t = 0$, as explained.

Next, we develop an example of a "shrinking maturity trade," i.e., a trading strategy relying on contracts with different maturities but referenced to the same forward swap rate. Assume that at time $t \equiv 0$, the term structure of the IRV swap rate $\mathbb{P}_{\mathrm{var},n}(0,T)$ is increasing in T for some fixed tenor, but that a view is held that this term structure is about to flatten soon. For example, we might expect that at some

Fig. 3.6 Shrinking maturity trade

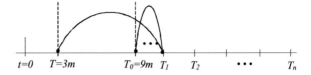

point $\tau < 3m$,

$$\mathbb{P}_{\text{var},n}(\tau, 3m) > \mathbb{P}_{\text{var},n}(0, 3m) \quad \text{and} \quad \mathbb{P}_{\text{var},n}(\tau, 9m) < \mathbb{P}_{\text{var},n}(0, 9m). \tag{3.32}$$

To express this view, one could go long a 3 month IRV swap struck at $\mathbb{P}_{\text{var},n}(0, 3m)$ and short a 9-month IRV swap struck at $\mathbb{P}_{\text{var},n}(0, 9m)$. Note that this trading strategy relies on IRV swaps for forward swap rates that are 9 months into $T_n - 9m$ years, at time $t = 0$, as Fig. 3.6 illustrates.

The payoffs to these two IRV swaps are as follows:

(i) The payoff of the 3 month IRV swap at $T = 3m$ is $V_n(0, 3m) \times \text{PVBP}_{3m}(T_1, \dots, T_n)$, where $\text{PVBP}_{3m}(T_1, \dots, T_n)$ is the value at T of the annuity beginning at T_1.

(ii) The payoff of the 9-month IRV swap at $T_0 = 9m$ is $V_n(0, 9m) \times \text{PVBP}_{9m}(T_1, \dots, T_n)$, where $\text{PVBP}_{9m}(T_1, \dots, T_n)$ is the value at T_0 of the annuity beginning at T_1.

The value of the 9-month contract relies on the price (at $t = 0$) of OTM swaptions that are 9 months into $T_n - 9m$ years. Instead, the value of the 3-month contract depends on the price of OTM swaptions with expiry $T = 3m$ and underlying swap starting at $T_0 = 9m$. Typically, swaption markets have maturities T equal to T_0. Theoretically, though, the pricing framework in this section is still valid even when $T < T_0$, contrary to the case of the 3-month contract in Fig. 3.6.

Note that markets may not exist (or, at best, be illiquid) for swaptions needed to price IRV contracts expiring strictly before the beginning of the tenor period, i.e. at $T < T_0$. In this case, pricing these contracts cannot be model-free. The simplest solution to price the illiquid swaptions could rely on the Black (1976) formula: one could use the implied volatilities for maturity $T_0 - t$ to price swaptions with time to maturity $T - t$. It is a rough approximation. Better approximations rely on local volatility (see Eq. (B.23) in Appendix B.2) or the SABR model (Hagan et al. 2002).

The strategy depicted in Fig. 3.6 eliminates the risk related to the fluctuations of realized interest rate swap volatility. Come time τ, its value is

$$\text{M-Var}_n(0, \tau, 3m) - \text{M-Var}_n(0, \tau, 9m)$$

$$= V_n(0, \tau) \times \text{PVBP}_\tau(T_1, \dots, T_n) + P_\tau(3m)\big[\mathbb{P}_{\text{var},n}(\tau, 3m) - \mathbb{P}_{\text{var},n}(0, 3m)\big]$$

$$- \big(V_n(0, \tau) \times \text{PVBP}_\tau(T_1, \dots, T_n) + P_\tau(9m)\big[\mathbb{P}_{\text{var},n}(0, 9m) - \mathbb{P}_{\text{var},n}(\tau, 9m)\big]\big)$$

$$> 0,$$

where we have made use of Eq. (3.23), and the last inequality follows if the view in (3.32) is correct. Note that this strategy eliminates the cumulative realized variance exposure $V_n(0, \tau) \times \text{PVBP}_\tau(T_1, \ldots, T_n)$. This property arises precisely because we are trading two IRV swaps referenced to the volatility of the same forward swap rate (9 months into $T_n - 9m$ years). Therefore, this strategy does not require additional positions in the IRV forward contract, as in the case of the constant tenor trades of Fig. 3.5.

3.4.2 Spot Trading Through Standardized IRV Swaps

The previous trading strategies rely on the IRV swap rate of Definition 3.2. Trading the standardized IRV swap rate of Definition 3.3 leads to comparable outcomes. For example, and in analogy with Eq. (3.29), suppose we hold the view that at some point $\tau < 3m$,

$$\mathbb{P}^*_{\text{var},\hat{n}}(\tau, 3m) > \mathbb{P}^*_{\text{var},\hat{n}}(0, 3m) \quad \text{and} \quad \mathbb{P}^*_{\text{var},n}(\tau, 9m) < \mathbb{P}^*_{\text{var},n}(0, 9m). \tag{3.33}$$

Consider a constant tenor trade of the type depicted in Fig. 3.5. That is, we implement the following strategy: (i) go long the $3m$ contract, (ii) short α units of the $9m$ contract, (iii) short the IRV forward contract of Definition 3.1 for the $3m$ maturity, with contract length equal to τ, and (iv) go long α units of the IRV forward contract for the $9m$ maturity, with contract length equal to τ.

Marking-to-market according to Eq. (3.24), we find that the value of this strategy at time τ is

$$\text{M-Var}^*_{\hat{n}}(0, \tau, 3m) - \alpha \cdot \text{M-Var}^*_n(0, \tau, 9m)$$

$$- V_{\hat{n}}(0, \tau) \times \text{PVBP}_\tau(\hat{T}_1, \ldots, \hat{T}_n) + \alpha \cdot V_n(0, \tau) \times \text{PVBP}_\tau(T_1, \ldots, T_n)$$

$$= \text{PVBP}_\tau(\hat{T}_1, \ldots, \hat{T}_n)\left[\mathbb{P}^*_{\text{var},\hat{n}}(\tau, 3m) - \mathbb{P}^*_{\text{var},\hat{n}}(0, 3m)\right]$$

$$+ \text{PVBP}_\tau(T_1, \ldots, T_n)\left[\mathbb{P}^*_{\text{var},n}(0, 9m) - \mathbb{P}^*_{\text{var},n}(\tau, 9m)\right]$$

$$> 0,$$

where α is as in Eq. (3.31) to ensure the portfolio is worthless at $t = 0$, and the inequality holds if the views in (3.33) are correct.

Likewise, and similar to Eq. (3.32), suppose that we hold the view that at some point $\tau < 3m$,

$$\mathbb{P}^*_{\text{var},n}(\tau, 3m) > \mathbb{P}^*_{\text{var},n}(0, 3m) \quad \text{and} \quad \mathbb{P}^*_{\text{var},n}(\tau, 9m) < \mathbb{P}^*_{\text{var},n}(0, 9m). \tag{3.34}$$

We now consider a shrinking maturity trade of the type depicted in Fig. 3.6. We go long the $3m$ and short the $9m$ standardized contracts. By Eq. (3.24), the value of this strategy at time τ is:

$$\text{M-Var}^*_n(0, \tau, 3m) - \text{M-Var}^*_n(0, \tau, 9m)$$

$$= \text{PVBP}_\tau(T_1, \ldots, T_n)$$

$$\times \left[\left(\mathbb{P}^*_{\text{var},n}(\tau, 3m) - \mathbb{P}^*_{\text{var},n}(0, 3m) \right) + \left(\mathbb{P}^*_{\text{var},n}(0, 9m) - \mathbb{P}^*_{\text{var},n}(\tau, 9m) \right) \right]$$
$$> 0,$$

where the inequality follows if the views in (3.34) are correct.

3.4.3 Forward Trading

Finally, we consider examples of trading strategies for expressing views on the implicit pricing that we can infer from current IRV swap rates. Suppose, for example, that at time $t = 0$, we hold the view that in one year's time, the IRV swap rate (Definition 3.2) for a contract expiring in a further year will be greater than the "implied forward" IRV swap rate, i.e. the rate implied by the current term structure of the IRV contracts, viz

$$\mathbb{P}_{\text{var},n}(1, 2) > \mathbb{P}_{\text{var},n}(0, 2) - \mathbb{P}_{\text{var},n}(0, 1). \tag{3.35}$$

The three IRV swap rates refer to contracts with tenors starting in 2 years' time from $t = 0$, and expiring at the same time T_n, similar to the contracts underlying the shrinking maturity trades of Fig. 3.6.

We synthesize the "cheap," implied IRV swap rate, by going long the following portfolio at $t = 0$:

(i) long a 2-year IRV swap, struck at $\mathbb{P}_{\text{var},n}(0, 2)$,

(ii) short a 1-year IRV swap, struck at $\mathbb{P}_{\text{var},n}(0, 1)$.   (3.P1)

This portfolio is costless at time zero, and if Eq. (3.35) holds true in 1 year, then in 1 year's time we close (i) and collect the payoff in (ii), securing a total payoff equal to:

$$\pi \equiv \text{M-Var}_n(0, 1, 2) + \left[\mathbb{P}_{\text{var},n}(0, 1) - V_n(0, 1) \times \text{PVBP}_1(T_1, \ldots, T_n) \right].$$

The first term is the value of the long position in (i) in one year's time, with M-Var$_n(0, 1, 2)$ as in Eq. (3.23). The second term is the payoff arising from (ii). Using Eq. (3.23), we obtain

$$\pi = \left[\mathbb{P}_{\text{var},n}(1, 2) - \mathbb{P}_{\text{var},n}(0, 2) \right] P_1(2) + \mathbb{P}_{\text{var},n}(0, 1)$$
$$\geq \left[\mathbb{P}_{\text{var},n}(1, 2) - \mathbb{P}_{\text{var},n}(0, 2) + \mathbb{P}_{\text{var},n}(0, 1) \right] P_1(2)$$
$$> 0,$$

where the first inequality follows by $P_1(2) < 1$, and the second follows whenever the views in (3.35) are correct.

Likewise, we can implement trading strategies consistent with views on implied standardized IRV swap rates (Definition 3.3). Suppose we anticipate that

$$\mathbb{P}^*_{\mathrm{var},n}(1,2) > \mathbb{P}^*_{\mathrm{var},n}(0,2) - \mathbb{P}^*_{\mathrm{var},n}(0,1), \tag{3.36}$$

and consider a shrinking maturity trade of the type depicted in Fig. 3.6. We can construct the same portfolio as in (3.P1), with standardized IRV contracts replacing IRV contracts. In one year's time, the value of this portfolio is

$$\text{M-Var}^*_n(0,1,2) + \left[\mathbb{P}^*_{\mathrm{var},n}(0,1) - V_n(0,1)\right] \times \text{PVBP}_1(T_1,\ldots,T_n)$$
$$= \left[\mathbb{P}^*_{\mathrm{var},n}(1,2) + \mathbb{P}^*_{\mathrm{var},n}(0,1) - \mathbb{P}^*_{\mathrm{var},n}(0,2)\right] \times \text{PVBP}_1(T_1,\ldots,T_n)$$
$$> 0,$$

where we have used the marking-to-market calculation in Eq. (3.24), and the inequality follows if the views in (3.36) are correct.

3.5 Interest Rate Swap Volatility Indexes

3.5.1 Basis Point Volatility Index

The BP standardized IRV swap rate in Definition 3.III can be normalized by the length of the contract, $T - t$, and quoted in terms of the current swap's PVBP, leading to the following forward-looking index of BP volatility:

$$\text{IRS-VI}^{\mathrm{bp}}_n(t,T) = \sqrt{\frac{1}{T-t}\mathbb{P}^{*\mathrm{bp}}_{\mathrm{var},n}(t,T)}, \qquad \mathbb{P}^{*\mathrm{bp}}_{\mathrm{var},n}(t,T) = \frac{\mathbb{F}^{\mathrm{bp}}_{\mathrm{var},n}(t,T)}{\text{PVBP}_t(T_1,\ldots,T_n)} \tag{3.37}$$

where $\mathbb{F}^{\mathrm{bp}}_{\mathrm{var},n}(t,T)$ is the value of the IRV forward agreement in Eq. (3.19). In Appendix B.2, we show that the expression in Eq. (3.38) can be simplified, so that the index can be calculated through the Black (1976) volatilities of the OTM swaption skew (see Eq. (B.20)).

Interestingly, our model-free BP volatility index squared and rescaled by $\frac{1}{T-t}$ coincides with the conditional second moment of the forward swap rate, as it turns out by comparing our formula in Eq. (3.37) with expressions in Trolle and Schwartz (2013). It is an interesting statistical property, which complements the asset pricing foundations laid down in Sect. 3.3, pertaining to security designs, hedging, replication, marking-to-market, and those relating to trading strategies in Sect. 3.4. In Sect. 3.5.4 we shall explain that our BP index has the additional property of being resilient to jumps: its value remains the same, and hence model-free, even in the presence of jumps.

3.5.2 Percentage Volatility Index

Similarly, the percentage counterpart to the IRS-VI index in Eq. (3.37) is the annualized square root of the percentage standardized IRV swap rate in Definition 3.3:

$$\text{IRS-VI}_n(t, T) = \sqrt{\frac{1}{T - t} \mathbb{P}^*_{\text{var},n}(t, T)} \tag{3.38}$$

where $\mathbb{P}^*_{\text{var},n}(t, T)$ is the standardized IRV swap rate in the second of Eqs. (3.22). We label the index in Eq. (3.38) "IRS-VI." Similar to the basis point index in Eq. (3.37), the percentage index in Eq. (3.38) can be calculated by feeding the Black (1976) formula with the swaption skew (see Eq. (B.19) in Appendix B.2).

3.5.3 Experiments

How do the previous interest rate swap volatility indexes compare to volatility expected in a risk-neutral market? Consider, for example, the percentage volatility index. If $V_n(t, T)$ and the path of the short-term rate in the time interval $[t, T]$ were independent, the IRS-VI index in Eq. (3.38) would collapse to the risk-neutral expectation of the future realized variance, defined as

$$\text{E-Vol}_n(t, T) \equiv \sqrt{\frac{1}{T - t} \mathbb{E}_t \big[V_n(t, T)\big]}, \tag{3.39}$$

where, as usual, $\mathbb{E}_t$ denotes the risk-neutral expectation conditional on time-t information.[5]

However, $V_n(t, T)$ and interest rates are likely to be correlated. For example, if the short-term rate is generated by the Vasicek (1977) model, the correlation between $V_n(t, T)$ and $\text{PVBP}_T(T_1, \ldots, T_n)$ is positive (a finding arising as a by-product of experiments reported in Appendix B.5). Despite this dependence, the model predicts that Eq. (3.38) is quite close to the risk-neutral expectation of future realized volatility, Eq. (3.39).

Figure 3.7 plots the relation between expected BP volatility and the forward rate arising within the Vasicek market. We consider a maturity $T - t = 1$ month and tenor $n = 5$ years (top panels), a maturity $T - t = 1$ year and tenor $n = 10$ years (bottom panels), and regular quarterly reset dates. The left panels plot the relation between the forward swap rate and: (i) the $\text{IRS-VI}_n^{\text{bp}}(t, T)$ index in Eq. (3.37), and (ii) the risk-neutral expectation of future realized BP volatility, calculated through Eq. (B.42) in Appendix B.5. The right panels plot the relation between the forward

---

[5]This claim follows from the definition of the Radon–Nikodym derivative of Q_{sw} with respect to Q (Eq. (3.11)) and Eq. (B.14) in Appendix B.

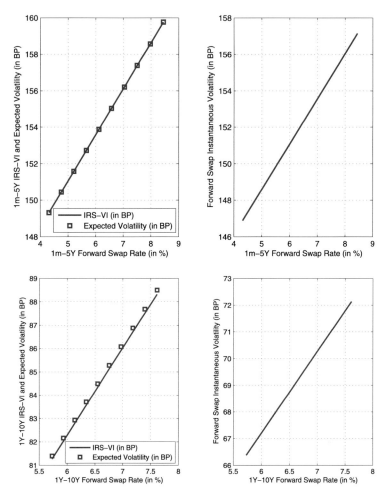

Fig. 3.7 *Left panels*: The IRS-VI index and expected volatility (both in Basis Points) in a risk-neutral Vasicek market, as a function of the level of the forward swap rate. *Right panels*: the instantaneous forward swap rate volatility (in Basis Points) as a function of the forward swap rate. *Top panels* relate to a length of the variance contract equal to 1 month and tenor of 5 years. *Bottom panels* relate to a length of the variance contract equal to 1 year and tenor of 10 years

swap rate and the instantaneous BP volatility of the forward swap rate as defined in Appendix B.5, Eq. (B.40).

Figure 3.8 reports experiments that are the percentage counterparts to those in Fig. 3.7. That is, we compare percentage volatility as referenced by our index for 5- and 1-year tenors, with that arising in a risk-neutral Vasicek market, as well as the instantaneous percentage volatility of the forward rate as predicted by the Vasicek model.

Appendix B.5 provides more details regarding these calculations, as well as additional numerical results relating to a number of alternative combinations of ma-

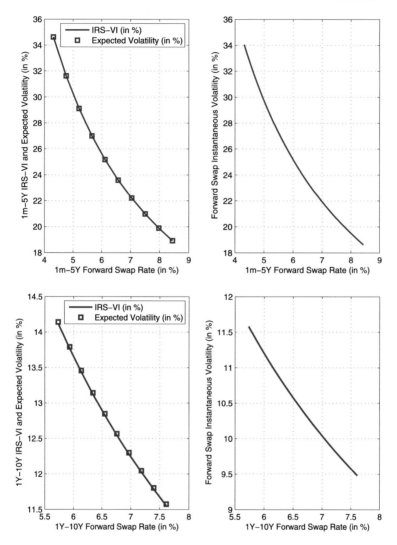

Fig. 3.8 *Left panels*: The IRS-VI index and expected volatility (both in percentage) in a risk-neutral Vasicek market, as a function of the level of the forward swap rate. *Right panels*: the instantaneous forward swap rate volatility (in percentage) as a function of the forward swap rate. *Top panels* relate to a length of the variance contract equal to 1 month and tenor of 5 years. *Bottom panels* relate to a length of the variance contract equal to 1 year and tenor of 10 years

turities and tenors (see Tables 3.7 and 3.8). In these examples, the IRS-VI indexes highly correlate with volatility expected under the risk-neutral probability, just as in Figs. 3.7 and 3.8. We find that both the IRS-VI and expected volatility respond in the same way to changes in market conditions, as summarized by movements in the forward swap rate, and deviate quite insignificantly from each other. Interestingly, in this Vasicek market, we observe the interest rate counterpart to the "leverage effect" observed in equity markets: low interest rates are associated with high interest rate volatility.

3.5.4 Jumps

Chapter 2 Sect. 2.7 explains that in the presence of jumps, *basis point* variance swaps are model-free. That is, their value is the same as that in the absence of jumps, independently of any parametric assumption regarding the jump data generating process. In contrast, the value of *percentage* variance contracts depends on the assumptions we make about jumps, as illustrated by Eq. (2.44) in Chap. 2 (see, also, Remark 2.4).

3.6 Swap Versus Equity Variance Contracts and Indexes

Chapter 2 emphasizes that similar to equity variance swaps, the pricing of IRV contracts relies on spanning arguments such as those in Bakshi and Madan (2000), Carr and Madan (2001), and others. In the case of IRV contracts, we need to generalize these arguments to account for the specific numéraires applying to each market of interest. We illustrate this feature of IRV contracts in the interest rate swap case.

As explained in Sect. 3.2, the payoffs of IRV contracts have two sources of uncertainty: one related to the realized variance of the forward swap rate, and another related to the forward PVBP. These features of the contract design and the natural assumption that swap rates are stochastic distinguish our contract design from the standard equity case. For example, the fair value of a variance swap in the equity case is the square of the following index:

$$\text{VIX}(t, T)$$

$$= \sqrt{\frac{1}{T-t} \frac{2}{e^{-\bar{r}(T-t)}} \left(\sum_{i:K_i < F_t(T)} \frac{\text{Put}_t(K_i, T)}{K_i^2} \Delta K_i + \sum_{i:K_i \geq F_t(T)} \frac{\text{Call}_t(K_i, T)}{K_i^2} \Delta K_i \right)},$$

(3.40)

where $\bar{r}$ is the instantaneous interest rate, *assumed to be constant*, and $\text{Put}_t(K, T)$ and $\text{Call}_t(K, T)$ are the market prices at time t of OTM European put and call equity options with strike prices equal to K and time to maturity $T - t$, $F_t(T)$ is the time-t value of a forward contract on equity, and ΔK_i are as in (3.19). Note that due to constant discounting, the fair strike for an equity variance contract is the same as the index, up to time-rescaling and squaring, and is given by:

$$\mathbb{P}_{\text{equity}}(t, T) \equiv (T - t) \cdot \text{VIX}^2(t, T).$$

(3.41)

As for IRV contracts, note that the fair value of the standardized IRV swap rate, $\mathbb{P}^*_{\text{var},n}(t, T)$ in Eq. (3.18), is still the index, up to time-rescaling and squaring and, by Eq. (3.38), is given by

IRS-VI$_n(t, T)$

$$
= \sqrt{\frac{1}{T-t} \frac{2}{\text{PVBP}_t(T_1, \ldots, T_n)} \left(\sum_{i:K_i < R_t} \frac{\text{Swpn}_t^R(K_i, T; T_n)}{K_i^2} \Delta K_i + \sum_{i:K_i \geq R_t} \frac{\text{Swpn}_t^P(K_i, T; T_n)}{K_i^2} \Delta K_i \right)}.
$$

$$(3.42)$$

While the two indexes in Eqs. (3.40) and (3.42) aggregate prices of OTM derivatives using the same weights, they differ for a number of reasons. Some of these reasons have been discussed in Chap. 2, while others are more specific to interest rate swap markets.

First, the equity VIX in Eq. (3.40) is constructed by rescaling a weighted average of OTM option prices through the inverse of the price of a zero with the same expiry date as that of the variance contract in a market with constant interest rates, $e^{-\bar{r}(T-t)}$. Instead, the IRS-VI index in Eq. (3.42) rescales a weighted average of out-of-the-money swaption prices with the inverse of the price of a basket of bonds, $\text{PVBP}_t(T_1, \ldots, T_n)$, each of them expiring at the end-points of the swap's fixed payment dates.

Second, the derivatives involved in the definition of the two indexes differ. The VIX(t, T) index is calculated through European options, whereas the IRS-VI$_n(t, T)$ index is fed by swaptions.

Third, swap rate volatility has one additional dimension, relating to the length of the tenor underlying the forward swap rate, $T_n - T_0$.

Fourth, consider the current value of a forward for delivery of variance at time T in Sect. 3.3.2 (Definition 3.1), which by Eq. (3.22) is:

$$
\mathbb{F}_{\text{var},n}(t, T) = \text{PVBP}_t(T_1, \ldots, T_n) \cdot \mathbb{P}^*_{\text{var},n}(t, T). \tag{3.43}
$$

As for equity, the price to be paid at t, for delivery of equity volatility at T, is, instead

$$
\mathbb{F}_{\text{equity}}(t, T) \equiv P_t(T) \cdot \mathbb{P}_{\text{equity}}(t, T). \tag{3.44}
$$

The two prices, $\mathbb{F}_{\text{var},n}$ and $\mathbb{F}_{\text{equity}}$, scale up to the time t fair values of their respective notionals at time T: (i) the fair value $\text{PVBP}_t(T_1, \ldots, T_n)$ of the *random notional* $\text{PVBP}_T(T_1, \ldots, T_n)$, in Eq. (3.43); and (ii) the fair value $P_t(T)$ of the *deterministic notional* of one dollar, in Eq. (3.44). Note that this property holds in spite of the different assumptions underlying the two contracts: (i) *random* interest rates (for the IRV contracts), and (ii) *constant* interest rates (for the equity variance swaps).

Fifth, consider the BP IRV contracts in Sect. 3.3.2. In Chap. 2, we provide motivation, explanations and properties regarding a basis point index, some of which are repeated here for convenience. By the expression in Eq. (3.19) for the fair value of the BP standardized IRV swap rate, the BP index in Eq. (3.37) can be written as:

$$\text{IRS-VI}_n^{\text{bp}}(t, T)$$

$$= \sqrt{\frac{1}{T-t} \frac{2}{\text{PVBP}_t(T_1, \ldots, T_n)} \left(\sum_{i:K_i < R_t} \text{Swpn}_t^{\text{R}}(K_i, T; T_n) \Delta K_i + \sum_{i:K_i \geq R_t} \text{Swpn}_t^{\text{P}}(K_i, T; T_n) \Delta K_i \right)}. \tag{3.45}$$

For the index of BP volatility in Eq. (3.45), the weights to be given to the OTM swaption prices are not inversely proportional to the square of the strike, as in the case for the equity volatility index in Eq. (3.40). In the theoretical case with a continuum of strikes, the BP index would actually be an *equally-weighted* average of out-of-the-money swaption prices, as revealed by the exact expression for the BP-standardized IRV swap rate in Appendix B.2 (see Eq. (B.18)), and consistent with the constant Gaussian vega explanations in Chap. 2 (see Eq. (2.21) in Sect. 2.3.4).[6]

Therefore, the BP interest rate volatility index and contracts differ from their equity counterparts, not only due to the aspects pointed out for "percentage" volatility, but also for the particular weighting each swaption price enters into the definition of the index and contracts. Note that these different weightings reflect different hedging strategies. As explained in Sect. 3.3.5, replicating a variance contract referencing "percentage" volatility requires "log-contracts," whereas replicating a variance contract referencing "basis point" realized variance requires "quadratic contracts."

3.7 Index Implementation

The index in Eq. (3.37) is the basis for the CBOE SRVIX index of BP interest rate swap volatility launched by the Chicago Board Options Exchange in June 2012. The index references a maturity of 1 year and tenor of 10 years. The reader is referred to the CBOE website, http://www.cboe.com/micro/srvix/default.aspx, for additional details. Figure 3.9 depicts the times-series of the index, along with additional maturities of 3 months and 2 years. The volatility of the index increases as the maturity shrinks. Mele et al. (2015a) provide an empirical analysis of the CBOE SRVIX and a comparison with the dynamics of the equity volatility as measured by CBOE VIX.

This section provides examples of a numerical implementation of the index. Section 3.7.1 is a step-by-step illustration of the index calculations. Section 3.7.2 contains examples of the index behavior around particular dates. We focus on the percentage volatility index, as this analysis complements other analyses regarding the basis point version (see Mele et al. 2015a).

---

[6]However, note that due to discretization and truncations, the weights in Eq. (3.45), ΔK_i, might not be exactly the same.

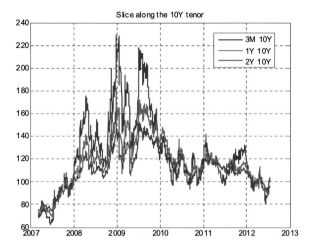

Fig. 3.9 Estimates of the CBOE-SRVIX calculated over different maturities and for a given tenor equal to 10 years

3.7.1 A Numerical Example

The following is a numerical illustration of the main steps involved in the construction of the IRS-VI indexes IRS-VI$_n^{bp}(t, T)$ and IRS-VI$_n(t, T)$ in Eq. (3.37) and Eq. (3.38). We utilize hypothetical data for implied volatilities expressed in percentage terms for swaptions maturing in 1 month and tenor equal to 5 years reflecting hypothetical market conditions on February 12, 2010. The first two columns of Table 3.5 report strike rates, K, and the "skew," i.e. the percentage implied volatilities for each strike, denoted as IV(K) ("Percentage Implied Vol"). For reference, the third column provides BP implied volatilities for each strike K, denoted as IV$^{bp}(K)$, and calculated as

$$\text{IV}^{bp}(K) = \text{IV}(K) \cdot R, \tag{3.46}$$

where R denotes the current forward swap rate ("Basis Point Implied Vol"). For example, from Table 3.5, $R = 2.7352\%$ and IV$(K)|_{K=R} = 35.80\%$ and, then, IV$^{bp}(K)|_{K=R} = 2.7352\% \times 35.80\% = 0.979202\%$, which is 97.9202 BP volatility. BP volatilities are not needed to calculate the indexes of this section.

The two IRS-VI indexes in Eq. (3.37) and Eq. (3.38) are implemented through the Black (1976) formula, as explained in Appendix B.2 (see Eq. (B.19) and Eq. (B.20)).

First, we plug the skew IV(K) into the Black (1976) formulae,

$$\hat{Z}\big(R, T, K_i; (T-t)\text{IV}^2(K_i)\big) = Z\big(R, T, K_i; (T-t)\text{IV}^2(K_i)\big) + K - R, \tag{3.47}$$

$$Z(R, T, K; V) = R\Phi(d) - K\Phi(d - \sqrt{V}), \quad d = \frac{\ln\frac{R}{K} + \frac{1}{2}V}{\sqrt{V}}, \tag{3.48}$$

where Φ denotes the cumulative standard normal distribution, and $\hat{Z}(\cdot)$ and $Z(\cdot)$ are the Black prices, i.e. swaption prices (receivers, for $\hat{Z}(\cdot)$; and payers, for $Z(\cdot)$), divided by the PVBP. Second, we use Eqs. (3.47)–(3.48), and calculate two indexes

Table 3.5 Implied volatilities and Black's prices

Strike Rate (%)	Percentage Implied Vol	Basis Point Implied Vol	Black's prices	
			Receiver Swaption ($\hat{Z}$)	Payer Swaption Z
1.7352	36.1900	98.9869	≈ 0	$10.0000 \cdot 10^{-3}$
1.9852	36.1900	98.9869	$0.0007 \cdot 10^{-3}$	$7.5007 \cdot 10^{-3}$
2.2352	36.1200	98.7954	$0.0259 \cdot 10^{-3}$	$5.0259 \cdot 10^{-3}$
2.4352	35.9900	98.4398	$0.1773 \cdot 10^{-3}$	$3.1773 \cdot 10^{-3}$
2.5352	35.9300	98.2757	$0.3692 \cdot 10^{-3}$	$2.3692 \cdot 10^{-3}$
2.6352	35.8600	98.0843	$0.6793 \cdot 10^{-3}$	$1.6793 \cdot 10^{-3}$
2.6852	35.8300	98.0022	$0.8855 \cdot 10^{-3}$	$1.3855 \cdot 10^{-3}$
2.7352 (ATM)	35.8000	97.9202	$1.1272 \cdot 10^{-3}$	$1.1272 \cdot 10^{-3}$
2.7852	35.7600	97.8108	$1.4037 \cdot 10^{-3}$	$0.9037 \cdot 10^{-3}$
2.8352	35.7300	97.7287	$1.7142 \cdot 10^{-3}$	$0.7142 \cdot 10^{-3}$
2.9352	35.6700	97.5646	$2.4270 \cdot 10^{-3}$	$0.4270 \cdot 10^{-3}$
3.0352	35.6000	97.3731	$3.2406 \cdot 10^{-3}$	$0.2406 \cdot 10^{-3}$
3.2352	35.4700	97.0175	$5.0644 \cdot 10^{-3}$	$0.0644 \cdot 10^{-3}$
3.4852	35.3100	96.5799	$7.5092 \cdot 10^{-3}$	$0.0092 \cdot 10^{-3}$
3.7352	35.1400	96.1149	$10.0010 \cdot 10^{-3}$	$0.0010 \cdot 10^{-3}$

in Eq. (3.37) and Eq. (3.38):

IRS-VI$_n^{\text{bp}}(t, T)$

$$= 100^2 \times \sqrt{\frac{2}{T - t} \left(\sum_{i:K_i < R_t} \hat{Z}\big(R, T, K_i; (T - t)\text{IV}^2(K_i)\big) \Delta K_i + \sum_{i:K_i \geq R_t} Z\big(R, T, K_i; (T - t)\text{IV}^2(K_i)\big) \Delta K_i \right)}, \tag{3.49}$$

and

IRS-VI$_n(t, T)$

$$= 100 \times \sqrt{\frac{2}{T - t} \left(\sum_{i:K_i < R_t} \frac{\hat{Z}\big(R, T, K_i; (T - t)\text{IV}^2(K_i)\big)}{K_i^2} \Delta K_i + \sum_{i:K_i \geq R_t} \frac{Z\big(R, T, K_i; (T - t)\text{IV}^2(K_i)\big)}{K_i^2} \Delta K_i \right)}. \tag{3.50}$$

The BP index in Eq. (3.49) is rescaled by 100^2. This rescaling parallels the market practice of expressing BP implied volatility as the product of rates times log-volatility, where both rates and log-volatility are multiplied by 100.[7]

---

[7]Consider, for example, the BP variance in Eq. (3.13), $V_n^{\text{bp}}(t, T)$, and suppose that the forward swap rate $R_\tau(T_1, \ldots, T_n)$ and its instantaneous volatility $\sqrt{\|\sigma_\tau(T_1, \ldots, T_n)\|^2}$ are both expressed in decimals, as in Table 3.5. Then, the *variance* $V_n^{\text{bp}}(t, T)$ is rescaled by $100^2 \times 100^2$, once

Table 3.6 Calculation of volatility indexes

Strike Rate (%)	Swaption Type	Price	Weights		Contributions to Strikes	
			Basis Point ΔK_i	Percentage $\Delta K_i / K_i^2$	Basis Point Contribution	Percentage Contribution
1.7352	Receiver	≈ 0	0.0025	8.3031	≈ 0	≈ 0
1.9852	Receiver	$0.0007 \cdot 10^{-3}$	0.0025	6.3435	$0.0018 \cdot 10^{-6}$	$0.0046 \cdot 10^{-3}$
2.2352	Receiver	$0.0259 \cdot 10^{-3}$	0.0022	4.5035	$0.0583 \cdot 10^{-6}$	$0.1167 \cdot 10^{-3}$
2.4352	Receiver	$0.1773 \cdot 10^{-3}$	0.0015	2.5294	$0.2660 \cdot 10^{-6}$	$0.4485 \cdot 10^{-3}$
2.5352	Receiver	$0.3692 \cdot 10^{-3}$	0.0010	1.5559	$0.3692 \cdot 10^{-6}$	$0.5744 \cdot 10^{-3}$
2.6352	Receiver	$0.6793 \cdot 10^{-3}$	0.0008	1.0800	$0.5095 \cdot 10^{-6}$	$0.7337 \cdot 10^{-3}$
2.6852	Receiver	$0.8855 \cdot 10^{-3}$	0.0005	0.6935	$0.4428 \cdot 10^{-6}$	$0.6141 \cdot 10^{-3}$
2.7352	ATM	$1.1272 \cdot 10^{-3}$	0.0005	0.6683	$0.5636 \cdot 10^{-6}$	$0.7533 \cdot 10^{-3}$
2.7852	Payer	$0.9037 \cdot 10^{-3}$	0.0005	0.6446	$0.4518 \cdot 10^{-6}$	$0.5825 \cdot 10^{-3}$
2.8352	Payer	$0.7142 \cdot 10^{-3}$	0.0007	0.9330	$0.5357 \cdot 10^{-6}$	$0.6664 \cdot 10^{-3}$
2.9352	Payer	$0.4270 \cdot 10^{-3}$	0.0010	1.1607	$0.4270 \cdot 10^{-6}$	$0.4956 \cdot 10^{-3}$
3.0352	Payer	$0.2406 \cdot 10^{-3}$	0.0015	1.6282	$0.3609 \cdot 10^{-6}$	$0.3917 \cdot 10^{-3}$
3.2352	Payer	$0.0644 \cdot 10^{-3}$	0.0023	2.1497	$0.1448 \cdot 10^{-6}$	$0.1384 \cdot 10^{-3}$
3.4852	Payer	$0.0092 \cdot 10^{-3}$	0.0025	2.0582	$0.0229 \cdot 10^{-6}$	$0.0188 \cdot 10^{-3}$
3.7352	Payer	$0.0010 \cdot 10^{-3}$	0.0025	1.7919	$0.0024 \cdot 10^{-6}$	$0.0017 \cdot 10^{-3}$
SUMS					$4.1567 \cdot 10^{-6}$	$5.5405 \cdot 10^{-3}$

The fourth column of Table 3.5 provides the values of $\hat{Z}$ and Z ("Black prices") for each strike rate.

Table 3.6 provides details regarding the calculation of the indexes IRS-VI$_n^{\mathrm{bp}}(t, T)$ and IRS-VI$_n(t, T)$ through Eqs. (3.49) and (3.50): the second column displays the type of OTM swaption entering into the calculation; the third column has the corresponding Black prices; the fourth and fifth columns report the weights each price bears towards the final calculation of the index before the final rescaling of $\frac{2}{T-t}$; finally, the sixth and seventh columns report each OTM swaption price multiplied by the appropriate weight (i.e. third column multiplied by the fourth column for the "Basis Point Contribution," and third column multiplied by the fifth column for the "Percentage Contribution").

The two indexes are calculated as

$$\text{IRS-VI}_n(t, T) = 100 \times \sqrt{\frac{2}{12^{-1}} \times 5.5405 \cdot 10^{-3}} = 36.4653,$$

$$\text{IRS-VI}_n^{\mathrm{bp}}(t, T) = 100^2 \times \sqrt{\frac{2}{12^{-1}} \times 4.1567 \cdot 10^{-6}} = 99.8803.$$

$R_\tau(T_1, \ldots, T_n)$ and $\sqrt{\|\sigma_\tau(T_1, \ldots, T_n)\|^2}$ are both rescaled by 100, leading to the scaling factor 100×100 for the *volatility* index in Eq. (3.49).

Fig. 3.10 Term structure of
the Interest Rate Swap
Volatility Index
(IRS-VI)—February 16, 2007

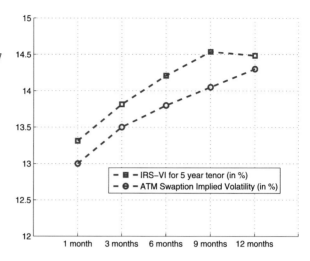

In comparison, ATM implied volatilities are $\mathrm{IV}(R) = 35.8000$ and $\mathrm{IV}^{\mathrm{bp}}(R) = 97.9202$.

3.7.2 Historical Performance

This section documents the performance of the IRS-VI index over selected days
based on data provided by a major interdealer broker. First, we consider the index behavior at four selected dates: (i) February 16, 2007, (ii) February 15, 2008,
(iii) February 13, 2009, and (iv) February 12, 2010. Second, we document the index
behavior over Lehman Brothers' collapse on September 2008. We only report experiments relating to the percentage index as defined in Eq. (3.38), $\mathrm{IRS\text{-}VI}_n(t, T)$. We
use ATM swaptions and OTM (OTM, henceforth) swaptions, with moneyness equal
to ∓200, ∓100, ∓50, and, finally, ∓25 basis points away from the current forward
swap rate. The index is calculated using "constant tenors" (similar to the trading
strategy depicted in Fig. 3.5). That is, for each time-to-maturity T, $\mathrm{IRS\text{-}VI}_n(t, T)$ is
the volatility index for a tenor period equal to $T_n - T$ and all considered n.

Figures 3.10, 3.11, 3.12, 3.13 compare the IRS-VI index with the implied volatility for ATM swaptions with 5-year tenors as a function of the maturity of the variance contract (1, 3, 6, 9, and 12 months) over the selected dates. Figures 3.14, 3.15,
3.16, 3.17 plot the IRS-VI "surface" for the same dates; that is, the plot of the IRS-
VI index against (i) the maturity of the variance contract and (ii) the tenor of the
forward rate.

The IRS-VI index deviates from the corresponding ATM volatilities, being at
times higher and other times lower, depending on the specific date of the variance
contract horizon. For example, on February 15, 2008, the IRS-VI index is higher
than ATM volatilities for short maturities, and lower otherwise. The reason for
these differences is that the IRS-VI aggregates information relating to both ATM

Fig. 3.11 Term structure of
the Interest Rate Swap
Volatility Index
(IRS-VI)—February 15, 2008

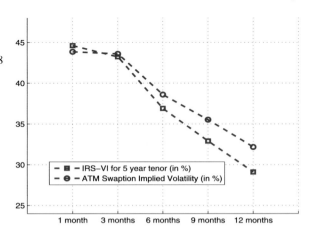

Fig. 3.12 Term structure of
the Interest Rate Swap
Volatility Index
(IRS-VI)—February 13, 2009

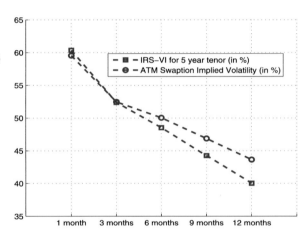

Fig. 3.13 Term structure of
the Interest Rate Swap
Volatility Index
(IRS-VI)—February 12, 2010

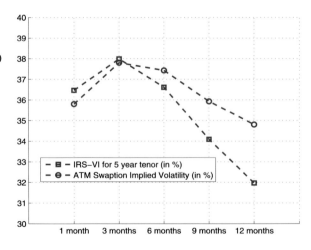

Fig. 3.14 Interest Rate Swap
Volatility Index
(IRS-VI)—February 16, 2007

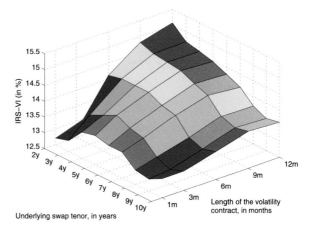

Fig. 3.15 Interest Rate Swap
Volatility Index
(IRS-VI)—February 15, 2008

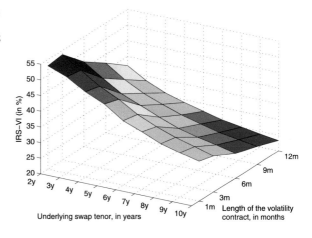

and OTM volatilities, in such a way as to isolate expected volatility from other
factors, such as changes in the underlying forward rates. On February 12, 2010,
OTM volatilities were overall higher (resp. lower) than ATM volatilities, for shorter
(longer) maturities. This pattern is captured by the IRS-VI, which is higher than
ATM volatilities for maturities up to 3 months, and lower than ATM volatilities for
maturities of 6, 9, and 12 months.

In these empirical case studies, the pattern of the IRS-VI index across different
maturities displays more pronounced characteristics than those of ATM volatilities.
On two particular days during "bad times" (February 15, 2008 and February 13,
2009), the IRS-VI slopes more sharply downwards than ATM volatilities do, thereby
signaling the market's expectation that the bad times will be followed by resolution
of uncertainty about interest rates. In good times (February 16, 2007), the IRS-VI
index and ATM volatilities both slope up, albeit moderately, reflecting market concerns that these good times, while persistent, will become slightly more uncertain
in the future. Even in this simple case, the IRS-VI index exhibits a richer behavior

Fig. 3.16 Interest Rate Swap
Volatility Index
(IRS-VI)—February 13, 2009

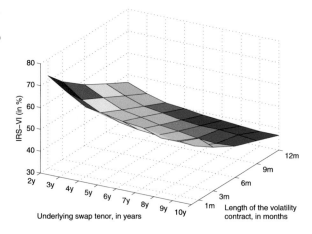

Fig. 3.17 Interest Rate Swap
Volatility Index
(IRS-VI)—February 12, 2010

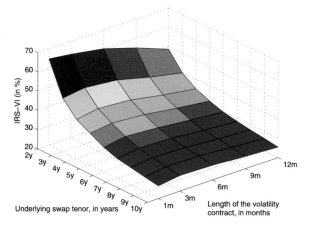

than ATM volatilities, flattening out for maturities greater than 9 months. Finally, the hump shape of the IRS-VI index and ATM volatilities occurring on February 12, 2010 reflects market expectations that uncertainty about interest rates would rise in the short term and subside thereafter.

These case studies reveal that the IRS-VI index and ATM volatilities do not display the same pattern, both quantitatively and qualitatively. Moreover, the IRV contracts in Sect. 3.3 cannot be priced using solely ATM volatilities, but require the IRS-VI index. Figures 3.14 through 3.17 further illustrate that the IRS-VI index can be used to express views on interest rate swap volatility. In fact, one of the added features of the IRS-VI index and contract, compared to the VIX index for equities, is the possibility that market participants may choose the tenor of the forward swap rate on top of the length of time over which to express volatility views. This added dimension arises quite naturally as a result of the increased complexity of fixed income derivatives. For example, the trading strategies summarized by Fig. 3.5 can be implemented on a particular tenor $T_n - T$ and another one of the tenors in Figs. 3.14–3.17 where the mispricing is thought to be occurring.

Fig. 3.18 Interest Rate Swap
Volatility Index (IRS-VI)
around the Lehman's collapse
days: I

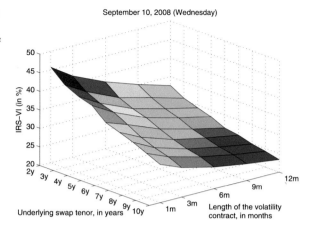

Fig. 3.19 Interest Rate Swap
Volatility Index (IRS-VI)
around the Lehman's collapse
days: II

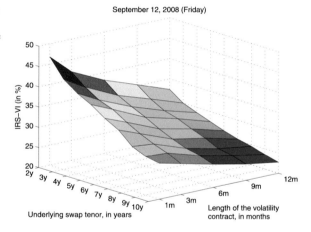

Finally, Figs. 3.18, 3.19, 3.20, 3.21 and 3.22 depict the behavior of the IRS-VI index around the Lehman collapse on September 15, 2008. A few days before the collapse, the index was downward sloping with respect to maturity, reflecting market expectations of possibly imminent negative tail events. These expectations continued to strengthen through the bankruptcy event, only to partially weaken after a week's time.

Appendix B: Appendix on Interest Rate Swap Markets

B.1 P&L of Option-Based Volatility Trading

We provide details regarding P&L calculations associated with swaption-based volatility trading. First, we provide basic definitions related to the Black (1976)

Fig. 3.20 Interest Rate Swap
Volatility Index (IRS-VI)
around the Lehman's collapse
days: III

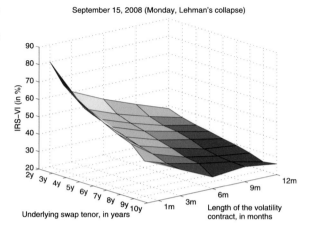

Fig. 3.21 Interest Rate Swap
Volatility Index (IRS-VI)
around the Lehman's collapse
days: IV

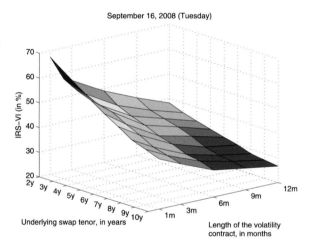

Fig. 3.22 Interest Rate Swap
Volatility Index (IRS-VI)
around the Lehman's collapse
days: V

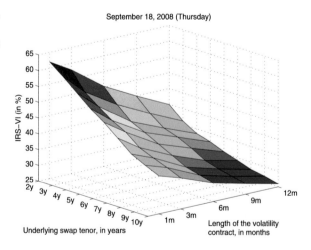

formula. Second, we derive the approximations in Eqs. (3.8) and (3.9). Third, we provide details regarding the experiments in Sect. 3.2.2 on directional volatility trading.

DEFINITIONS. The price of a payer swaption in Eq. (3.14) is known in closed-form when the volatility σ_t in Eq. (3.12), and hence the integrated variance $V_n(t, T)$ in Eq. (3.13), are deterministic. It is given by the Black (1976) formula

$$\text{Swpn\_Bl}_t^P\left(R_t^n, \text{PVBP}_t, K, T; \bar{V}\right) = \text{PVBP}_t \cdot Z_t\left(R_t^n, T, K; \bar{V}\right), \qquad (B.1)$$

where

$$Z_t(R_t; T, K; \bar{V}) = R_t \Phi(d_t) - K \Phi(d_t - \sqrt{\bar{V}}), \quad d_t = \frac{\ln \frac{R_t}{K} + \frac{1}{2} \bar{V}}{\sqrt{\bar{V}}},$$

Φ denotes the cumulative standard normal distribution, and to alleviate notation, $R_t^n \equiv R_t(T_1, \ldots, T_n)$, $\text{PVBP}_t \equiv \text{PVBP}_t(T_1, \ldots, T_n)$, and $\bar{V}$ is the constant value of $V_n(t, T)$ in Eq. (3.13). By the definition of the forward swap rate in Eq. (3.5), we have

$$\text{Swpn\_Bl}_t^P\left(R_t^n, \text{PVBP}_t, K, T; \bar{V}\right)$$

$$\equiv \left(P_t(T) - P_t(T_n)\right)\Phi(d_t) - \text{PVBP}_t K \Phi(d_t - \sqrt{\bar{V}}). \qquad (B.2)$$

Equation (B.2) suggests that the swaption can be hedged through portfolios of zero-coupon bonds: (i) long $\Phi(d_t)$ units of a portfolio which is long one zero expiring at T and short one zero expiring at T_n, and (ii) short $K \Phi(d_t - \sqrt{\bar{V}})$ units of the PVBP_t basket.

Next, and against the assumption underlying Eq. (B.2), we assume that the forward swap rate has stochastic volatility, σ_t, as in Eq. (3.12) of the main text. Suppose, further, that we have a view that volatility will rise compared to the current implied volatility, defined as the value

$$\text{IV}_t = \sqrt{\frac{\bar{V}}{T - t}}, \qquad (B.3)$$

so that, once re-normalized again by $T - t$ and inserted into Eq. (B.2), it delivers the swaption market price.

VOLATILITY TRADING BASED ON DELTA-HEDGED SWAPTIONS. Consider the strategy of buying the swaption and hedging it with the portfolio underlying Eq. (B.2). We derive the P&L of this strategy. Let υ denote the value of a self-financed hedging portfolio that includes (i) long a zero expiring at T and short a zero expiring at T_n, (ii) a basket of zeros with value equal to the PVBP. The value of this portfolio satisfies:

$$\upsilon_\tau = a_\tau\left(P_\tau(T) - P_\tau(T_n)\right) + b_\tau \text{PVBP}_\tau, \quad \tau \in [t, T],$$

where, denoting for simplicity $R_t \equiv R_t^n$,

$$v_t = \text{Swpn\_Bl}_t^P\big(R_t, \text{PVBP}_t, K, T, (T-t)\text{IV}_t^2\big),$$

$$a_\tau = \Phi\left(\frac{\ln\frac{R_\tau}{K} + \frac{1}{2}(T-t)\text{IV}_t^2}{\sqrt{T-t}\text{IV}_t}\right) \quad \text{and} \quad b_\tau = -K\Phi\left(\frac{\ln\frac{R_\tau}{K} - \frac{1}{2}(T-t)\text{IV}_t^2}{\sqrt{T-t}\text{IV}_t}\right).$$
(B.4)

Because this portfolio is self-financed, $dv_\tau = a_\tau d[P_\tau(T) - P_\tau(T_n)] + b_\tau d\text{PVBP}_\tau$, and, hence

$$dv_\tau = \big[\mu_\tau^b v_\tau + \big(\mu_\tau^{\Delta P} - \mu_\tau^b\big)a_\tau\big(P_\tau(T) - P_\tau(T_n)\big)\big]d\tau$$
$$+ \big[\sigma_\tau^b v_\tau + \big(\sigma_\tau^{\Delta P} - \sigma_\tau^b\big)a_\tau\big(P_\tau(T) - P_\tau(T_n)\big)\big]dW_\tau,$$
(B.5)

where W_τ is a Wiener process under the physical probability and, accordingly, μ_τ^b and σ_τ^b denote the drift and instantaneous volatility of $\frac{d\text{PVBP}_\tau}{\text{PVBP}_\tau}$, and $\mu_\tau^{\Delta P}$ and $\sigma_\tau^{\Delta P}$ denote the drift and instantaneous volatility of $\frac{d(P_\tau(T) - P_\tau(T_n))}{P_\tau(T) - P_\tau(T_n)}$.

Next, consider the swaption price in Eq. (B.1) with $\bar{V}$ replaced by $(T-t)\text{IV}_t^2$, as in Eq. (B.3), i.e. $\text{Swpn\_Bl}_\tau^P \equiv \text{Swpn\_Bl}_\tau^P(R_\tau, \text{PVBP}_\tau, K, T, (T-t)\text{IV}_t^2)$. By Itô's lemma, the definition of the forward swap rate in Eq. (3.5), and the partial differential equation satisfied by the pricing function $Z_t(R, T, K; \bar{V})$, the swaption price satisfies

$$d\text{Swpn\_Bl}_\tau^P$$
$$= \text{PVBP}_\tau dZ_\tau + Z_\tau d\text{PVBP}_\tau + dZ_\tau d\text{PVBP}_\tau$$
$$= \text{PVBP}_\tau \underbrace{\left(\frac{\partial Z_\tau}{\partial \tau} + \frac{1}{2}\frac{\partial^2 Z_\tau}{\partial R^2}R_\tau^2\text{IV}_t^2\right)}_{=0}d\tau$$
$$+ \text{PVBP}_\tau\left(\frac{1}{2}\frac{\partial^2 Z_\tau}{\partial R^2}R_\tau^2\big(\sigma_\tau^2 - \text{IV}_t^2\big) + \frac{\partial Z_\tau}{\partial R}\mu_\tau^R R_\tau\right)d\tau$$
$$+ \left(\mu_\tau^b\text{Swpn\_Bl}_\tau^P + \frac{\partial Z_\tau}{\partial R}\sigma_\tau^b\sigma_\tau\big(P_\tau(T) - P_\tau(T_n)\big)\right)d\tau$$
$$+ \left(\frac{\partial Z_\tau}{\partial R}\sigma_\tau\big(P_\tau(T) - P_\tau(T_n)\big) + \sigma_\tau^b\text{Swpn\_Bl}_\tau^P\right)dW_\tau,$$
(B.6)

where μ_τ^R is the drift of $\frac{dR_\tau}{R_\tau}$ under the physical probability, and equals

$$\mu_\tau^R = \mu_\tau^{\Delta p} - \mu_\tau^b - \sigma_\tau\sigma_\tau^b, \quad \sigma_\tau = \sigma_\tau^{\Delta p} - \sigma_\tau^b,$$
(B.7)

and the second equality in Eqs. (B.7) follows by Itô's lemma. By the definition of the forward swap rate in Eq. (3.5), Eqs. (B.7), and the relation $a_\tau = \frac{\partial Z_\tau}{\partial R}$, we can integrate the difference between $d\text{Swpn\_Bl}_\tau^P$ in Eq. (B.6) and dv_τ in Eq. (B.5) to obtain the P&L of the hedged swaption at the swaption maturity:

$$\text{Swpn\_Bl}_T^P - \upsilon_T = \text{PVBP}_T \cdot [R_T - K]^+ - \upsilon_T$$

$$= \frac{1}{2} \int_t^T \frac{\partial^2 Z_\tau}{\partial R^2} R_\tau^2 (\sigma_\tau^2 - \text{IV}_t^2)(\Lambda_{\tau,T} \text{PVBP}_\tau) d\tau$$

$$+ \int_t^T \Lambda_{\tau,T} \sigma_\tau^b (\text{Swpn\_Bl}_\tau^P - \upsilon_\tau) dW_\tau, \qquad (\text{B.8})$$

where we have used the first relation in Eqs. (B.4), $\text{Swpn\_Bl}_\tau^P = \upsilon_\tau$ for $\tau = t$, and defined $\Lambda_{\tau,T} = e^{\int_\tau^T \mu_s^b ds}$. The approximation in Eq. (3.8), which merely aims to simplify the presentation, relies on: (i) $\text{PVBP}_T \approx \Lambda_{\tau,T} \text{PVBP}_\tau$; (ii) $\frac{\widetilde{\Delta R_t}}{R_t} \approx dW_t$; and (iii) disregarding the term $\Lambda_{\tau,T}$ inside the stochastic integral in Eq. (B.8).

VOLATILITY TRADING BASED ON STRADDLES. Here we derive the P&L regarding the straddle. It is useful in the sequel to note that the parity for payer and receiver swaptions is

$$\text{Swpn}_t^P(K, T; T_1, \ldots, T_n) = \text{IRS}_t(K; T_1, \ldots, T_n) + \text{Swpn}_t^R(K, T; T_1, \ldots, T_n). \qquad (\text{B.9})$$

Consider the value of a straddle, $\text{Straddle}_\tau = \text{Swpn}_\tau^P + \text{Swpn}_\tau^R$. By the Black (1976) formula, the price of the receiver swaption is

$$\text{Swpn\_Bl}_t^R(R_t^n, \text{PVBP}_t, K, T; \bar{V}) = \text{PVBP}_t \cdot \hat{Z}_t(R_t^n, T, K; \bar{V}),$$

$$\hat{Z}_t(R_t, T, K; \bar{V}) = K(1 - \Phi(d_t - \sqrt{\bar{V}})) - R_t(1 - \Phi(d_t)), \quad d_t = \frac{\ln \frac{R_t}{K} + \frac{1}{2}\bar{V}}{\sqrt{\bar{V}}}. \qquad (\text{B.10})$$

Therefore, the dynamics of Swpn_Bl_τ^R are the same as Swpn_Bl_τ^P in Eq. (B.6), but with $\hat{Z}_\tau$ replacing Z_t. Moreover, we have, by Eq. (B.9) and Eq. (3.6) of the main text,

$$\frac{\partial \text{Straddle}_\tau}{\partial R} = \text{PVBP}_\tau \left(1 + 2\frac{\partial \hat{Z}_\tau}{\partial R}\right), \quad \frac{\partial \hat{Z}_\tau}{\partial R} = \frac{\partial Z_\tau}{\partial R} - 1,$$

where Z_τ is as in Eq. (B.1). Assuming the straddle delta is sufficiently small, which it is by the previous equation when $2\frac{\partial Z_\tau}{\partial R} \approx 1$, the value of the straddle is, by Eq. (B.6) and the previous arguments concerning the dynamics of Swpn_Bl_τ^R,

$$\text{Straddle}_T - \text{Straddle}_t$$

$$= \int_t^T \frac{\partial^2 Z_\tau}{\partial R^2} R_\tau^2 (\sigma_\tau^2 - \text{IV}_t^2)(\Lambda_{\tau,T} \text{PVBP}_\tau) d\tau + \int_t^T \Lambda_{\tau,T} \sigma_\tau^b \text{Straddle}_\tau dW_\tau. \qquad (\text{B.11})$$

The approximation in Eq. (3.9) relies on the same arguments leading to Eq. (3.8). Note that the approximation $2\frac{\partial Z_\tau}{\partial R} \approx 1$ is quite inadequate as the forward swap rate drifts away from ATM—a very well-known feature discussed, for example, by Mele (2014, Chap. 10) in the case of equity straddles.

DIRECTIONAL VOLATILITY TRADES. The P&L displayed in Fig. 3.3 are calculated using daily data from January 1998 to December 2009, and comprise yield curve data as well as implied volatilities for swaptions. Yield curve data are needed to calculate forward swap rates for a fixed 5-year tenor and their realized volatilities, and implied volatilities are needed to compute P&Ls.

As for the straddle, the strategy is to go long an ATM swaption straddle. Let t denote the beginning of the holding period (1 month or 3 months). The terminal P&L of the straddle, say as of time T, is

$$
\text{PVBP}_T(T_1,\ldots,T_n)\big(\big[R_T(T_1,\ldots,T_n)-K\big]^+ + \big[K-R_T(T_1,\ldots,T_n)\big]^+\big)
$$
$$
-\text{Straddle}_t(T_1,\ldots,T_n),
$$

where $\text{Straddle}_t(T_1,\ldots,T_n)$ is the cost of the straddle at t, and $T-t$ is either 1 month (as in the top panel of Fig. 3.3) or 3 months (as in the bottom panel).

Instead, the terminal P&L of the variance swap contract is calculated consistently with Definition 3.2 and Eq. (3.22) as

$$
\frac{1}{T-t}\left(\text{PVBP}_T(T_1,\ldots,T_n)V_n(t,T) - \frac{\mathbb{F}^*_{\text{var},n}(t,T)}{P_t(T)}\right),
$$

where $\mathbb{F}^*_{\text{var},n}(t,T)$ approximates the unnormalized strike of the contract to be entered at t. Its exact value, $\mathbb{F}_{\text{var},n}(t,T)$, is that in Eq. (3.21). By Eq. (3.38),

$$
\frac{1}{T-t}\mathbb{F}_{\text{var},n}(t,T) = \text{PVBP}_t(T_1,\ldots,T_n)\cdot\text{IRS-VI}_n^2(t,T),
$$

which we approximate with

$$
\frac{1}{T-t}\mathbb{F}^*_{\text{var},n}(t,T) \equiv \text{PVBP}_t(T_1,\ldots,T_n)\cdot\text{ATM}_n^2(t,T),
$$

where $\text{ATM}_n(t,T)$ is the ATM implied volatility.

Finally, the realized variance rescaled by $T-t$, $\frac{1}{T-t}V_n(t,T)$, is calculated as

$$
\overline{\text{RV}}_t^m = \left(\frac{21\cdot m}{252}\right)^{-1}\text{RV}_t^m, \quad \text{where } \text{RV}_t^m = \sum_{i=t-21\cdot m}^{t}\ln^2\frac{R_i(T_1,\ldots,T_n)}{R_{i-1}(T_1,\ldots,T_n)},
$$

with $m\in\{1,3\}$ and $T_n-t=5$. The variance risk premium is defined as $\overline{\text{RV}}_t^m - \text{ATM}_n^2(t-21\cdot m,T)$, where $T-t$ is either $\frac{1}{12}$ (for 1 month) or $\frac{3}{12}$ (for 3 months). Finally, to make the P&Ls of the volatility swap contract and the straddle line up to the same order of magnitude, the values in Fig. 3.3 are obtained by re-multiplying the P&Ls of the variance contracts by $\frac{1}{12}$ (top panel) and $\frac{3}{12}$ (bottom panel).

B.2 Spanning IRS Variance Contracts

The pricing results of this chapter are a special case of those in Chap. 2. However, we derive them in the context of the interest rate swap market of this chapter for three reasons. First, these are the results that we first derived in Mele and Obayashi (2012); second, the derivations in this appendix make this chapter self-contained; third, some of the derivations and notation in the following proof reveal useful for hedging variance swaps.

Because the reader may be more familiar with the pricing of *percentage* variance swaps, we first provide details for these contracts, and then details regarding *basis point* variance swaps.

PRICING I: PERCENTAGE. To alleviate notation, we set $R_t \equiv R_t(T_1, \ldots, T_n)$. By the usual Taylor expansion with remainder,

$$
\ln \frac{R_T}{R_t} = \frac{1}{R_t}(R_T - R_t) - \int_0^{R_t} (K - R_T)^+ \frac{1}{K^2} dK - \int_{R_t}^{\infty} (R_T - K)^+ \frac{1}{K^2} dK.
$$
(B.12)

Multiplying both sides of the previous equation by $\mathrm{PVBP}_t(T_1, \ldots, T_n)$, and taking expectations under the annuity probability Q_{sw} (see Eq. (3.11) in the main text):

$$
\mathrm{PVBP}_t(T_1, \ldots, T_n) \mathbb{E}_t^{Q_{\mathrm{sw}}} \left(\ln \frac{R_T}{R_t} \right)
$$
$$
= - \int_0^{R_t} \frac{\mathrm{Swpn}_t^{\mathrm{R}}(K, T; T_n)}{K^2} dK - \int_{R_t}^{\infty} \frac{\mathrm{Swpn}_t^{\mathrm{P}}(K, T; T_n)}{K^2} dK,
$$
(B.13)

where we have used (i) the fact that by Eq. (3.12) of the main text, the forward swap rate is a martingale under Q_{sw}, and (ii) the pricing Equations (3.14) and (3.15) of the main text.

By Eq. (3.12) and a change of measure obtained through Eq. (3.11),

$$
-2\mathbb{E}_t^{Q_{\mathrm{sw}}} \left(\ln \frac{R_T}{R_t} \right)
$$
$$
= \mathbb{E}_t^{Q_{\mathrm{sw}}} \left(\int_t^T \sigma_s^2(T_1, \ldots, T_n) ds \right)
$$
$$
= \frac{1}{\mathrm{PVBP}_t(T_1, \ldots, T_n)}
$$
$$
\times \mathbb{E}_t \left[e^{-\int_t^T r_s ds} \left(\mathrm{PVBP}_T(T_1, \ldots, T_n) \int_t^T \sigma_s^2(T_1, \ldots, T_n) ds \right) \right],
$$
(B.14)

where $\mathbb{E}_t$ denotes the risk-neutral expectation conditional on information available at time-t. Combining Eq. (B.13) and Eq. (B.14) gives:

$$\mathbb{E}_t\left[e^{-\int_t^T r_s ds}\left(\mathrm{PVBP}_T(T_1,\ldots,T_n)\int_t^T \sigma_s^2(T_1,\ldots,T_n)ds\right)\right]$$
$$=2\left(\int_0^{R_t}\frac{\mathrm{Swpn}_t^{\mathrm{R}}(K,T;T_1,\ldots,T_n)}{K^2}dK+\int_{R_t}^\infty\frac{\mathrm{Swpn}_t^{\mathrm{P}}(K,T;T_1,\ldots,T_n)}{K^2}dK\right).$$
$$(\mathrm{B.15})$$

The L.H.S. of the previous equation is the value of the IRV forward agreement in Definition 3.1 (in percentage), and Eq. (3.21) is its approximation.

To derive the IRV swap rate $\mathbb{P}_{\mathrm{var},n}(t,T)$ in the first of Eqs. (3.22), note that by Eq. (3.16), it solves

$$0=\mathbb{E}_t\left(e^{-\int_t^T r_s ds}\left(V_n(t,T)\times \mathrm{PVBP}_T(T_1,\ldots,T_n)-\mathbb{P}_{\mathrm{var},n}(t,T)\right)\right),$$

which yields the expression of $\mathbb{P}_{\mathrm{var},n}(t,T)$ in the first of Eqs. (3.22) after rearranging terms. As for the standardized IRV swap rate $\mathbb{P}_{\mathrm{var},n}^*(t,T)$ in the second of Eqs. (3.22), note that by Eq. (3.18), it satisfies:

$$0=\mathbb{E}_t\left(e^{-\int_t^T r_s ds}\left(V_n(t,T)-\mathbb{P}_{\mathrm{var},n}^*(t,T)\right)\times \mathrm{PVBP}_T(T_1,\ldots,T_n)\right),$$

which yields the expression of $\mathbb{P}_{\mathrm{var},n}^*(t,T)$ in the second of Eqs. (3.22) by the definition of the Radon–Nikodym derivative in Eq. (3.11) of the main text after rearranging terms.

PRICING II: BASIS POINT. We price contracts for which the variance payoff is given by

$$V_n^{\mathrm{bp}}(t,T)\times \mathrm{PVBP}_T(T_1,\ldots,T_n),$$

where $V_n^{\mathrm{bp}}(t,T)$ is the realized variance of the forward swap rate changes during the interval $[t,T]$ (see the first of Eqs. (3.13)).

By a Taylor expansion with remainder, we have

$$R_T^2=R_t^2+2R_t(R_T-R_t)+2\left(\int_0^{R_t}(K-R_T)^+dK+\int_{R_t}^\infty(R_T-K)^+dK\right).$$
$$(\mathrm{B.16})$$

Multiplying both sides of this equation by $\mathrm{PVBP}_t(T_1,\ldots,T_n)$ and taking expectations under Q_{sw} leads to

$$\mathrm{PVBP}_t(T_1,\ldots,T_n)\mathbb{E}_t^{Q_{\mathrm{sw}}}\left(R_T^2-R_t^2\right)$$
$$=2\left(\int_0^{R_t}\mathrm{Swpn}_t^{\mathrm{R}}(K,T;T_1,\ldots,T_n)dK+\int_{R_t}^\infty\mathrm{Swpn}_t^{\mathrm{P}}(K,T;T_1,\ldots,T_n)dK\right).$$
$$(\mathrm{B.17})$$

Moreover, by Eq. (3.12) and Itô's lemma,

$$\mathbb{E}_t^{Q_{\mathrm{sw}}}\left(R_T^2 - R_t^2\right) = \mathbb{E}_t^{Q_{\mathrm{sw}}}\left(V_n^{\mathrm{bp}}(t,T)\right).$$

Substituting this expression into Eq. (B.17) yields

$$2\left(\int_0^{R_t} \mathrm{Swpn}_t^{\mathrm{R}}(K,T;T_1,\ldots,T_n)dK + \int_{R_t}^{\infty} \mathrm{Swpn}_t^{\mathrm{P}}(K,T;T_1,\ldots,T_n)dK\right)$$

$$= \mathrm{PVBP}_t(T_1,\ldots,T_n)\mathbb{E}_t^{Q_{\mathrm{sw}}}\left(R_T^2 - R_t^2\right)$$

$$= \mathrm{PVBP}_t(T_1,\ldots,T_n)\mathbb{E}_t^{Q_{\mathrm{sw}}}\left(V_n^{\mathrm{bp}}(t,T)\right)$$

$$= \mathbb{E}_t\left[e^{-\int_t^T r_s ds}\left(\mathrm{PVBP}_T(T_1,\ldots,T_n)V_n^{\mathrm{bp}}(t,T)\right)\right], \qquad (\mathrm{B}.18)$$

where the last line follows by the Radon–Nikodym derivative defined in Eq. (3.11) of the main text. The R.H.S. of this equation is the value of the IRV forward agreement in Definition 3.1 in basis points.

VOLATILITY INDEXES AND BLACK'S IMPLIED VOLATILITIES. The percentage standardized IRV swap rate in Definition 3.3 leads to an index of interest rate swap volatility, Eq. (3.38) (Sect. 3.5). This expression can be simplified by the Black (1976) formula, as claimed in the main text. Indeed, substitute Eq. (3.21) into Eq. (3.38), and use the Black formulae in Eqs. (B.1) and (B.10). The result is

IRS-VI$_n(t,T)$

$$= \sqrt{\frac{2}{T-t}\left(\sum_{i:K_i<R_t} \frac{\hat{Z}_t(R_t^n,T,K_i;(T-t)\cdot \mathrm{IV}_{i,t}^2)}{K_i^2}\Delta K_i + \sum_{i:K_i\geq R_t} \frac{Z_t(R_t^n,T,K_i;(T-t)\cdot \mathrm{IV}_{i,t}^2)}{K_i^2}\Delta K_i\right)},$$
$$(\mathrm{B}.19)$$

where the expressions for Z_t and $\hat{Z}_t$ are given in Eqs. (B.1) and (B.10), R_t^n is the current forward swap rate for maturity T and tenor length $T_n - T$, $R_t^n \equiv R_t(T_1,\ldots,T_n)$, and, finally, $\mathrm{IV}_{i,t}$ denotes the time t implied percentage volatility for swaptions with strike equal to K_i.

Similarly, the BP volatility index in Eq. (3.37) can be simplified while calculating the L.H.S. of Eq. (B.18) as follows:

IRS-VI$_n^{\mathrm{bp}}(t,T)$

$$= \sqrt{\frac{2}{T-t}\left(\sum_{i:K_i<R_t} \hat{Z}_t\left(R_t^n,T,K_i;(T-t)\cdot \mathrm{IV}_{i,t}^2\right)\Delta K_i + \sum_{i:K_i\geq R_t} Z_t\left(R_t^n,T,K_i;(T-t)\cdot \mathrm{IV}_{i,t}^2\right)\Delta K_i\right)}.$$
$$(\mathrm{B}.20)$$

CONNECTIONS WITH LOCAL VOLATILITY SURFACES. We develop arguments that hinge upon those Dumas (1995) and Britten-Jones and Neuberger (2000) put forth for the equity case. Consider the price of the payer swaption in Eq. (3.14) of the main text, $\text{Swpn}_t^P(K, T; T_n)$. For simplicity, let $R_t \equiv R_t(T_1, \ldots, T_n)$. We have

$$\frac{\partial \text{Swpn}_t^P}{\partial T} = \frac{\partial \ln \text{PVBP}_t}{\partial T} \cdot \text{Swpn}_t^P + \text{PVBP}_t \cdot \frac{d\mathbb{E}_t^{Q_{sw}}(R_T - K)^+}{dT}. \tag{B.21}$$

Since the forward swap rate is a martingale under Q_{sw}, with instantaneous volatility $\sigma_\tau(\cdot)$ as in Eq. (3.12) of the main text, we have

$$\frac{d\mathbb{E}_t^{Q_{sw}}(R_T - K)^+}{dT} = \frac{1}{2}\mathbb{E}_t^{Q_{sw}}\left[\delta(R_T - K)R_T^2\sigma_T^2\right], \tag{B.22}$$

where $\delta(\cdot)$ is the Dirac delta. We can elaborate on the R.H.S. of Eq. (B.22), obtaining

$$\mathbb{E}_t^{Q_{sw}}\left[\delta(R_T - K)R_T^2\sigma_T^2\right] = \iint \delta(R_T - K)R_T^2\sigma_T^2 \underbrace{\phi_T^c(\sigma_T|R_T)\phi_T^m(R_T)}_{\equiv \text{joint density of } (\sigma_T, R_T)}dR_T d\sigma_T$$

$$= K^2\phi_T^m(K)\mathbb{E}_t^{Q_{sw}}\left(\sigma_T^2|R_T = K\right)$$

$$= K^2\frac{\frac{\partial^2 \text{Swpn}_t^P}{\partial K^2}}{\text{PVBP}_t}\mathbb{E}_t^{Q_{sw}}\left(\sigma_T^2|R_T = K\right),$$

where $\phi_T^c(\sigma_T|R_T)$ denotes the conditional density of σ_T given R_T under Q_{sw}, $\phi_T^m(R_T)$ denotes the marginal density of R_T under Q_{sw}, and the third line follows by a well-known property of option-like prices. By substituting this result into Eq. (B.22) and then into Eq. (B.21), we obtain

$$\frac{\partial \text{Swpn}_t^P}{\partial T} = \frac{\partial \ln \text{PVBP}_t}{\partial T} \cdot \text{Swpn}_t^P + \frac{1}{2}K^2\frac{\partial^2 \text{Swpn}_t^P}{\partial K^2}\mathbb{E}_t^{Q_{sw}}\left(\sigma_T^2|R_T = K\right).$$

Rearranging this equation, we obtain

$$\sigma_{\text{loc}}^2(K, T) \equiv \mathbb{E}_t^{Q_{sw}}\left(\sigma_T^2|R_T = K\right) = 2\frac{\frac{\partial \text{Swpn}_t^P}{\partial T} - \frac{\partial \ln \text{PVBP}_t}{\partial T}\text{Swpn}_t^P}{K^2\frac{\partial^2 \text{Swpn}_t^P}{\partial K^2}}. \tag{B.23}$$

Next, assume that the volatility of the forward swap rate in Eq. (3.12) of the main text is only a function of the forward swap rate and calendar time, $\sigma_s = \sigma(R_s, s)$. Define

$$\frac{dR_\tau}{R_\tau} = \sigma_{\text{loc}}(R_\tau, \tau)dW_\tau^{sw}, \quad \tau \in [t, T],$$

where $\sigma_{\text{loc}}(R_\tau, \tau)$ is as in Eq. (B.23). This model can then match the cross-section of swaptions prices (without errors) and then be used to price all of the non-traded swaptions in Eq. (3.21) through Monte Carlo integration.

MARKING TO MARKET. First, we derive the updates in Eq. (3.23). For a given $\tau \in (t, T)$, we need to derive the following conditional expectation of the payoff

Var-Swap$_n(t, T)$ in Eq. (3.16):

$$\mathbb{E}_\tau^Q\left(e^{-\int_\tau^T r_u du}\text{Var-Swap}_n(t, T)\right)$$

$$= \mathbb{E}_\tau^Q\left[e^{-\int_\tau^T r_u du}\left(V_n(t, \tau) + V_n(\tau, T)\right) \times \text{PVBP}_T(T_1, \ldots, T_n)\right]$$

$$- P_\tau(T)\mathbb{P}_{\text{var},n}(t, T)$$

$$= V_n(t, \tau)\text{PVBP}_\tau(T_1, \ldots, T_n) + \mathbb{F}_{\text{var},n}(\tau, T) - P_\tau(T)\mathbb{P}_{\text{var},n}(t, T), \quad \text{(B.24)}$$

where we have used the definition of $\mathbb{F}_{\text{var},n}(\cdot, T)$ in Definition 3.1. Equation (3.23) follows after plugging the expression for $\mathbb{P}_{\text{var},n}(t, T)$ in the first of Eqs. (3.22) into Eq. (B.24).

Next, we derive Eq. (3.24). Utilizing the expression of Var-Swap$_n^*(t, T)$ in Eq. (3.18), we obtain

$$\mathbb{E}_\tau^Q\left(e^{-\int_\tau^T r_u du}\text{Var-Swap}_n^*(t, T)\right)$$

$$= \mathbb{E}_\tau^Q\left[e^{-\int_\tau^T r_u du}\left(V_n(t, \tau) + V_n(\tau, T) - \mathbb{P}_{\text{var},n}^*(t, T)\right) \times \text{PVBP}_T(T_1, \ldots, T_n)\right]$$

$$= \text{PVBP}_\tau(T_1, \ldots, T_n)\left(V_n(t, \tau) + \mathbb{P}_{\text{var},n}^*(\tau, T) - \mathbb{P}_{\text{var},n}^*(t, T)\right),$$

where the second equality follows by the definition of $\mathbb{F}_{\text{var},n}(\cdot, T)$ in Definition 3.1 and the expression of $\mathbb{P}_{\text{var},n}^*(t, T)$ in the second of Eqs. (3.22).

B.3 Hedging

We provide details relating to hedging IRV contracts. One element in these proofs involves the construction of a portfolio that replicates the forward swap rate. Accordingly, we first clarify how this portfolio is constructed, and proceed with the proofs in the second step.

REPLICATION OF THE FORWARD SWAP RATE. We provide details regarding how to replicate the forward swap rate. Consider the forward swap rate at t, as defined in Eqs. (3.5) and (3.3), which we repeat for the reader's convenience:

$$R_t = \frac{P_t(T_0) - P_t(T_n)}{\text{PVBP}_t}, \quad \text{PVBP}_t = \sum_{i=1}^n \delta_{i-1} P_t(T_i),$$

where $R_t \equiv R_t(T_1, \ldots, T_n)$ and $\text{PVBP}_t \equiv \text{PVBP}_t(T_1, \ldots, T_n)$ to simplify notation.

We aim to set up a portfolio that replicates the forward swap rate, R_t. Note that R_t is a function of three freely tradable assets:

$$R_t = \varphi\left(P_t(T_0), P_t(T_n), \text{PVBP}_t\right), \quad \varphi(x_1, x_2, x_3) \equiv \frac{x_1 - x_2}{x_3}, \quad \text{(B.25)}$$

suggesting a portfolio that comprises these three assets. Let V_t be the value of this portfolio, which satisfies

$$dV_t = \theta_{1t} d P_t(T_0) + \theta_{2t} d P_t(T_n) + \theta_{3t} d\text{PVBP}_t + \theta_{4t} d M_t, \qquad \text{(B.26)}$$

where M_t denotes the value of a money market account at time t, and $\theta_{\cdot t}$ are the units of assets in the portfolio at time t. Next, note that by Itô's lemma, and Eq. (B.25), the forward swap rate is the solution to

$$
\begin{aligned}
d R_t = {} & \varphi_1\big(P_t(T_0), P_t(T_n), \text{PVBP}_t\big) d P_t(T_0) \\
& + \varphi_2\big(P_t(T_0), P_t(T_n), \text{PVBP}_t\big) d P_t(T_n) \\
& + \varphi_3\big(P_t(T_0), P_t(T_n), \text{PVBP}_t\big) d\text{PVBP}_t \\
& + \text{VC}_t dt,
\end{aligned}
\qquad \text{(B.27)}
$$

where subscripts denote partial derivatives,

$$
\begin{aligned}
\text{VC}_t \equiv {} & \frac{1}{2}\varphi_{11}\big(P_t(T_0), P_t(T_n), \text{PVBP}_t\big)\big\|\sigma_{P_t}(T_0)\big\|^2 \\
& + \frac{1}{2}\varphi_{22}\big(P_t(T_0), P_t(T_n), \text{PVBP}_t\big)\big\|\sigma_{P_t}(T_n)\big\|^2 \\
& + \frac{1}{2}\varphi_{33}\big(P_t(T_0), P_t(T_n), \text{PVBP}_t\big)\big\|\sigma_{\text{PVBP}_t}(T_0)\big\|^2 \\
& + \varphi_{12}\big(P_t(T_0), P_t(T_n), \text{PVBP}_t\big)\sigma_{P_t}(T_0) \cdot \sigma_{P_t}(T_n) \\
& + \varphi_{13}\big(P_t(T_0), P_t(T_n), \text{PVBP}_t\big)\sigma_{P_t}(T_0) \cdot \sigma_{\text{PVBP}_t} \\
& + \varphi_{23}\big(P_t(T_0), P_t(T_n), \text{PVBP}_t\big)\sigma_{P_t}(T_n) \cdot \sigma_{\text{PVBP}_t},
\end{aligned}
$$

with $\sigma_{P_t}(T_j)$ and σ_{PVBP_t} denoting the instantaneous volatility vector of $P_t(T_j)$ and PVBP_t.

We have by Eqs. (B.26) and (B.27):

$$
\begin{aligned}
dV_t - d R_t = {} & \big(\theta_{1t} - \varphi_1\big(P_t(T_0), P_t(T_n), \text{PVBP}_t\big)\big) d P_t(T_0) \\
& + \big(\theta_{2t} - \varphi_2\big(P_t(T_0), P_t(T_n), \text{PVBP}_t\big)\big) d P_t(T_n) \\
& + \big(\theta_{3t} - \varphi_3\big(P_t(T_0), P_t(T_n), \text{PVBP}_t\big)\big) d\text{PVBP}_t \\
& + (r_t \theta_{4t} M_t - \text{VC}_t) dt
\end{aligned}
$$

where we have used the dynamics of the money market account, $dM_t = r_t M_t dt$, with r_t denoting the instantaneous short-term rate. Therefore, we replicate the forward swap rate through a portfolio comprising the following proportions of assets

$$\theta_{1t} = \varphi_1\big(P_t(T_0), P_t(T_n), \text{PVBP}_t\big) = \frac{1}{\text{PVBP}_t},$$

$$\theta_{2t} = \varphi_2\big(P_t(T_0), P_t(T_n), \mathrm{PVBP}_t\big) = -\frac{1}{\mathrm{PVBP}_t},$$

$$\theta_{3t} = \varphi_3\big(P_t(T_0), P_t(T_n), \mathrm{PVBP}_t\big) = -\frac{R_t}{\mathrm{PVBP}_t}.$$

For the portfolio to replicate the forward swap rate, one chooses

$$\theta_{4t} M_t = R_t - \theta_{1t} P_t(T_0) - \theta_{2t} P_t(T_n) - \theta_{3t}\mathrm{PVBP}_t = R_t.$$

By construction, the portfolio replicates the forward swap rate R_t, although it gives rise to a hedging cost $\varepsilon_t \equiv R_t - V_t$ at time t, which satisfies: $d\varepsilon_t = (\mathrm{VC}_t - r_t R_t)dt$.

HEDGING PERCENTAGE VOLATILITY CONTRACTS. We provide the details leading to Table 3.1; those for Table 3.2 are nearly identical. By Itô's lemma:

$$\mathrm{PVBP}_T(T_1, \ldots, T_n) \int_t^T \sigma_s^2(T_1, \ldots, T_n)ds$$

$$= 2\mathrm{PVBP}_T(T_1, \ldots, T_n) \int_t^T \frac{dR_s(T_1, \ldots, T_n)}{R_s(T_1, \ldots, T_n)}$$

$$- 2\mathrm{PVBP}_T(T_1, \ldots, T_n) \ln \frac{R_T(T_1, \ldots, T_n)}{R_t(T_1, \ldots, T_n)}. \tag{B.28}$$

By Eq. (B.12), the value of the interest rate swap in Eq. (3.7) and the swaption premiums in Eqs. (3.14)–(3.15), the second term on the R.H.S. is the payoff at T of a portfolio set up at t, which is two times: (a) short $1/R_t(T_1, \ldots, T_n)$ units of a fixed interest payer swap struck at $R_t(T_1, \ldots, T_n)$, and (b) long a continuum of OTM swaptions with weights $K^{-2}dK$. This portfolio is the static position (ii) in Table 3.1. By previous results in this appendix (see Eqs. (B.12)–(B.15)), its cost is $\mathbb{F}_{\mathrm{var},n}(t, T)$. We borrow $\mathbb{F}_{\mathrm{var},n}(t, T)$ at time t, and repay it back at time T, as in row (iii) of Table 3.1.

Next, we derive the portfolio of zero-coupon bonds (i) in Table 3.1. This portfolio needs to be worthless at time t and replicate the first term on the R.H.S. of Eq. (B.28). First, we replicate the forward swap through a portfolio of zero-coupon bonds, as in derivations following Eq. (B.25). Note that this replication entails a hedging cost as explained above. For simplicity, we refer to this portfolio as the "forward swap rate." Then, we consider a self-financed strategy in (a) this portfolio and (b) a money market account (MMA, in the sequel) worth $M_\tau \equiv e^{\int_t^\tau r_s ds}$ at time $\tau \geq t$. The value of this portfolio is

$$v_\tau = \theta_\tau R_\tau(T_1, \ldots, T_n) + \psi_\tau M_\tau,$$

where θ_τ are the units in the forward swap rate and ψ_τ are the units in the MMA. Consider the following portfolio:

$$\hat{\theta}_\tau R_\tau(T_1, \ldots, T_n) = \mathrm{PVBP}_\tau(T_1, \ldots, T_n),$$

$$\hat{\psi}_\tau M_\tau = \mathrm{PVBP}_\tau(T_1, \ldots, T_n)\left(\int_t^\tau \frac{dR_s(T_1, \ldots, T_n)}{R_s(T_1, \ldots, T_n)} - 1\right). \tag{B.29}$$

We have that $\hat{\upsilon}_\tau = \hat{\theta}_\tau R_\tau + \hat{\psi}_\tau M_\tau$ satisfies

$$\hat{\upsilon}_\tau = \text{PVBP}_\tau(T_1, \ldots, T_n) \int_t^\tau \frac{dR_s(T_1, \ldots, T_n)}{R_s(T_1, \ldots, T_n)}, \qquad (\text{B.30})$$

so that

$$\hat{\upsilon}_t = 0, \quad \text{and} \quad \hat{\upsilon}_T = \text{PVBP}_T(T_1, \ldots, T_n) \int_t^T \frac{dR_s(T_1, \ldots, T_n)}{R_s(T_1, \ldots, T_n)}.$$

Therefore, by going long two portfolios $(\hat{\theta}_\tau, \hat{\psi}_\tau)$, the first term on the R.H.S. of Eq. (B.28)—and by the previous results, the entire R.H.S. of Eq. (B.28)—can be replicated, provided $(\hat{\theta}_\tau, \hat{\psi}_\tau)$ is self-financed. A hedging cost is only incurred in the replication of the forward swap rate. We are only left to show that $(\hat{\theta}_\tau, \hat{\psi}_\tau)$ is self-financed. We have

$$d\hat{\upsilon}_\tau = \hat{\theta}_\tau R_\tau(T_1, \ldots, T_n) \frac{dR_\tau(T_1, \ldots, T_n)}{R_\tau(T_1, \ldots, T_n)} + \hat{\psi}_\tau M_\tau \frac{dM_\tau}{M_\tau}$$

$$= \text{PVBP}_\tau(T_1, \ldots, T_n) \left(\frac{dR_\tau(T_1, \ldots, T_n)}{R_\tau(T_1, \ldots, T_n)} - r_\tau d\tau \right)$$

$$+ \left(\text{PVBP}_\tau(T_1, \ldots, T_n) \int_t^\tau \frac{dR_s(T_1, \ldots, T_n)}{R_s(T_1, \ldots, T_n)} \right) r_\tau d\tau$$

$$= \text{PVBP}_\tau(T_1, \ldots, T_n) \left(\frac{dR_\tau(T_1, \ldots, T_n)}{R_\tau(T_1, \ldots, T_n)} - r_\tau d\tau \right) + r_\tau \hat{\upsilon}_\tau d\tau$$

$$= \hat{\theta}_\tau R_\tau(T_1, \ldots, T_n) \left(\frac{dR_\tau(T_1, \ldots, T_n)}{R_\tau(T_1, \ldots, T_n)} - r_\tau d\tau \right) + r_\tau \hat{\upsilon}_\tau d\tau, \qquad (\text{B.31})$$

where the second line follows by the expressions for $(\hat{\theta}_\tau, \hat{\psi}_\tau)$ in Eqs. (B.29), the third line holds by Eq. (B.30), and the fourth follows, again, by the expression for $\hat{\theta}_\tau R_\tau$ in Eq. (B.29). It is easy to check that the dynamics of $\hat{\upsilon}_\tau$ in Eq. (B.31) are those of a self-financed strategy.

HEDGING BASIS POINT VOLATILITY CONTRACTS. We provide the details leading to Table 3.3 only, as those for Table 3.4 are nearly identical. Itô's lemma gives us

$$\text{PVBP}_T(T_1, \ldots, T_n) V_n^{\text{bp}}(t, T)$$

$$= -2\text{PVBP}_T(T_1, \ldots, T_n) \int_t^T R_s(T_1, \ldots, T_n) dR_s(T_1, \ldots, T_n)$$

$$+ \text{PVBP}_T(T_1, \ldots, T_n) \left(R_T^2(T_1, \ldots, T_n) - R_t^2(T_1, \ldots, T_n) \right). \qquad (\text{B.32})$$

By Eq. (B.16), the second term on the R.H.S. of Eq. (B.32) is the payoff at T of a portfolio set up at t, which is: (a) long $2R_t(T_1, \ldots, T_n)$ units a fixed interest payer swap struck at $R_t(T_1, \ldots, T_n)$, and (b) long a continuum of OTM swaptions with weights $2dK$. It is the static position (ii) in Table 3.3. By previous results in this appendix (see Eqs. (B.16)–(B.18)), its cost is $\mathbb{F}^{\text{bp}}_{\text{var},n}(t, T)$, which we borrow at t and repay back at T, as in row (iii) of Table 3.3. The portfolio to be short in row (i) of Table 3.3 is obtained similarly as the portfolio in row (i) of Table 3.1, but with

$$\hat{\psi}_\tau M_\tau = \text{PVBP}_\tau(T_1, \ldots, T_n) \left(\int_t^\tau 2R_s(T_1, \ldots, T_n) dR_s(T_1, \ldots, T_n) - 1 \right),$$

replacing the percentage counterparts in Eq. (B.29).

B.4 Constant Maturity Swaps

Consider the fair value of a single payment of a Constant Maturity Swap (CMS) (see Sect. 3.3.6), say the first one occurring at time $T_0 + \kappa$. To simplify notation, set $S(T_0) \equiv R_{T_0}(T_1, \ldots, T_n)$. The current value of $S(T_0)$ to be paid at time $T_0 + \kappa$ is the same as the current value of $S(T_0) P_{T_0}(T_0 + \kappa)$ to be paid at T_0, so that

$$\text{cms}(t, T_0 + \kappa) \equiv \mathbb{E}_t \left(e^{-\int_t^{T_0} r_u du} S(T_0) P_{T_0}(T_0 + \kappa) \right) = P_t(T_0 + \kappa) \mathbb{E}_t^{Q_{\text{sw}}} \left(S(T_0) \frac{\mathcal{G}_{T_0}}{\mathcal{G}_t} \right),$$
$$(B.33)$$

where $\mathcal{G}_\tau \equiv \frac{P_\tau(T_0+\kappa)}{\text{PVBP}_\tau(T_1,\ldots,T_n)}$. We determine $\mathcal{G}_\tau$, by discounting through the reset times, $T_i - T_0$ using the flat rate formula, $P_{T_0}(T_i) = (1 + \delta S(T_0))^{-i}$, where δ is the year fraction between the reset times, so that

$$\mathcal{G}_\tau \equiv \frac{P_\tau(T_0 + \kappa)}{\text{PVBP}_\tau(T_1, \ldots, T_n)} \approx \frac{(1 + \delta S(\tau))^{-(T_0+\kappa-\tau)}}{\delta \sum_{i=1}^n (1 + \delta S(\tau))^{-(T_i-\tau)}} = \frac{S(\tau)(1 + \delta S(\tau))^{-\kappa}}{1 - \frac{1}{(1+\delta S(\tau))^n}}$$

$$\equiv G(S(\tau)). \qquad\qquad\qquad\qquad\qquad\qquad\qquad\qquad\qquad (B.34)$$

The previous derivations closely follow those in Hagan (2003), and are provided for completeness. We now make the connection between the value $\text{cms}(t, T_0 + \kappa)$ in Eq. (B.33) and the value of the IRV forward contract in BP of Definition 3.1. Replacing the approximation in Eq. (B.34) into Eq. (B.33) leaves

$$\text{cms}(t, T_0 + \kappa)$$

$$= P_t(T_0 + \kappa) \mathbb{E}_t^{Q_{\text{sw}}} \left(S(T_0) \frac{G(S(T_0))}{G(R_t)} \right)$$

$$\approx P_t(T_0 + \kappa) \mathbb{E}_t^{Q_{\text{sw}}} \left[S(T_0) \left(1 + \frac{G'(R_t)}{G(R_t)} (S(T_0) - R_t) \right) \right]$$

$$= P_t(T_0 + \kappa) \mathbb{E}_t^{Q_{\text{sw}}} (S(T_0))$$

$$+ \text{PVBP}_t(T_1, \ldots, T_n) G'(R_t) \mathbb{E}_t^{Q_{\text{sw}}} \left[S(T_0)(S(T_0) - R_t) \right]$$

$$= P_t(T_0 + \kappa)R_t + G'(R_t)\text{PVBP}_t(T_1, \ldots, T_n)\mathbb{E}_t^{Q_{\text{sw}}}\left(S^2(T_0) - R_t^2\right)$$

$$= P_t(T_0 + \kappa)R_t + G'(R_t)\mathbb{F}_{\text{var},n}^{\text{bp}}(t, T_0),$$

where the second line follows by a first order Taylor approximation of the function G around R_t, the third from the definition of $G(R_t)$, the fourth from the martingale property of the forward swap rate under Q_{sw}, $\mathbb{E}_t^{Q_{\text{sw}}}(S(T_0)) = R_t$ and the fifth line from Eqs. (B.17)–(B.18) and the definition of the value of the IRV forward agreement (see Definition 3.1), $\mathbb{F}_{\text{var},n}^{\text{bp}}(t, T)$. Equation (3.25) of the main text follows by summing every single CMS payment, cms$(t, T_j + \kappa)$ say, for $j = 0, \ldots, N - 1$.

As mentioned in the main text, it has been well known since at least Hagan (2003) and Mercurio and Pallavicini (2006) that a CMS is linked to the entire skew. However, our representation of the fair value of a CMS in Eq. (3.25) differs non-trivially from previous ones. For example, Mercurio and Pallavicini (2006) utilize spanning arguments different from ours, which we now explain.

Consider a Taylor expansion with remainder of the function $f(R_T) \equiv R_T^2$ around some point R_o:

$$R_T^2 = R_o^2 + 2R_o(R_T - R_o) + 2\left(\int_0^{R_o} (K - R_T)^+ dK + \int_{R_o}^{\infty} (R_T - K)^+ dK\right).$$
$$\text{(B.35)}$$

Mercurio and Pallavicini (2006) consider the point $R_o = 0$, in which case Eq. (B.35) collapses to

$$R_T^2 = 2\int_0^{\infty} (R_T - K)^+ dK,$$

so that

$$\mathbb{E}_t^{Q_{\text{sw}}}\left(R_T^2\right) = \frac{2}{\text{PVBP}_t(T_1, \ldots, T_n)} \int_0^{\infty} \text{Swpn}_t^{\text{P}}(K, T; T_n) dK.$$

Our approach differs as we take $R_o = R_t$ in Eq. (B.35) (see Eq. (B.16)), so that

$$\mathbb{E}_t^{Q_{\text{sw}}}\left(R_T^2 - R_t^2\right) = 2\mathbb{E}_t^{Q_{\text{sw}}}\left(\int_0^{R_t} (K - R_T)^+ dK + \int_{R_t}^{\infty} (R_T - K)^+ dK\right)$$

$$= \frac{2}{\text{PVBP}_t(T_1, \ldots, T_n)}$$

$$\times \left(\int_0^{R_t} \text{Swpn}_t^{\text{R}}(K, T; T_n) dK + \int_{R_t}^{\infty} \text{Swpn}_t^{\text{P}}(K, T; T_n) dK\right).$$

B.5 The Contract and Index in the Vasicek Market

We use the Vasicek (1977) model as the basis for the experiments in Sect. 3.5.3 by assuming that the short-term rate follows a Gaussian process:

$$dr_t = \kappa(\bar{r} - r_t)dt + \sigma_v d\hat{W}_t, \qquad (B.36)$$

where $\hat{W}_t$ is a Wiener process under the risk-neutral probability, and κ, $\bar{r}$ and σ_v are three positive constants. The price of a zero-coupon bond predicted by this model is $P_t(r_t; T) = A(T - t)e^{-B(T-t)r_t}$ for two functions A and B, with obvious notation.

The model predicts that the forward swap rate is

$$R_t(r_t; T_1, \ldots, T_n) = \frac{P_t(r_t; T_0) - P_t(r_t; T_n)}{\text{PVBP}_t(r_t; T_1, \ldots, T_n)},$$

$$\text{PVBP}_t(r_t; T_1, \ldots, T_n) = \sum_{i=1}^{n} \delta_{i-1} P_t(r_t; T_i), \qquad (B.37)$$

the counterpart to $R_t(T_1, \ldots, T_n)$, Eq. (3.10).

The instantaneous percentage bond return volatility is

$$\text{Vol-P}_t(r_t; T) \equiv -B(T - t)\sigma_v. \qquad (B.38)$$

Finally, the instantaneous percentage forward swap rate volatility is

$$\sigma_t(r_t; T_1, \ldots, T_n) = \frac{\text{Vol-P}_t(r_t; T_0)P_t(r_t; T_0) - \text{Vol-P}_t(r_t; T_n)P_t(r_t; T_n)}{P_t(r_t; T_0) - P_t(r_t; T_n)}$$

$$- \frac{\sum_{i=1}^{n} \delta_{i-1} P_t(r_t; T_i)\text{Vol-P}_t(r_t; T_i)}{\text{PVBP}_t(r_t; T_1, \ldots, T_n)}, \qquad (B.39)$$

the counterpart to $\sigma_t(T_1, \ldots, T_n)$, Eq. (3.12).

We simulate Eq. (B.36) using a Milstein approximation method with initial values of r in the interval $[0.01, 0.10]$ and the following parameter values: $\kappa = 0.3807$, $\bar{r} = 0.072$, taken from Veronesi (2010, Chap. 15), and $\sigma_v = \sqrt{0.02} \times 0.2341$. We generate values of $\sigma_t(r_t; T_1, \ldots, T_n)$ by plugging in the simulated values of r_t. The value of σ_v implies that the instantaneous BP volatility is $100 \times \sigma_v \approx 3$ bps, which is consistent with standard average estimates reported in the literature (see, e.g., Chap. 12 in Mele 2014).

The instantaneous BP forward swap rate volatility is

$$\sigma_{\text{bp},t}(r_t; T_1, \ldots, T_n) = R_t(r_t; T_1, \ldots, T_n)\sigma_t(r_t; T_1, \ldots, T_n). \qquad (B.40)$$

The two expectations,

$$\mathbb{E}_t^{Q_{sw}}\left(V_n^{bp}(r;t,T)\right)$$

$$\equiv \frac{1}{\text{PVBP}_t(r_T;T_1,\ldots,T_n)}$$

$$\times \mathbb{E}_t\left[e^{-\int_t^T r_s ds}\left(\text{PVBP}_T(r_T;T_1,\ldots,T_n)\int_t^T \sigma_{bp,s}^2(r_s;T_1,\ldots,T_n)ds\right)\right] \tag{B.41}$$

and

$$\mathbb{E}_t\left(V_n^{bp}(r;t,T)\right) \equiv \mathbb{E}_t\left(\int_t^T \sigma_{bp,s}^2(r_s;T_1,\ldots,T_n)ds\right), \tag{B.42}$$

are estimated through Monte Carlo integration. The expectations regarding percentage volatility are also estimated through Monte Carlo integration. The forward swap rates in Figs. 3.7 and 3.8 (and in Tables 3.7 and 3.8 below) are obtained by plugging the initial values of the short-term rate into Eq. (B.37). We assume reset dates are quarterly and set $\delta_i = \frac{1}{4}$.

Regarding basis point variance, we calculate the IRS-VI index in Eq. (3.45) as $\sqrt{\frac{1}{T-t}\mathbb{E}_t^{Q_{sw}}(V_n^{bp}(r_t;t,T))}$ and future expected volatility in a risk-neutral market as $\sqrt{\frac{1}{T-t}\mathbb{E}_t(V_n^{bp}(r_t;t,T))}$ (see Eqs. (B.41)–(B.42)). Percentage variance is dealt with similarly.

The left panel of Fig. 3.7 and Table 3.7 report results from experiments regarding the calculation of BP volatility corresponding to the index IRS-VI$_n^{bp}(t,T)$ in Eq. (3.45), as well as future expected volatility in a risk-neutral market. The left panel of Fig. 3.8 and Table 3.8 report results from experiments regarding the percentage case. The right panels of Figs. 3.7 and 3.8 depict the instantaneous forward swap rate volatilities, calculated through Eqs. (B.40) and (B.39), respectively.

Table 3.7 The model prediction of the BP volatility index, IRS-VI$_n^{bp}(t, T)$ in Eq. (3.37) (labeled IRS-VI^{bp}), and future expected volatility in a risk-neutral market, $\sqrt{\frac{1}{T-t}\mathbb{E}_t(V_n^{bp}(r_t; t, T))}$, where $V_n^{bp}(r_t; t, T)$ denotes the model-implied realized BP variance (labeled E-Vol^{bp}), for (i) time-to maturity $T - t$ equal to 1, 6 and 9 months, and 1, 2 and 3 years, and (ii) tenor length n equal to 1, 2, 3 and 5 years, and for time-to maturity $T - t$ equal to 1, 6 and 12 months, and (iii) tenor length n equal to 10, 20 and 30 years. The index and expected volatilities are computed under the assumption the short-term interest rate is as in the Vasicek model of Eq. (B.36), with parameter values $\kappa = 0.3807$, $\bar{r} = 0.072$ and $\sigma_v = 0.0331$

	Maturity of the variance contract						
Tenor = 1 year	1 month				6 months		
Short Term Rate	Forward Swap Rate	IRS-VI^{bp}	E-Vol^{bp}		Forward Swap Rate	IRS-VI^{bp}	E-Vol^{bp}
1.00	2.18	272.43	272.43		2.90	252.92	252.97
3.00	3.80	273.79	273.79		4.28	253.99	254.05
5.00	5.44	275.15	275.16		5.68	255.07	255.13
7.00	7.08	276.52	276.54		7.08	256.15	256.22
9.00	8.73	277.90	277.92		8.49	257.23	257.32
10.00	9.56	278.59	278.62		9.19	257.78	257.87
Tenor = 2 years	1 month				6 months		
Short Term Rate	Forward Swap Rate	IRS-VI^{bp}	E-Vol^{bp}		Forward Swap Rate	IRS-VI^{bp}	E-Vol^{bp}
1.00	2.93	229.60	229.60		3.53	213.30	213.37
3.00	4.31	231.15	231.15		4.70	214.53	214.60
5.00	5.68	232.71	232.72		5.88	215.76	215.84
7.00	7.07	234.29	234.30		7.07	217.00	217.09
9.00	8.47	235.88	235.89		8.26	218.25	218.35
10.00	9.18	236.68	236.69		8.86	218.87	218.99
Tenor = 3 years	1 month				6 months		
Short Term Rate	Forward Swap Rate	IRS-VI^{bp}	E-Vol^{bp}		Forward Swap Rate	IRS-VI^{bp}	E-Vol^{bp}
1.00	3.51	196.13	196.14		4.02	182.41	182.48
3.00	4.68	197.97	197.97		5.02	183.85	183.93
5.00	5.87	199.82	199.82		6.03	185.31	185.40
7.00	7.06	201.68	201.69		7.05	186.78	186.88
9.00	8.27	203.57	203.58		8.08	188.26	188.38
10.00	8.87	204.52	204.53		8.60	189.01	189.13

Table 3.7 (continued)

	Maturity of the variance contract					
Tenor = 5 years	1 month			6 months		
Short Term Rate	Forward Swap Rate	IRS-VI[bp]	E-Vol[bp]	Forward Swap Rate	IRS-VI[bp]	E-Vol[bp]
1.00	4.31	149.31	149.31	4.69	139.19	139.27
3.00	5.20	151.57	151.58	5.46	140.98	141.07
5.00	6.11	153.87	153.87	6.23	142.79	142.90
7.00	7.03	156.20	156.21	7.02	144.63	144.75
9.00	7.97	158.56	158.57	7.82	146.49	146.61
10.00	8.44	159.76	159.76	8.22	147.43	147.56
Tenor = 1 year	9 months			1 year		
Short Term Rate	Forward Swap Rate	IRS-VI[bp]	E-Vol[bp]	Forward Swap Rate	IRS-VI[bp]	E-Vol[bp]
1.00	3.27	242.33	242.41	3.61	232.52	232.62
3.00	4.53	243.26	243.36	4.76	233.32	233.44
5.00	5.80	244.19	244.30	5.91	234.13	234.27
7.00	7.08	245.13	245.26	7.07	234.95	235.10
9.00	8.36	246.07	246.21	8.23	235.77	235.93
10.00	9.00	246.54	246.69	8.82	236.18	236.35
Tenor = 2 years	9 months			1 year		
Short Term Rate	Forward Swap Rate	IRS-VI[bp]	E-Vol[bp]	Forward Swap Rate	IRS-VI[bp]	E-Vol[bp]
1.00	3.84	204.45	204.55	4.13	196.23	196.36
3.00	4.91	205.51	205.63	5.10	197.16	197.30
5.00	5.98	206.57	206.72	6.07	198.09	198.25
7.00	7.06	207.65	207.81	7.05	199.03	199.20
9.00	8.15	208.73	208.90	8.04	199.98	200.16
10.00	8.69	209.28	209.46	8.53	200.45	200.64
Tenor = 3 years	9 months			1 year		
Short Term Rate	Forward Swap Rate	IRS-VI[bp]	E-Vol[bp]	Forward Swap Rate	IRS-VI[bp]	E-Vol[bp]
1.00	4.28	174.93	175.05	4.52	167.99	168.13
3.00	5.20	176.18	176.32	5.36	169.09	169.24
5.00	6.12	177.45	177.60	6.19	170.19	170.36
7.00	7.05	178.72	178.90	7.04	171.31	171.49
9.00	7.98	180.01	180.20	7.89	172.43	172.63
10.00	8.45	180.66	180.85	8.31	172.99	173.20

Table 3.7 (continued)

	Maturity of the variance contract					
Tenor = 5 years	9 months			1 year		
Short Term Rate	Forward Swap Rate	IRS-VI^{bp}	E-Vol^{bp}	Forward Swap Rate	IRS-VI^{bp}	E-Vol^{bp}
1.00	4.89	133.66	133.79	5.07	128.51	128.66
3.00	5.59	135.22	135.36	5.71	129.88	130.04
5.00	6.30	136.80	136.96	6.36	131.25	131.43
7.00	7.01	138.39	138.57	7.01	132.65	132.84
9.00	7.74	140.01	140.20	7.67	134.06	134.26
10.00	8.11	140.82	141.02	8.00	134.77	134.97
Tenor = 1 year	2 years			3 years		
Short Term Rate	Forward Swap Rate	IRS-VI^{bp}	E-Vol^{bp}	Forward Swap Rate	IRS-VI^{bp}	E-Vol^{bp}
1.00	4.67	199.65	199.91	5.38	174.98	175.33
3.00	5.45	200.12	200.39	5.92	175.25	175.62
5.00	6.24	200.58	200.88	6.46	175.53	175.91
7.00	7.04	201.05	201.36	7.00	175.81	176.21
9.00	7.83	201.53	201.85	7.54	176.09	176.50
10.00	8.23	201.76	202.10	7.81	176.23	176.64
Tenor = 2 years	2 years			3 years		
Short Term Rate	Forward Swap Rate	IRS-VI^{bp}	E-Vol^{bp}	Forward Swap Rate	IRS-VI^{bp}	E-Vol^{bp}
1.00	5.01	168.67	168.95	5.61	147.95	148.30
3.00	5.68	169.21	169.51	6.07	148.26	148.64
5.00	6.35	169.74	170.07	6.52	148.58	148.97
7.00	7.02	170.29	170.62	6.98	148.91	149.30
9.00	7.69	170.83	171.19	7.44	149.23	149.64
10.00	8.03	171.11	171.47	7.67	149.39	149.81
Tenor = 3 years	2 years			3 years		
Short Term Rate	Forward Swap Rate	IRS-VI^{bp}	E-Vol^{bp}	Forward Swap Rate	IRS-VI^{bp}	E-Vol^{bp}
1.00	5.28	144.62	144.91	5.79	127.00	127.35
3.00	5.85	145.26	145.57	6.18	127.38	127.75
5.00	6.42	145.90	146.23	6.57	127.76	128.14
7.00	7.01	146.55	146.89	6.97	128.14	128.54
9.00	7.58	147.20	147.56	7.36	128.53	128.93
10.00	7.87	147.53	147.89	7.56	128.73	129.13

Table 3.7 (continued)

	Maturity of the variance contract					
Tenor = 5 years	2 years			3 years		
Short Term Rate	Forward Swap Rate	IRS-VI^{bp}	E-Vol^{bp}	Forward Swap Rate	IRS-VI^{bp}	E-Vol^{bp}
1.00	5.65	111.04	111.33	6.04	97.75	98.09
3.00	6.09	111.84	112.14	6.34	98.23	98.58
5.00	6.53	112.64	112.96	6.64	98.71	99.07
7.00	6.98	113.45	113.79	6.95	99.20	99.56
9.00	7.43	114.27	114.62	7.26	99.69	100.06
10.00	7.65	114.68	115.04	7.41	99.93	100.31
Tenor = 10 years	1 month			6 months		
Short Term Rate	Forward Swap Rate	IRS-VI^{bp}	E-Vol^{bp}	Forward Swap Rate	IRS-VI^{bp}	E-Vol^{bp}
1.00	5.25	93.49	93.50	5.48	87.58	87.65
3.00	5.81	96.00	96.00	5.97	89.57	89.65
5.00	6.39	98.56	98.56	6.47	91.60	91.69
7.00	6.98	101.18	101.19	6.97	93.67	93.77
9.00	7.59	103.86	103.87	7.49	95.78	95.89
10.00	7.90	105.23	105.23	7.76	96.85	96.96
				12 months		
Short Term Rate				Forward Swap Rate	IRS-VI^{bp}	E-Vol^{bp}
1.00				5.7299	81.26	81.39
3.00				6.1350	82.78	82.93
5.00				6.5477	84.33	84.49
7.00				6.9682	85.91	86.07
9.00				7.3965	87.51	87.68
10.00				7.6137	88.32	88.49
Tenor = 20 years	1 month			6 months		
Short Term Rate	Forward Swap Rate	IRS-VI^{bp}	E-Vol^{bp}	Forward Swap Rate	IRS-VI^{bp}	E-Vol^{bp}
1.00	5.7998	61.21	61.21	5.95	57.56	57.62
3.00	6.1705	63.43	63.43	6.27	59.33	59.40
5.00	6.5546	65.72	65.72	6.60	61.15	61.23
7.00	6.9526	68.09	68.09	6.94	63.02	63.10
9.00	7.3649	70.53	70.53	7.29	64.94	65.03
10.00	7.5766	71.77	71.78	7.47	65.92	66.01

Table 3.7 (continued)

	Maturity of the variance contract					
				12 months		
Short Term Rate				Forward Swap Rate	IRS-VI^{bp}	E-Vol^{bp}
1.00				6.1146	53.62	53.72
3.00				6.3827	54.99	55.10
5.00				6.6577	56.38	56.50
7.00				6.9396	57.80	57.93
9.00				7.2287	59.25	59.39
10.00				7.3759	59.99	60.13
Tenor = 30 years	1 month			6 months		
Short Term Rate	Forward Swap Rate	IRS-VI^{bp}	E-Vol^{bp}	Forward Swap Rate	IRS-VI^{bp}	E-Vol^{bp}
1.00	5.96	51.91	51.91	6.09	48.88	48.93
3.00	6.27	53.95	53.95	6.36	50.51	50.57
5.00	6.60	56.06	56.07	6.64	52.19	52.25
7.00	6.94	58.25	58.25	6.93	53.91	53.98
9.00	7.29	60.51	60.51	7.23	55.69	55.77
10.00	7.47	61.66	61.66	7.39	56.59	56.68
				12 months		
Short Term Rate				Forward Swap Rate	IRS-VI^{bp}	E-Vol^{bp}
1.00				6.22	45.59	45.68
3.00				6.45	46.85	46.94
5.00				6.69	48.13	48.24
7.00				6.93	49.45	49.56
9.00				7.17	50.79	50.91
10.00				7.30	51.47	51.60

Table 3.8 The model prediction of the percentage volatility index, IRS-VI$_n(t, T)$ in Eq. (3.38) (labeled IRS-VI), and future expected volatility in a risk-neutral market, $\sqrt{\frac{1}{T-t}\mathbb{E}_t(V_n(r_t; t, T))}$, where $V_n(r_t; t, T)$ denotes the model-implied realized percentage variance (labeled E-Vol), for (i) time-to maturity $T - t$ equal to 1, 6 and 9 months, and 1, 2 and 3 years, and (ii) tenor length n equal to 1, 2, 3 and 5 years, and for time-to maturity $T - t$ equal to 1, 6 and 12 months, and (iii) tenor length n equal to 10, 20 and 30 years. The index and expected volatilities are computed under the assumption the short-term interest rate is as in the Vasicek model of Eq. (B.36), with parameter values $\kappa = 0.3807$, $\bar{r} = 0.072$ and $\sigma_v = 0.0331$

	Maturity of the variance contract					
Tenor = 1 year	1 month			6 months		
Short Term Rate	Forward Swap Rate	IRS-VI	E-Vol	Forward Swap Rate	IRS-VI	E-Vol
1.00	2.18	124.83	124.82	2.48	106.53	106.53
3.00	3.80	71.97	71.97	4.01	66.39	66.38
5.00	5.44	50.62	50.62	5.54	48.25	48.25
7.00	7.08	39.08	39.08	7.08	37.93	37.93
9.00	8.73	31.85	31.86	8.63	31.27	31.27
10.00	9.56	29.17	29.17	9.41	28.78	28.75
Tenor = 2 years	1 month			6 months		
Short Term Rate	Forward Swap Rate	IRS-VI	E-Vol	Forward Swap Rate	IRS-VI	E-Vol
1.00	2.93	78.12	78.12	3.53	60.11	60.12
3.00	4.31	53.66	53.66	4.70	45.49	45.50
5.00	5.68	40.93	40.93	5.88	36.62	36.62
7.00	7.07	33.12	33.12	7.07	30.66	30.67
9.00	8.47	27.84	27.84	8.26	26.39	26.40
10.00	9.18	25.80	25.80	8.86	24.68	24.69
Tenor = 3 years	1 month			6 months		
Short Term Rate	Forward Swap Rate	IRS-VI	E-Vol	Forward Swap Rate	IRS-VI	E-Vol
1.00	3.51	55.76	55.76	4.02	45.18	45.20
3.00	4.68	42.23	42.23	5.02	36.50	36.51
5.00	5.87	34.04	34.04	6.04	30.64	30.65
7.00	7.06	28.55	28.55	7.06	26.44	26.45
9.00	8.27	24.62	24.62	8.08	23.26	23.27
10.00	8.87	23.04	23.04	8.60	21.95	21.96

Table 3.8 (continued)

	Maturity of the variance contract					
Tenor = 5 years	1 month			6 months		
Short Term Rate	Forward Swap Rate	IRS-VI	E-Vol	Forward Swap Rate	IRS-VI	E-Vol
1.00	4.31	34.63	34.62	4.69	29.56	29.58
3.00	5.20	29.11	29.11	5.46	25.75	25.77
5.00	6.11	25.17	25.17	6.23	22.84	22.86
7.00	7.04	22.20	22.20	7.02	20.55	20.56
9.00	7.97	19.89	19.89	7.82	18.70	18.71
10.00	8.44	18.92	18.92	8.22	17.90	17.91
Tenor = 1 year	9 months			1 year		
Short Term Rate	Forward Swap Rate	IRS-VI	E-Vol	Forward Swap Rate	IRS-VI	E-Vol
1.00	3.27	73.50	73.52	3.61	64.25	64.26
3.00	4.53	53.39	53.40	4.76	49.10	49.10
5.00	5.80	41.93	41.94	5.91	39.72	39.72
7.00	7.08	34.54	34.55	7.07	33.35	33.35
9.00	8.36	29.38	29.38	8.23	28.75	28.75
10.00	9.00	27.34	27.35	8.82	26.90	26.90
Tenor = 2 years	9 months			1 year		
Short Term Rate	Forward Swap Rate	IRS-VI	E-Vol	Forward Swap Rate	IRS-VI	E-Vol
1.00	3.84	52.74	52.76	4.13	47.30	47.32
3.00	4.91	41.59	41.61	5.10	38.61	38.62
5.00	5.98	34.36	34.37	6.07	32.62	32.62
7.00	7.06	29.29	29.30	7.05	28.25	28.25
9.00	8.15	25.53	25.55	8.04	24.92	24.92
10.00	8.69	24.01	24.02	8.53	23.53	23.54
Tenor = 3 years	9 months			1 year		
Short Term Rate	Forward Swap Rate	IRS-VI	E-Vol	Forward Swap Rate	IRS-VI	E-Vol
1.00	4.28	40.52	40.54	4.52	36.92	36.94
3.00	5.20	33.69	33.71	5.36	31.47	31.49
5.00	6.12	28.85	28.87	6.19	27.44	27.45
7.00	7.05	25.25	25.27	7.04	24.33	24.34
9.00	7.98	22.47	22.49	7.89	21.87	21.88
10.00	8.45	21.30	21.32	8.31	20.82	20.83

Table 3.8 (continued)

	Maturity of the variance contract					
Tenor = 5 years	9 months			1 year		
Short Term Rate	Forward Swap Rate	IRS-VI	E-Vol	Forward Swap Rate	IRS-VI	E-Vol
1.00	4.89	27.16	27.18	5.07	25.19	25.22
3.00	5.59	24.05	24.07	5.71	22.66	22.68
5.00	6.30	21.61	21.63	6.36	20.60	20.61
7.00	7.01	19.64	19.66	7.01	18.90	18.91
9.00	7.74	18.01	18.03	7.67	17.47	17.48
10.00	8.11	17.30	17.32	8.00	16.83	16.84
Tenor = 1 year	2 years			3 years		
Short Term Rate	Forward Swap Rate	IRS-VI	E-Vol	Forward Swap Rate	IRS-VI	E-Vol
1.00	4.67	41.87	41.92	5.38	31.66	31.72
3.00	5.45	36.06	36.10	5.92	28.90	28.96
5.00	6.24	31.66	31.69	6.46	26.58	26.63
7.00	7.04	28.22	28.25	7.00	24.61	24.65
9.00	7.83	25.46	25.48	7.54	22.91	22.95
10.00	8.23	24.27	24.29	7.81	22.15	22.19
Tenor = 2 years	2 years			3 years		
Short Term Rate	Forward Swap Rate	IRS-VI	E-Vol	Forward Swap Rate	IRS-VI	E-Vol
1.00	5.01	32.95	33.01	5.61	25.69	25.76
3.00	5.68	29.27	29.32	6.07	23.86	23.92
5.00	6.35	26.34	26.38	6.52	22.27	22.33
7.00	7.02	23.94	23.97	6.98	20.89	20.94
9.00	7.69	21.94	21.97	7.44	19.66	19.71
10.00	8.03	21.07	21.10	7.67	19.11	19.15
Tenor = 3 years	2 years			3 years		
Short Term Rate	Forward Swap Rate	IRS-VI	E-Vol	Forward Swap Rate	IRS-VI	E-Vol
1.00	5.28	26.87	26.93	5.79	21.41	21.48
3.00	5.85	24.41	24.46	6.18	20.14	20.21
5.00	6.42	22.37	22.41	6.57	19.02	19.08
7.00	7.00	20.64	20.69	6.97	18.02	18.08
9.00	7.58	19.17	19.21	7.36	17.12	17.17
10.00	7.87	18.51	18.55	7.56	16.70	16.76

Table 3.8 (continued)

	Maturity of the variance contract					
Tenor = 5 years	2 years			3 years		
Short Term Rate	Forward Swap Rate	IRS-VI	E-Vol	Forward Swap Rate	IRS-VI	E-Vol
1.00	5.65	19.35	19.41	6.04	15.86	15.93
3.00	6.09	18.10	18.16	6.34	15.19	15.26
5.00	6.53	17.01	17.06	6.64	14.58	14.64
7.00	6.98	16.05	16.10	6.95	14.02	14.08
9.00	7.43	15.20	15.25	7.26	13.50	13.56
10.00	7.65	14.81	14.86	7.41	13.26	13.31
Tenor = 10 years	1 month			6 months		
Short Term Rate	Forward Swap Rate	IRS-VI	E-Vol	Forward Swap Rate	IRS-VI	E-Vol
1.00	5.25	17.80	17.80	5.48	15.92	15.93
3.00	5.81	16.51	16.51	5.97	14.96	14.97
5.00	6.39	15.42	15.41	6.47	14.13	14.14
7.00	6.98	14.48	14.48	6.97	13.40	13.41
9.00	7.59	13.67	13.67	7.49	12.75	12.77
10.00	7.91	13.31	13.31	7.76	12.46	12.47
				12 months		
Short Term Rate				Forward Swap Rate	IRS-VI	E-Vol
1.00				5.72	14.12	14.14
3.00				6.13	13.43	13.45
5.00				6.54	12.82	12.84
7.00				6.96	12.27	12.29
9.00				7.39	11.78	11.80
10.00				7.61	11.55	11.57
Tenor = 20 years	1 month			6 months		
Short Term Rate	Forward Swap Rate	IRS-VI	E-Vol	Forward Swap Rate	IRS-VI	E-Vol
1.00	5.79	10.55	10.55	5.9561	9.65	9.66
3.00	6.17	10.27	10.27	6.2765	9.44	9.45
5.00	6.55	10.03	10.03	6.6067	9.24	9.26
7.00	6.95	9.79	9.79	6.9470	9.06	9.07
9.00	7.36	9.57	9.57	7.2978	8.88	8.90
10.00	7.57	9.47	9.47	7.4771	8.81	8.82

Table 3.8 (continued)

	Maturity of the variance contract					
				12 months		
Short Term Rate				Forward Swap Rate	IRS-VI	E-Vol
1.00				6.11	8.74	8.76
3.00				6.38	8.59	8.60
5.00				6.65	8.44	8.46
7.00				6.93	8.30	8.32
9.00				7.22	8.17	8.18
10.00				7.37	8.10	8.12
Tenor = 30 years	1 month			6 months		
Short Term Rate	Forward Swap Rate	IRS-VI	E-Vol	Forward Swap Rate	IRS-VI	E-Vol
1.00	5.96	8.70	8.70	6.0928	8.01	8.02
3.00	6.27	8.59	8.59	6.3651	7.93	7.93
5.00	6.60	8.49	8.49	6.6466	7.84	7.85
7.00	6.94	8.39	8.39	6.9374	7.76	7.77
9.00	7.29	8.29	8.29	7.2378	7.68	7.69
10.00	7.47	8.24	8.24	7.3917	7.65	7.66
				12 months		
Short Term Rate				Forward Swap Rate	IRS-VI	E-Vol
1.00				6.22	7.30	7.31
3.00				6.45	7.23	7.25
5.00				6.69	7.17	7.19
7.00				6.93	7.11	7.13
9.00				7.17	7.05	7.07
10.00				7.30	7.02	7.04

Chapter 4
Government Bonds and Time-Deposits

4.1 Introduction

The fragility of the US banking sector and stagnant economic growth caused by the global financial crisis culminating in Lehman Brothers' collapse on 15 September 2008 prompted the Federal Reserve and the US Treasury to implement large-scale operations and policy changes designed to stabilize financial institutions and provide economic stimulus. These government actions and market forces, such as flight-to-quality in times of distress and an ever evolving outlook on the global macroeconomic landscape, have been accompanied by large movements in interest rates and pronounced bouts of volatility. Figure 4.1 shows the yield on on-the-run 10-year government bonds rallying from above 5 % at the onset of the crisis to around 2 % at its depth with periods of increased volatility. A similarly dramatic drop and increase in volatility can be seen in the Eurodollar futures-implied LIBOR in Fig. 4.2.

While Figs. 4.1 and 4.2 suggest equity volatility shares some common trends and spikes with both Treasury and Eurodollar volatilities, Fig. 4.3 demonstrates that equity volatility is hardly a proxy for government bond or time deposit volatility. The correlation estimates in this figure are those in the 2nd and 3rd row of the 1st column of Fig. 1.2 in Chap. 1. They are estimates of correlations between 20-day realized volatility for SPY (an ETF tracking the S&P 500 Index) and both 10-year Treasuries and Eurodollars. The estimates change significantly over time. Before the financial crisis, they were close to zero or negative. At the beginning of the crisis, they were high, probably driven by global concerns about disorderly tail events. Even during the crisis, we see periods of marked divergences between the two volatilities, for example from 2010 and 2012 when the market's focus shifted from the private sector to the public, and the correlation between equity volatility and both Treasury and Eurodollar reached levels as low as 30 %. Note, finally, how Eurodollar volatility decouples from equity volatility over the last years in the sample, plummeting to negative values of almost −40 %.

Despite these facts and the wild success of CBOE's VIX options and futures, no counterpart had existed for government bond or time deposit volatility until the

© Springer International Publishing Switzerland 2015
A. Mele, Y. Obayashi, *The Price of Fixed Income Market Volatility*, Springer Finance,
DOI 10.1007/978-3-319-26523-0_4

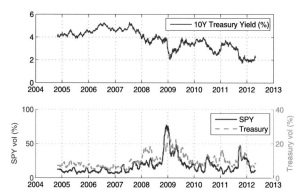

Fig. 4.1 *Top panel*: The yield on on-the-run ten year government bonds, annualized, percent. *Bottom panel*: estimates of 20 day realized volatility of the SPY daily returns (*solid line*), and bond price returns, annualized, percent, $100\sqrt{12\sum_{t=1}^{21}\ln^2\frac{S_{t-i+1}}{S_{t-i}}}$, where S_t denotes either the SPY or the Treasury bond price corresponding to the yield in the *top panel*. The sample includes daily data from 27 September 2004 to March 27, 2012, for a total of 1881 observations

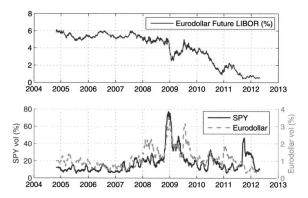

Fig. 4.2 *Top panel*: The Eurodollar Future three-month LIBOR for a one year maturity, annualized, percent. *Bottom panel*: estimates of 20 day realized volatility of the SPY and Eurodollar Future returns, annualized, percent, $100\sqrt{12\sum_{t=1}^{21}\ln^2\frac{S_{t-i+1}}{S_{t-i}}}$, where S_t denotes either the SPY or the Eurodollar Future defined as $\tilde{Z}_t = 100 \times (1 - \tilde{z}_t)$, where $\tilde{z}_t$ is the Eurodollar Future three-month LIBOR for a one year maturity. The sample includes daily data from 27 September 2004 to March 27, 2012, for a total of 1881 observations

2013 launch of TYVIX, the CBOE/CBOT 10-Year US Treasury Note Volatility Index based on work of the present chapter. Subsequently, S&P Dow Jones Indices and Japan Exchange Group collaborated in applying the methodology to Japanese Government Bond (JGB) markets to launch the S&P/JPX JGB VIX in 2015 based on Osaka Exchange's JGB future options.

These facts provide the motivation behind the innovation of this chapter. We aim to develop model-free designs of implied volatility indexes for these asset classes

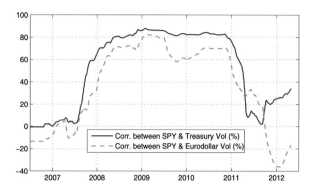

Fig. 4.3 Moving average estimates of the correlation between: (i) the 20 day realized volatilities of SPY and Treasury in the bottom panel of Fig. 4.1 (*solid line*), and (ii) the 20 day realized volatilities of SPY and Eurodollar in the bottom panel of Fig. 4.2 (*dashed line*). Each correlation estimate is calculated over the previous two years of data, and the sample includes daily data from 27 September 2004 to March 27, 2012, for a total of 1881 observations

and related derivatives, relying on the general theory in Chap. 2, and propose indexes of government bond and time deposit volatility that link to the fair value of variance swap contract designs tailored to asset classes.

Government bonds and time deposits pose a common methodological challenge in this context. Their market value has volatility that obviously depends on interest rate changes and yet, to price this volatility, we need to discount it through the very same interest rates. This issue is similar to that arising in swap markets (see Chap. 3, and Mele and Obayashi 2012). However, the context we study in this chapter is distinct from that in Chap. 3.

First, the probability associated with the market numéraire is the forward, not the annuity. Second, the risks involved in these markets may require model-dependent adjustments. For example, we cannot construct a model-free, forward-looking index of interest rate or price volatility based on the price of, say, one-month out-of-the-money options written on three month forwards. A model-dependent adjustment is needed to account for the fact that one-month options are priced under the one-month forward probability, whereas three-month futures are martingales under the three-month forward probability. We point to precise situations in which we could rely on model-free expressions for the price of future volatility for both government bonds and time deposits, and construct option-based designs for volatility indexes arising therefrom. Finally, we illustrate pitfalls arising whilst applying a naïve methodology to price these variance contracts based on what is known from the literature on equity volatility.

We consider both *percentage* and *basis point* volatility index formulations to match the market practice of expressing implied volatility of fixed income derivatives in both log-normal and normal terms. But we also introduce new indexes that match the market practice of expressing expected volatility in terms of *yield* volatility and *duration-based* volatility. Finally, we show how American-style options on

Treasury or Eurodollar futures may be used as inputs for the calculation of the indexes. While our exposition will, at times, be centered around US Treasuries and the Eurodollar futures, the proposed methodology applies to any bond market in which equivalent bond futures and futures options are traded, including but not limited to those currently traded on Eurex for European Government bonds and on the Osaka Exchange for Japanese Government Bonds.

The plan of this chapter is as follows. Section 4.2 deals with the evaluation of government bond *price* volatility. First, we shall explain some issues arising while pricing this volatility in a model-free format in the case where the underlying assets have a finite maturity date (e.g., Treasury bonds). Our model-free approach relies on "spanning arguments" such as those in Bakshi and Madan (2000) and Carr and Madan (2001). However, the payoffs and contract designs we consider are quite distinct and pose new methodological issues. We shall develop methodology relating to government bond volatility, which includes pricing, marking to market, hedging variance contracts, and an appropriate use of American option data as inputs to index calculations, numerical examples of the index calculation and the construction of model-free measures of *basis point yield volatility*; finally we shall also consider measures of *duration-based volatility*, obtained using as inputs both the percentage and the basis point *price* volatility indexes.

Section 4.3 deals with the pricing of time deposit volatility. It is more concise than Sect. 4.2, because security design and model-free pricing of government bond and time deposit volatility involve the same numéraire: the forward probability. Note that the contract designs and pricing in Sect. 4.3 are not an extension of those in Sect. 4.2; rather, the issues pertaining to the pricing of government bond and time deposit volatility have common origins, although we choose to organize this chapter by beginning with a presentation of the government bond case.

Section 4.4 deals with problems arising when the maturity of the options underlying volatility indexes differs from that of the risks underlying the options. The pricing of interest rate volatility is model-dependent in these cases. It is an interesting instance of a methodological disconnect between the pricing of equity volatility (leading to CBOE's VIX index) and the complex nature of fixed income volatility, where many additional dimensions come into play. Section 4.5 explains instances of how to deal with situations in which the necessary options data are generated based on specific listing cycles. It proposes a judicious use of indexes referenced to different horizons, leading to volatility indexes with constant horizons. Finally, Appendix C contains results omitted from the main text.

Before diving in, a word on the presentation of this chapter. We are about to deal with a number of ramifications and departures from the standard evaluation framework in the previous chapters. To alleviate the presentation, this chapter shall not provide index formulae based on discretizations of Riemann integrals, thereby relying on the assumption of a continuum of derivatives. These formulae can be made operational through the same discretizations explained in the previous chapter.

4.2 Government Bonds

This section develops variance swap contract designs for government bonds and corresponding indexes relying on the fair price of these contracts by applying the general framework of Chap. 2 to government bond derivatives.

We begin with an introductory section that explains why volatility cannot be priced in a model-free fashion when the underlying asset has a fixed maturity date. The remaining sections provide details pertaining to variance swaps, their fair value, marking to market, replication, and, finally, ensuing government bond volatility indexes expressed in a model-free format.[1] While these indexes relate to *prices*, not *yields*, we also discuss one method for converting our model-free price volatility indexes into an equivalent model-free yield volatility index. We also propose algorithms that make consistent use of American future option prices to feed the indexes.

In Sect. 4.4, we provide extensions to deal with cases in which the maturities of a few key derivatives underlying the index might differ. These extensions require model-dependent adjustments as alluded to in the introduction of the present chapter.

4.2.1 Pricing Spot Volatility

This section explains some preliminary points arising while attempting to price government bond volatility in a model-free manner. The major hindrance we encounter relates to the pricing of volatility on any asset with a finite maturity.

This pricing problem cannot be handled in a model-free fashion when the horizon of the variance swap is the same as the time-to-maturity of the of "spanning assets." Consider, for example, a zero expiring in 3 months, and the volatility of its returns realized over the next 3 months. Intuition gained over the equity case suggests that to price this volatility, we could use options on the zero maturing in 3 months. However, at maturity, these options have a trivial payoff, since the 3 month zero is at par in three months with certainty assuming no default risk. Naturally, these "degenerate" options do not exist in the first place. Our main point is that, in this case, the price of volatility cannot be expressed in a model-free fashion. We develop detailed explanations of these facts, to pave the way for the government bond variance contracts and indexes developed in the subsequent sections. We identify two potential issues, which we need to address while pricing government bond volatility in a model-free format.

---

[1]Mueller et al. (2012) make use of Britten-Jones and Neuberger (2000) formula for equity volatility to calculate an index for Treasury bonds, thereby relying on the assumptions that: (i) interest rates are constant, and (ii) European options are available for trading, not American, as is typically the case in Treasury markets. This chapter develops a methodology which (i) is internally consistent, allowing for random interest rates and, hence, strictly positive interest rate volatility, and (ii) incorporates the early exercise premiums embedded by American options.

First, the horizon set to price volatility has to be lower than the expiration of the asset. Consider, for example, a 3 month zero. Under certain conditions, we can only price the return volatility of this zero in a model-free manner over, say, the next month or two, or more generally up to but not including 3 months.

Second, prices of variance swaps are model-free when these contracts reference to *forwards*, not *spots*. Naturally, forwards and spots have the same volatility when interest rates are constant. For example, the VIX, while spanned by options on futures, is sometimes described as an index on spots. We do not usually pay attention to the fact the equity VIX is an index relying on forwards because a standard assumption underlying the index methodology is that interest rates are constant.[2] In contrast, this nuance is fundamental when it comes to pricing the volatility of government bonds.

4.2.1.1 Government Bond Prices and Volatility

Our focus in this chapter is the pricing of volatility of coupon-bearing bonds. To simplify the exposition of this introductory section, we only rely on zero-coupon bonds. Let $P_\tau(T)$ be the time τ price of a pure discount bond, or a zero, expiring at time T. We assume that information is driven by Brownian motions. Under the risk-neutral probability Q,

$$\frac{dP_\tau(T)}{P_\tau(T)} = r_\tau d\tau + \sigma_\tau(T) \cdot dW_\tau, \quad \tau \in [t, T], \tag{4.1}$$

where W_τ is a multidimensional Brownian motion under Q, r_τ is the short-term rate process, adapted to W_τ, and $\sigma_\tau(T)$ is the vector of instantaneous volatilities, also adapted to W_τ. Consider a variance swap with payoff occurring precisely at the expiration of the zero, T,

$$\int_t^T \|\sigma_\tau(T)\|^2 d\tau - \mathbb{P}(t, T),$$

where the fair value of the strike is:

$$\mathbb{P}(t, T) = \frac{1}{P_t(T)} \mathbb{E}_t \left(e^{-\int_t^T r_\tau d\tau} \int_t^T \|\sigma_\tau(T)\|^2 d\tau \right),$$

and $\mathbb{E}_t$ denotes the expectation under Q, conditional on the information at t.

4.2.1.2 The Two Steps to Variance Swap Evaluation

Pricing variance contracts in a model-free fashion takes two steps as explained in Chap. 2. In the first step, we link the price of volatility to that of other contracts

---

[2]See Appendix C.1 for a generalization to the case in which this assumption is removed.

such as a quadratic or a log-contract. In the second step, we cast the price of these contracts in a model-free fashion by using "spanning" arguments. For the case we consider in this section, this second step does not lead to model-free solutions, as we now discuss.

First Step: Pricing Volatility Through Risk-Neutral Evaluation

To characterize the strike of a variance swap contract, we elaborate on the dynamics of the asset price that underlies the contract. For example, we characterize $\mathbb{P}(t, T)$ for the government zeros of this section by elaborating on Eq. (4.1), using Itô's lemma, obtaining

$$\mathbb{P}(t, T) = -2\mathbb{E}_t \left(\ln \frac{P_T(T)}{P_t(T)} \right) + \frac{2}{P_t(T)} \mathbb{E}_t \left(e^{-\int_t^T r_\tau d\tau} \int_t^T \frac{d P_\tau(T)}{P_\tau(T)} \right), \qquad (4.2)$$

where $P_T(T) = 1$ by the boundary condition satisfied by the zero.

The fair value of the variance swap, $\mathbb{P}(t, T)$, is made up of two components. The first component is the negative of the return of the zero, which is deterministic as $P_T(T) = 1$. The second component is proportional to the discounted risk-neutral expectation of the total realized returns on the zero until maturity. Under regularity conditions, this second term can also be written as: $2 \cdot \mathbb{E}_t^{Q_{FT}} (\int_t^T \frac{d P_\tau(T)}{P_\tau(T)})$, where Q_{FT} is the forward probability introduced in Eq. (4.13) below.

While pricing existing variance contracts, P_τ is usually a forward price as explained in Chap. 2, in which case the second term drops out (see Sect. 4.2.2). For example, the strike leading to the VIX (Demeterfi et al. 1999a, 1999b) and that leading to the CBOE SRVIX index of interest rate swap volatility (see Chap. 3, and Mele and Obayashi 2012) are made up of a single component: the expectation under the market probability of the asset return (or the change in the forward swap rate) underlying the variance contract. The next section contains similar representations applied to government bonds. Note that for a government bond contract, this expected return is, in fact, *the* (deterministic) return, due to the boundary condition $P_T(T) = 1$,

$$\mathbb{E}_t \left(\ln \frac{P_T(T)}{P_t(T)} \right) = -\ln P_t(T). \qquad (4.3)$$

It appears that if it were not for the second term in Eq. (4.2), we would be done, as the strike would then simply equal $2 \cdot \ln P_t(T)$. However, note that this hypothetical strike is negative, and therefore the second term on the R.H.S. of Eq. (4.2) is crucial. By contrast, note that the expected log-return under Q is negative on assets with random payoffs such as stock indexes or government bonds trading before expiry (see the following sections). It is equal to the negative of the value of a portfolio of out-of-the-money European options, as discussed below.

Second Step: Matching Volatility Through Spanning Contracts

The second step involves characterizing the expected return on the asset underlying the contract by using spanning arguments. We can use these arguments and achieve a decomposition of the deterministic return in Eq. (4.3). Note that

$$\ln \frac{P_T(T)}{P_t(T)} = \frac{P_T(T) - P_t(T)}{P_t(T)} - \int_0^{P_t(T)} (K - P_T(T))^+ \frac{1}{K^2} dK$$
$$- \int_{P_t(T)}^{\infty} (P_T(T) - K)^+ \frac{1}{K^2} dK.$$

By the boundary condition, $P_T(T) = 1$, and the obvious fact that $P_t(T) \leq 1$, the previous equation collapses to one where the deterministic return on the zero, $-\ln P_t(T)$, is decomposed into two constituents:

$$\mathcal{R}_{t,T} \equiv -\ln P_t(T) = \frac{1 - P_t(T)}{P_t(T)} - \int_{P_t(T)}^{1} (1 - K) \frac{1}{K^2} dK, \qquad (4.4)$$

so that we can write Eq. (4.2) as follows:

$$\mathbb{P}(t, T) = -2\mathcal{R}_{t,T} + \frac{2}{P_t(T)} \mathbb{E}_t \left(e^{-\int_t^T r_\tau d\tau} \int_t^T \frac{d P_\tau(T)}{P_\tau(T)} \right). \qquad (4.5)$$

The expected log-return under Q is negative on assets with random payoffs. Instead, the deterministic return on the zero, $\mathcal{R}_{t,T}$, is positive, and Eq. (4.4) provides a decomposition of it. The first component of $\mathcal{R}_{t,T}$ is the arithmetic return on the zero-coupon bond, which is positive. The second component is responsible for the negative part of $\mathcal{R}_{t,T}$ and is less than the first component in absolute value.

Now consider the fair value of the variance swap, $\mathbb{P}(t, T)$, in Eq. (4.5). Because $\mathcal{R}_{t,T}$ is positive, the second term in Eq. (4.5) has a crucial economic content. Yet it cannot be expressed in a model-free fashion. In the remainder of this section, we aim to develop government bond contract designs to overcome these issues, which lead to indexes cast in a model-free fashion. Before this, we provide one additional instance where applying the standard equity methodology does not allow an interpretation of the resulting index as the fair value of a variance swap.

4.2.1.3 Naïve Model-Free Methodology and Pricing Biases

The previous section explains that we cannot price a government bond volatility in a model-free fashion when the bond and option maturities are the same. We now make a further step and deal with the following issues. Suppose a number of quotes relating to out-of-the-money (OTM) options is available, say bond options settled on cash, not futures, as is customary in OTC markets, and of course with maturity shorter than the underlying bond. Assume, further, that we aggregate these option prices to create a VIX-like index, relying on the standard equity methodology.

Would this index reflect the fair value of some notion of interest rate volatility? The answer is in the negative due to pricing biases, to which we now turn.

Suppose again we want to price a variance swap on bond prices, not on bond futures, and assume that bond forwards or futures are not available. We know from the previous section that a model-free solution does not exist when the maturities of the variance swap and the bond are the same. We now show that the same problem arises even when the maturity of the bond is longer than that of the variance swap.

Let T be the maturity of the interest rate variance swap, and $P_t(\mathbb{T})$ be the price of a zero-coupon bond expiring at time $\mathbb{T} > T$. We assume $P_t(\mathbb{T})$ is driven by a Brownian motion, just as $P_t(T)$ is in Eq. (4.1). We use Itô's lemma, and generalize Eq. (4.2) to

$$\mathbb{E}_t\left(e^{-\int_t^T r_\tau d\tau} \ln \frac{P_T(\mathbb{T})}{P_t(\mathbb{T})}\right) = P_t(T)\left(\int_t^T \mathbb{E}_t^{Q_{FT}}(r_\tau)d\tau - \frac{1}{2}\mathbb{P}(t, T, \mathbb{T})\right), \quad (4.6)$$

where $\mathbb{P}(t, T, \mathbb{T})$ denotes the strike of the variance swap on the bond volatility,

$$\mathbb{P}(t, T, \mathbb{T}) = \frac{1}{P_t(T)}\mathbb{E}_t\left(e^{-\int_t^T r_\tau d\tau}\int_t^T \left\|\sigma_\tau(\mathbb{T})\right\|^2 d\tau\right).$$

By the usual spanning arguments, we also have that

$$\mathbb{E}_t\left(e^{-\int_t^T r_\tau d\tau} \ln \frac{P_T(\mathbb{T})}{F_t^o(T, \mathbb{T})}\right) = \int_0^{F_t^o(T, \mathbb{T})} \text{Put}_t(K, T, \mathbb{T})\frac{1}{K^2}dK$$

$$+ \int_{F_t^o(T, \mathbb{T})}^\infty \text{Call}_t(K, T, \mathbb{T})\frac{1}{K^2}dK, \quad (4.7)$$

where $F_t^o(T, \mathbb{T}) \equiv \frac{P_t(\mathbb{T})}{P_t(T)}$ is the "shadow price" of a forward contract, and $\text{Put}_t(K, T, \mathbb{T})$ and $\text{Call}_t(K, T, \mathbb{T})$ are the prices of OTM put and call options struck at K and having the "bond forward" as the underlying.

Matching Eq. (4.6) to Eq. (4.7) leaves the following expression for the fair value of the strike:

$$\mathbb{P}(t, T, \mathbb{T}) = \mathbb{P}_{\text{VIX}}(t, T, \mathbb{T}) + \text{Bias}(t, T), \quad (4.8)$$

where

$$\mathbb{P}_{\text{VIX}}(t, T, \mathbb{T})$$

$$\equiv \frac{2}{P_t(T)}\left(\int_0^{F_t^o(T, \mathbb{T})} \text{Put}_t(K, T, \mathbb{T})\frac{1}{K^2}dK + \int_{F_t^o(T, \mathbb{T})}^\infty \text{Call}_t(K, T, \mathbb{T})\frac{1}{K^2}dK\right)$$

$$\text{Bias}(t, T) \equiv 2\left(\int_t^T \mathbb{E}_t^{Q_{FT}}(r_\tau)d\tau + \ln P_t(T)\right).$$

The motivation underlying this notation is that $\mathbb{P}_{\text{VIX}}(t, T, \mathbb{T})$ is the value of a variance swap determined through a naïve use of the standard equity VIX methodology, in which interest rates are constant. Indeed, it is easy to check that the bias term is zero in the hypothetical—and internally inconsistent—setting of constant interest rates.[3]

In the general case of random interest rates, the bias is different from zero and, hence, no model-free solution is available to price a variance swap. We explain this fact through an alternative representation of the bias term in Eq. (4.8). Denote the time t instantaneous forward rate for time τ with $f(t,\tau) \equiv -\frac{\partial}{\partial Z} \ln P_t(Z)|_{Z=\tau}$. By exploiting the relation, $f(t,\tau) = \mathbb{E}_t^{Q_{F\tau}}(r_\tau)$ (e.g., Mele 2014, Chap. 12), we determine the bias to be

$$\text{Bias}(t,T) = 2 \int_t^T \left(\mathbb{E}_t^{Q_{FT}}(r_\tau) - \mathbb{E}_t^{Q_{F\tau}}(r_\tau) \right) d\tau. \qquad (4.9)$$

Equation (4.9) shows that the pricing bias is twice the integrated difference between two expectations taken under the forward probability, the T-forward minus the τ-forward for all $\tau \in [t,T]$. These expectations are obviously *not* model-free, as they are dependent on the term-structure of the bond return volatility.

In other words, a naïve application of the standard VIX methodology to the pricing of a variance swap for rates leads to a bias. Appendix C.2 contains numerical experiments that illustrate these facts, whereby a VIX index is calculated based on Eq. (4.8), and assuming a market with interest rates as in the Vasicek (1977) model.

4.2.2 Basis Assets

We now study variance swaps in the government bond space in a more general context than the zero-coupon case, and focus on forward prices rather than spot prices.

Consider a coupon-bearing bond issued at time T_0, and paying coupons $\frac{C_i}{n}$ over a sequence of dates T_i, $i = 1, \dots, N$, where n is the frequency of coupon payments. For example $n = 2$ for semiannual coupon payments, in which case $T_i - T_{i-1} = \frac{1}{2}$; in general, $T_i - T_{i-1} = \frac{1}{n}$. For notational simplicity, we set $\mathbb{T} \equiv T_N$. The price of this bond may henceforth be interpreted as that of a specific bond or that of the "cheapest to deliver" from a set of deliverable bonds into a futures contract, and quoted in terms of either a traded flat price or an adjusted price based on some scalar "conversion factor." In the absence of arbitrage, the price of this bond is given by

$$B_t(\mathbb{T}) \equiv \sum_{i=1}^N \frac{C_i}{n} P_t(T_i) + P_t(\mathbb{T}), \quad t \equiv T_0. \qquad (4.10)$$

---

[3]Note that if $\mathbb{T} = T$, the fair value of a government bond variance swap, $\mathbb{P}(t,T)$ in Eq. (4.5), collapses to this bias, viz $\mathbb{P}(t,T) = \text{Bias}(t,T)$.

After T_0, the market value of the bond (the "dirty" or "invoice" price) is:

$$B_t(\mathbb{T}) \equiv \sum_{i=i_t}^{N} \frac{C_i}{n} P_t(T_i) + P_t(\mathbb{T}), \quad t \geq T_0, \tag{4.11}$$

where T_{i_t} is the first available coupon payment date a bondholder would have access to.

We are interested in developing a framework for pricing the volatility of this coupon-bearing bond price.

Let $F_t(T, \mathbb{T})$ be the forward price at t, for delivery at T, of the coupon-bearing bond expiring at $\mathbb{T}$. That is, come time T, the payoff for going long the forward is, $B_T(\mathbb{T}) - F_t(T, \mathbb{T})$, so that in the absence of arbitrage,

$$F_t(T, \mathbb{T}) = \frac{B_t(\mathbb{T})}{P_t(T)}. \tag{4.12}$$

It is well known (see, e.g., Mele 2014, Chap. 12) that $F_t(T, \mathbb{T})$ is a martingale under the so-called forward probability, Q_{FT}, defined through the Radon–Nikodym derivative,

$$\left. \frac{dQ_{FT}}{dQ} \right|_{\mathbb{F}_T} = \frac{e^{-\int_t^T r_\tau d\tau}}{P_t(T)}, \tag{4.13}$$

where $\mathbb{F}_T$ is the information set as of time T, so that $F_t(T, \mathbb{T})$ satisfies

$$\frac{dF_\tau(T, \mathbb{T})}{F_\tau(T, \mathbb{T})} = v_\tau(T, \mathbb{T}) \cdot dW_\tau^{F^T}, \quad \tau \in [t, T], \tag{4.14}$$

and $W_\tau^{F^T}$ is a multidimensional Brownian motion under Q_{FT}; $v_\tau(T, \mathbb{T}) \equiv \sigma_\tau^B(\mathbb{T}) - \sigma_\tau(T)$ is the vector of instantaneous volatilities, with $\sigma_\tau^B(\mathbb{T})$ denoting the vector of instantaneous volatilities of the coupon bearing bond; and, finally, $\sigma_\tau(T)$ the vector of instantaneous volatilities of a zero coupon bond expiring at T, as defined in Eq. (4.1).

In light of the discussion in the previous section, we shall price this volatility but only up to time $T \leq \mathbb{T}$. For example, we can consider "1m-10Y contracts," i.e., contracts for which the expiration for volatility delivery (T) is 1 month and volatility is referenced to returns on coupon bearing bonds expiring 10 years after T, at $\mathbb{T}$.

Our objective is to price the volatility of forward prices. In our continuous time setting, this volatility is approximately the same as the volatility of futures. Indeed, forwards are martingales under the forward probability, whereas futures are martingales under the risk-neutral probability. By Girsanov's theorem, their drifts differ but their realized volatilities are approximately the same.[4] Following Chap. 2, we consider two types of variance contracts and indexes: one based on percentage volatility (in Sect. 4.2.3), and another based on basis point volatility (in Sect. 4.2.4).

---

[4]Let the future price be $\tilde{F}_\tau(T, \mathbb{T}) = \mathbb{E}_\tau(B_T(\mathbb{T}))$ and the forward $F_\tau(T, \mathbb{T}) = \mathbb{E}_\tau^{Q_{FT}}(B_T(\mathbb{T}))$, where $\mathbb{E}_\tau(\cdot)$ denotes the conditional expectation under the risk-neutral probability, and $\mathbb{E}_\tau^{Q_{FT}}(\cdot)$

4.2.3 Percentage Price Volatility

Consider the following payoff at T of a government bond variance swap:

$$\pi(T, \mathbb{T}) \equiv V(t, T, \mathbb{T}) - \mathbb{P}(t, T, \mathbb{T}), \quad T \leq \mathbb{T}, \tag{4.15}$$

where

$$V(t, T, \mathbb{T}) \equiv \int_t^T \left\| v_\tau (T, \mathbb{T}) \right\|^2 d\tau, \tag{4.16}$$

and $\mathbb{P}(t, T, \mathbb{T})$ is the fair value of the strike.[5] It equals

$$\mathbb{P}(t, T, \mathbb{T}) = \frac{1}{P_t(T)} \mathbb{E}_t \left(e^{-\int_t^T r_\tau d\tau} V(t, T, \mathbb{T}) \right) = \mathbb{E}_t^{Q_{FT}} \left(V(t, T, \mathbb{T}) \right), \tag{4.17}$$

where $\mathbb{E}_t^{Q_{FT}}$ denotes the conditional expectation under the forward probability Q_{FT}, and the second equality follows by a change of probability.

4.2.3.1 Pricing

We are ready to cast this strike value in a model-free format through the two steps we discussed in Sect. 4.2.1. As for the first step, we apply Itô's lemma to $\ln F_\tau(T, \mathbb{T})$ in Eq. (4.14) and determine the value of a *log-contract*, i.e. a contract for delivery of $\ln F_T(T, \mathbb{T})$ at time T. This value is proportional to

$$-\mathbb{E}_t^{Q_{FT}} \left(\ln \frac{F_T(T, \mathbb{T})}{F_t(T, \mathbb{T})} \right) = \frac{1}{2} \mathbb{E}_t^{Q_{FT}} \left(V(t, T, \mathbb{T}) \right) = \frac{1}{2} \mathbb{P}(t, T, \mathbb{T}), \tag{4.18}$$

where we have used the definition of the swap variance rate in Eq. (4.17).

Note that because $F_\tau(T, \mathbb{T})$ is a martingale under Q_{FT}, Eq. (4.18) does not include terms involving the total realized return of the asset (i.e., the second term of the R.H.S. of Eq. (4.2)). Therefore, we only need to calculate the expected return on the forward, which is the second step in terms of our discussion in Sect. 4.2.1.

For this second step, we proceed with the usual spanning arguments. A log-contract for the forward price $F_T(T, \mathbb{T})$ is valued the same as a weighted average of

---

the conditional expectation under the forward. We have, $F_\tau(T, \mathbb{T}) = \mathbb{E}_\tau(\eta_\tau(T) B_T(\mathbb{T}))$, so that $F_\tau(T, \mathbb{T}) = \tilde{F}_\tau(T, \mathbb{T}) + h_\tau$, where $\eta_\tau(T) \equiv \frac{dQ_{FT}}{dQ}|_{\mathcal{F}_\tau}$ and $h_\tau \equiv \text{cov}_\tau^Q(\eta_\tau(T), B_T(\mathbb{T}))$, with obvious notation. The volatilities of the future and the forward are exactly the same when h_τ has bounded variation as, for example, in the Vasicek (1977) model.

[5] For simplicity, we use the same notation as that in Sect. 4.2.1.3 to denote the fair value of the strike.

a continuum of out-of-the-money options, viz

$$\mathbb{E}_t^{Q_{FT}}\left(\ln\frac{F_T(T,\mathbb{T})}{F_t(T,\mathbb{T})}\right)$$

$$= -\frac{1}{P_t(T)}\left(\int_0^{F_t(T,\mathbb{T})}\text{Put}_t(K,T,\mathbb{T})\frac{1}{K^2}dK + \int_{F_t(T,\mathbb{T})}^{\infty}\text{Call}_t(K,T,\mathbb{T})\frac{1}{K^2}dK\right),$$

$$(4.19)$$

where, for simplicity, we are using the same notation as that in Sect. 4.2.1.3 and referring to $\text{Put}_t(K,T,\mathbb{T})$ and $\text{Call}_t(K,T,\mathbb{T})$ as the prices of out-of-the-money European puts and calls struck at K, expiring at time T, and written at t on the coupon bearing bond maturing at $\mathbb{T}$.

Finally, we express the variance swap rate, $\mathbb{P}(t,T,\mathbb{T})$ in Eq. (4.17), in a model-free format. Combining Eq. (4.18) and Eq. (4.19) leads to:

$$\mathbb{P}(t,T,\mathbb{T})$$

$$= \frac{2}{P_t(T)}\left(\int_0^{F_t(T,\mathbb{T})}\text{Put}_t(K,T,\mathbb{T})\frac{1}{K^2}dK + \int_{F_t(T,\mathbb{T})}^{\infty}\text{Call}_t(K,T,\mathbb{T})\frac{1}{K^2}dK\right).$$

$$(4.20)$$

Accordingly, an index of percentage government bond volatility is,

$$\boxed{\text{GB-VI}(t,T,\mathbb{T}) \equiv 100 \times \sqrt{\frac{\mathbb{P}(t,T,\mathbb{T})}{T-t}}} \qquad (4.21)$$

4.2.3.2 Discussion

For the GB-VI index in Eqs. (4.20)–(4.21), a portfolio of out-of-the-money options is rescaled by the inverse of the price of a zero, $1/P_t(T)$. This formula for government bond volatility remarkably resembles that for the VIX, where a portfolio of out-of-the-money options is rescaled by $e^{\bar{r}(T-t)}$, and $\bar{r}$ is the constant short-term rate. If it wasn't that Eqs. (4.20)–(4.21) obviously relate to the volatility of interest rate instruments, one would naturally conclude that the GB-VI index extends the standard equity volatility index to the case where rates are stochastic. It is not a trivial statement. The rescaling property in Eq. (4.20) is, instead, remarkable because in Appendix C.1, we show that to achieve a model-free expression for the equity VIX in markets with random interest rates, we need to tilt the market probability from the risk-neutral to the forward, exactly as in this chapter.

The key to understanding these properties is that the government bond variance swaps in this section illustrate a specific instance of the theory in Chap. 2. That is, government bond variance swaps can be priced in a model-free fashion while making reference to the forward probability simply because zero coupon bonds are the market numéraires in the context of the present chapter. However, Eqs. (4.20) and (4.21) only provide us with an incomplete account of the complexity arising whilst pricing government bond volatility. As anticipated in the introduction of this chapter, we need to consider maturity mismatches that arise when the options maturity is shorter than the underlying. Section 4.4 explains how to deal with this additional complication.

4.2.4 Basis Point Price Volatility

Next, we consider a variance contract cast in basis point terms, that is, one in which the payoff is

$$\pi^{\text{bp}}(T, \mathbb{T}) \equiv V^{\text{bp}}(t, T, \mathbb{T}) - \mathbb{P}^{\text{bp}}(t, T, \mathbb{T}), \quad T \leq \mathbb{T}, \tag{4.22}$$

where

$$V^{\text{bp}}(t, T, \mathbb{T}) \equiv \int_t^T F_\tau^2(T, \mathbb{T}) \left\| v_\tau(T, \mathbb{T}) \right\|^2 d\tau, \tag{4.23}$$

and $\mathbb{P}^{\text{bp}}(t, T, \mathbb{T})$ is the fair value of the strike. In analogy with Eq. (4.17),

$$\mathbb{P}^{\text{bp}}(t, T, \mathbb{T}) = \frac{1}{P_t(T)} \mathbb{E}_t \left(e^{-\int_t^T r_\tau d\tau} V^{\text{bp}}(t, T, \mathbb{T}) \right) = \mathbb{E}^{Q_{FT}} \left(V^{\text{bp}}(t, T, \mathbb{T}) \right). \tag{4.24}$$

To implement the usual first step towards a model-free expression for the fair strike, we consider a *quadratic contract*, similarly as in the approach leading to the basis point volatility index for interest rate swaps in Chap. 3. By applying Itô's lemma to $F_\tau^2(T, \mathbb{T})$ in Eq. (4.14), we find that the price of a quadratic contract (i.e., one for delivery of $F_T^2(T, \mathbb{T})$ at time T) is

$$\mathbb{E}_t^{Q_{FT}} \left(F_T^2(T, \mathbb{T}) \right) - F_t^2(T, \mathbb{T}) = \mathbb{E}_t^{Q_{FT}} \left(V^{\text{bp}}(t, T, \mathbb{T}) \right) = \mathbb{P}^{\text{bp}}(t, T, \mathbb{T}), \tag{4.25}$$

where the last equality follows by Eq. (4.24).

The second step regards spanning the quadratic contract with options. By the usual arguments in Chap. 2, we have that,

$$F_T^2(T, \mathbb{T}) - F_t^2(T, \mathbb{T})$$
$$= 2F_t(T, \mathbb{T}) \left(F_T(T, \mathbb{T}) - F_t(T, \mathbb{T}) \right)$$
$$+ 2 \left(\int_0^{F_t(T, \mathbb{T})} (K - F_T(T, \mathbb{T}))^+ dK + \int_{F_t(T, \mathbb{T})}^\infty (F_T(T, \mathbb{T}) - K)^+ dK \right),$$

so that, by taking expectation under the forward probability,

$$\mathbb{E}_t^{Q_{FT}} \left(F_T^2(T, \mathbb{T}) \right) - F_t^2(T, \mathbb{T})$$
$$= 2 \left(\int_0^{F_t(T, \mathbb{T})} \mathbb{E}_t^{Q_{FT}} (K - F_T(T, \mathbb{T}))^+ dK \right.$$

$$+ \int_{F_t(T,\mathbb{T})}^{\infty} \mathbb{E}_t^{Q_{FT}} \left(F_T(T,\mathbb{T}) - K\right)^+ dK\bigg)$$

$$= \frac{2}{P_t(T)} \left(\int_0^{F_t(T,\mathbb{T})} \mathrm{Put}_t(K,T,\mathbb{T})dK + \int_{F_t(T,\mathbb{T})}^{\infty} \mathrm{Call}_t(K,T,\mathbb{T})dK \right), \quad (4.26)$$

where the second equality follows by a change in probability. Comparing Eq. (4.25) with Eq. (4.26) leaves

$$\mathbb{P}^{\mathrm{bp}}(t,T,\mathbb{T}) = \frac{2}{P_t(T)} \left(\int_0^{F_t(T,\mathbb{T})} \mathrm{Put}_t(K,T,\mathbb{T})dK + \int_{F_t(T,\mathbb{T})}^{\infty} \mathrm{Call}_t(K,T,\mathbb{T})dK \right).$$
$$(4.27)$$

The basis point version of the government bond volatility index is

$$\boxed{\mathrm{GB\text{-}VI}^{\mathrm{bp}}(t,T,\mathbb{T}) \equiv 100^2 \times \sqrt{\frac{\mathbb{P}^{\mathrm{bp}}(t,T,\mathbb{T})}{T-t}}} \qquad (4.28)$$

4.2.5 Marking to Market

In Appendix C.3, we show that the marking-to-market value of the percentage government bond variance swap is

$$\text{M-Var}_\tau(T,\mathbb{T}) \equiv P_\tau(T)\big[V(t,\tau,\mathbb{T}) + \mathbb{P}(\tau,T,\mathbb{T}) - \mathbb{P}(t,T,\mathbb{T})\big], \qquad (4.29)$$

and that of the Basis Point is the same as M-Var$_\tau(T,\mathbb{T})$, but with $V^{\mathrm{bp}}(t,T,\mathbb{T})$ replacing $V(t,T,\mathbb{T})$.

4.2.6 Replication

4.2.6.1 Percentage Variance Swaps

We provide details on how to replicate the payoff of the variance swap $\pi(T,\mathbb{T})$ in Eq. (4.15) through forwards and out-of-the-money options. In Appendix C.4, we show that to hedge against this payoff we need two identical zero cost portfolios, constructed with the following positions:

(i) A dynamic position in government bond forwards aiming to replicate the cumulative increments of the government bond forward returns over $[t,T]$, which equals $\int_t^T \frac{dF_s(T,\mathbb{T})}{F_s(T,\mathbb{T})} = \frac{1}{2}V(,T,\mathbb{T}) + \ln\frac{F_T(T,\mathbb{T})}{F_t(T,\mathbb{T})}$.

(ii) Static positions in forwards expiring at time T and out-of-the-money options expiring at time T, aiming to replicate the *log-contract* for the coupon bearing bond forward, $F_t(T,\mathbb{T})$. The positions are as follows:

Table 4.1 Replication of the percentage government bond variance contract payoff $\pi(T, \mathbb{T})$ in Eq. (4.15), $\pi(T, \mathbb{T}) = V(t, T, \mathbb{T}) - \mathbb{P}(t, T, \mathbb{T})$, $T \leq \mathbb{T}$

Portfolio	Value at t	Value at T
(i) long self-financed portfolio of forwards	0	$V(t, T, \mathbb{T}) + 2 \ln \frac{F_T(T, \mathbb{T})}{F_t(T, \mathbb{T})}$
(ii) short forwards and long OTM options	$-\mathbb{P}(t, T, \mathbb{T}) P_t(T)$	$-2 \ln \frac{F_T(T, \mathbb{T})}{F_t(T, \mathbb{T})}$
(iii) borrow $\mathbb{P}(t, T, \mathbb{T}) P_t(T)$	$+\mathbb{P}(t, T, \mathbb{T}) P_t(T)$	$-\mathbb{P}(t, T, \mathbb{T})$
Net cash flows	0	$V(t, T, \mathbb{T}) - \mathbb{P}(t, T, \mathbb{T})$

    (ii.1) Short $1/F_t(T, \mathbb{T})$ units of a forward expiring at T, and struck at the current forward value, $F_t(T, \mathbb{T})$.

    (ii.2) Long out-of-the-money options, each of them carrying a weight equal to $\frac{1}{K^2}$.

(iii) A static borrowing position aimed to finance the out-of-the-money option positions in (ii.2).

    The details for the self-financed portfolio in (i) are given in Appendix C.4. Table 4.1 summarizes the costs of each of these trades at t, as well as the payoffs at T, which sum up to be the same as the payoff $\pi(T, \mathbb{T})$ in Eq. (4.15), as required to honour the contract.

4.2.6.2 Basis Point Variance Swaps

Next, we explain how to replicate basis point variance contracts, that is, those contracts with a payoff equal to $\pi^{bp}(T, \mathbb{T})$ in Eq. (4.22). We need positioning into *quadratic* contracts (i.e., contracts delivering $F_T^2(T, \mathbb{T}) - F_t^2(T, \mathbb{T})$), rather than on the *log-contracts* underlying row (ii) of Table 4.1. We utilize the following portfolios:

  (i) A dynamic position in government bond forwards, aiming to replicate the increments of the government bond forwards over $[t, T]$, weighted by the very same government bond forwards, which equals $-\int_t^T F_s(T, \mathbb{T}) dF_s(T, \mathbb{T}) = \frac{1}{2}[V^{bp}(t, T, \mathbb{T}) - (F_T^2(T, \mathbb{T}) - F_t^2(T, \mathbb{T}))]$.

  (ii) Static positions in forwards expiring at T and out-of-the-money options expiring at T, aiming to replicate the payoff of the *quadratic contract* on the forward, as follows:

    (ii.1) Long $2F_t(T, \mathbb{T})$ units of a forward struck at $F_t(T, \mathbb{T})$.

    (ii.2) Long out-of-the-money options, each of them carrying a weight equal to $2\Delta K$.

(iii) A static borrowing position aimed to finance the out-of-the-money option positions in (ii.2).

    Table 4.2 provides the value of this portfolio at time t and time T, resulting in a perfect replication of the payoff $\pi^{bp}(T, \mathbb{T})$ in Eq. (4.22).

Table 4.2 Replication of the basis point government bond variance contract $\pi^{\mathrm{bp}}(T, \mathbb{T})$ in Eq. (4.22), $\pi^{\mathrm{bp}}(T, \mathbb{T}) = V^{\mathrm{bp}}(t, T, \mathbb{T}) - \mathbb{P}^{\mathrm{bp}}(t, T, \mathbb{T}), T \leq \mathbb{T}$

Portfolio	Value at t	Value at T
(i) short self-financed forwards	0	$V^{\mathrm{bp}}(t, T, \mathbb{T}) - (F_T^2(T, \mathbb{T}) - F_t^2(T, \mathbb{T}))$
(ii) long forwards, long OTM options	$-\mathbb{P}^{\mathrm{bp}}(t, T, \mathbb{T}) P_t(T)$	$F_T^2(T, \mathbb{T}) - F_t^2(T, \mathbb{T})$
(iii) borrow $\mathbb{P}^{\mathrm{bp}}(t, T, \mathbb{T}) P_t(T)$	$\mathbb{P}^{\mathrm{bp}}(t, T, \mathbb{T}) P_t(T)$	$-\mathbb{P}^{\mathrm{bp}}(t, T, \mathbb{T})$
Net cash flows	0	$V^{\mathrm{bp}}(t, T, \mathbb{T}) - \mathbb{P}^{\mathrm{bp}}(t, T, \mathbb{T})$

4.2.7 Forward Price Adjustments

In absence of price quotes for ATM options, we can approximate the theoretical value of the indexes through a device similar to that used in CBOE's calculations of VIX (see Chicago Board Options Exchange 2009). Let K_0 denote the first strike traded below the current forward price $F_t(T, \mathbb{T})$, where in case $F_t(T, \mathbb{T})$ is not observed, it is approximated by the strike price at which the absolute difference between the call and put prices is smallest. (Naturally, if there is an option struck at F_t, then $K_0 = F_t$.)

In Appendix C.5, we show that the percentage index and the basis point index are approximated by the following expressions:

$$
\begin{aligned}
\text{GB-VI}_o(t, T, \mathbb{T}) &\equiv 100 \times \sqrt{\frac{\mathbb{P}_o(t, T, \mathbb{T})}{T - t}}, \\
\text{GB-VI}_o^{\mathrm{bp}}(t, T, \mathbb{T}) &\equiv 100^2 \times \sqrt{\frac{\mathbb{P}_o^{\mathrm{bp}}(t, T, \mathbb{T})}{T - t}}
\end{aligned}
$$

where

$$
\begin{aligned}
\mathbb{P}_o(t, T, \mathbb{T}) &\equiv \frac{2}{P_t(T)} \left(\int_0^{K_0} \mathrm{Put}_t(K, T, \mathbb{T}) \frac{1}{K^2} dK + \int_{K_0}^{\infty} \mathrm{Call}_t(K, T, \mathbb{T}) \frac{1}{K^2} dK \right) \\
&\quad - \left(\frac{F_t(T, \mathbb{T}) - K_0}{K_0} \right)^2,
\end{aligned} \tag{4.30}
$$

and

$$
\begin{aligned}
\mathbb{P}_o^{\mathrm{bp}}(t, T, \mathbb{T}) &\equiv \frac{2}{P_t(T)} \left(\int_0^{K_0} \mathrm{Put}_t(K, T, \mathbb{T}) dK + \int_{K_0}^{\infty} \mathrm{Call}_t(K, T, \mathbb{T}) dK \right) \\
&\quad - \left(F_t(T, \mathbb{T}) - K_0 \right)^2. \tag{4.31}
\end{aligned}
$$

4.2.8 Model-Free Measures of Basis Point Yield Volatility

It is market practice to publish model-dependent measures of "basis point yield volatility," defined as the level of volatility of yields consistent with a given ATM option price. This section provides a model-free approach to measuring basis point yield volatility, which arises from a combination of the price volatility indexes in Eq. (4.21) (percentage) and Eq. (4.28) (basis point), or on those based on the forward approximations in Eq. (4.30) and Eq. (4.31). We begin by introducing a notion of certainty equivalent prices.

4.2.8.1 Certainty Equivalent Bond Prices

By Eq. (4.18) and Eq. (4.25), we can express the two volatility indexes as follows:

$$\text{GB-VI}(t, T, \mathbb{T}) = \sqrt{\frac{1}{T-t}\frac{1}{P_t(T)}\mathbb{E}_t\left(e^{-\int_t^T r_\tau d\tau}\int_t^T \left\|v_\tau(T, \mathbb{T})\right\|^2 d\tau\right)}, \quad (4.32)$$

and

$$\text{GB-VI}^{\text{bp}}(t, T, \mathbb{T}) = \sqrt{\frac{1}{T-t}\frac{1}{P_t(T)}\mathbb{E}_t\left(e^{-\int_t^T r_\tau d\tau}\int_t^T F_\tau^2(T, \mathbb{T})\left\|v_\tau(T, \mathbb{T})\right\|^2 d\tau\right)}. \quad (4.33)$$

We look for a certainty equivalent price for the coupon bearing bond that matches the two indexes $\text{GB-VI}(t, T, \mathbb{T})$ and $\text{GB-VI}^{\text{bp}}(t, T, \mathbb{T})$. In other words, we ask what is the guaranteed price of the coupon-bearing bond at time T, say $\mathcal{B}(t, T, \mathbb{T})$, such that the expected basis point volatility on the forward, $\text{GB-VI}^{\text{bp}}(t, T, \mathbb{T})$, is the same as the basis point volatility in a certainty equivalent market? Clearly, in such a hypothetical market, the forward would be constant and equal to $\mathcal{B}(t, T, \mathbb{T})$, so that

$$\mathcal{B}(t, T, \mathbb{T}) : \text{GB-VI}^{\text{bp}}(t, T, \mathbb{T})$$

$$= \sqrt{\frac{1}{T-t}\frac{1}{P_t(T)}\mathbb{E}_t\left(e^{-\int_t^T r_\tau d\tau}\int_t^T \mathcal{B}^2(t, T, \mathbb{T})\left\|v_\tau(T, \mathbb{T})\right\|^2 d\tau\right)}.$$

Then, by Eqs. (4.32)–(4.33), $\mathcal{B}(t, T, \mathbb{T})$ satisfies

$$\text{GB-VI}^{\text{bp}}(t, T, \mathbb{T}) = \mathcal{B}(t, T, \mathbb{T}) \times \text{GB-VI}(t, T, \mathbb{T}). \quad (4.34)$$

That is, the certainty equivalent for the coupon-bearing bond price, $\mathcal{B}(t, T, \mathbb{T})$, is simply the ratio of the basis point price volatility index to the percentage.

Market practice defines basis point volatility as the product of percentage ATM volatility times the underlying. Equation (4.34) is a model-free counterpart to this practice. Note that Eq. (4.34) does not rely on any assumption, except for those underlying the framework of previous sections, and is simply another way to characterize the basis point volatility index $\text{GB-VI}^{\text{bp}}(t, T, \mathbb{T})$ in Eq. (4.28).

How does Eq. (4.34) relate to the "skew factor" $\xi(t, T)$ of Proposition 2.3 in Chap. 2 (Sect. 2.4.3)? Note that

$$\text{GB-VI}^{\text{bp}}(t, T, \mathbb{T}) = F_t(T, \mathbb{T})\sqrt{(T - t)^{-1}\xi(t, T)} = \mathcal{B}(t, T, \mathbb{T})\text{GB-VI}(t, T, \mathbb{T}),$$

where the first equality follows by Eq. (2.28) in Chap. 2, and the second by Eq. (4.34). Therefore,

$$\sqrt{(T - t)^{-1}\xi(t, T)} = \frac{\mathcal{B}(t, T, \mathbb{T})}{F_t(T, \mathbb{T})}\text{GB -VI}(t, T, \mathbb{T}).$$

That is, the skew component of the basis point volatility index is a biased estimate of the percentage index, with the bias reducing to zero only when the certainty equivalent price collapses to the forward.

4.2.8.2 Yield Volatility

Yield volatility is a subtle concept in our context because government bond option trading relates to *prices* instead of *yields*, and market practice relies on model-based measures of yield volatility. We propose two model-free measures of yield volatility by mapping the new representation of basis point expected price volatility in Eq. (4.34) into its yield equivalent.

Define $y_\mathcal{B}(t, T, \mathbb{T})$ as the solution to the following equation:

$$y_\mathcal{B}(t, T, \mathbb{T}) : \mathcal{B}(t, T, \mathbb{T}) = \frac{\text{GB-VI}^{\text{bp}}(t, T, \mathbb{T})}{\text{GB-VI}(t, T, \mathbb{T})} = \hat{P}\big(y_\mathcal{B}(t, T, \mathbb{T})\big), \tag{4.35}$$

where,

$$\hat{P}(y) \equiv \sum_{i=1}^{N} \frac{C_i}{n}\left(1 + \frac{y}{n}\right)^{-i} + 100\left(1 + \frac{y}{n}\right)^{-N}, \tag{4.36}$$

and C_i are the coupons, defined as in Eq. (4.10). In Appendix C.6, we show that $y_\mathcal{B}(t, T, \mathbb{T})$ is well-defined, in that it exists and is positive. Our first model-free measure of volatility relies on the duration of the certainty equivalent price, $\mathcal{B}(t, T, \mathbb{T})$. A second relies on the model-free yield, $y_\mathcal{B}(t, T, \mathbb{T})$.

Duration-Based Yield Volatility

Define $D_\mathcal{B}(t, T, \mathbb{T})$ as the modified duration of the guaranteed price $\mathcal{B}(t, T, \mathbb{T})$,

$$D_\mathcal{B}(t, T, \mathbb{T}) \equiv \frac{1}{1 + \frac{y_\mathcal{B}(t,T,\mathbb{T})}{n}}\left(\sum_{i=1}^{N} \omega_i \frac{i}{n} + \hat{\omega}_N \frac{N}{n}\right)$$

$$\omega_i \equiv \frac{\frac{C_i}{n}/(1 + \frac{y_\mathcal{B}(t,T,\mathbb{T})}{n})^i}{\mathcal{B}(t, T, \mathbb{T})}, \qquad \hat{\omega}_N \equiv \frac{100/(1 + \frac{y_\mathcal{B}(t,T,\mathbb{T})}{n})^N}{\mathcal{B}(t, T, \mathbb{T})} \tag{4.37}$$

where $y_\mathcal{B}(t, T, \mathbb{T})$ and $\mathcal{B}(t, T, \mathbb{T})$ are defined in Eq. (4.35). A model-free gauge of duration-based yield volatility is

$$\text{GB-VI}_{\text{Yd}}^{\text{bp}}(t, T, \mathbb{T}) = 100 \times \frac{\text{GB-VI}(t, T, \mathbb{T})}{D_\mathcal{B}(t, T, \mathbb{T})}, \quad (4.38)$$

or, using the definitions of $\hat{P}$ in Eq. (4.36) and that of $D_\mathcal{B}$ in Eqs. (4.37),

$$\text{GB-VI}_{\text{Yd}}^{\text{bp}}(t, T, \mathbb{T})$$
$$= \frac{100 \times (1 + \frac{1}{n} \hat{P}^{-1}[\frac{\text{GB-VI}^{\text{bp}}(t,T,\mathbb{T})}{\text{GB-VI}(t,T,\mathbb{T})}]) \times \text{GB-VI}^{\text{bp}}(t,T,\mathbb{T})}{\sum_{i=1}^{N} \frac{C_i}{n}(1 + \frac{1}{n}\hat{P}^{-1}[\frac{\text{GB-VI}^{\text{bp}}(t,T,\mathbb{T})}{\text{GB-VI}(t,T,\mathbb{T})}])^{-i}\frac{i}{n} + 100(1 + \frac{1}{n}\hat{P}^{-1}[\frac{\text{GB-VI}^{\text{bp}}(t,T,\mathbb{T})}{\text{GB-VI}(t,T,\mathbb{T})}])^{-N}\frac{N}{n}} \quad (4.39)$$

where $\hat{P}^{-1}$ denotes the inverse function of $\hat{P}$ in Eq. (4.36). In Appendix C.6, we extend this volatility gauge to the "post-issuance case," where the maturity T of the forward is higher than the date of issuance of the bond (see Eq. (C.19)).

Equation (4.39) is a model-free index of yield volatility, as it relates to the duration of the certainty equivalent price for the coupon-bearing bond, $\mathcal{B}(t, T, \mathbb{T})$, which is model-free and recovered by means of Eq. (4.34), and through the market price of OTM options. Moreover, this measure of yield volatility can be spanned and hedged through a dedicated basis point variance swap, for the reason that $y_\mathcal{B}(t, T, \mathbb{T})$ and $\mathcal{B}(t, T, \mathbb{T})$ and, then, $D_\mathcal{B}(t, T, \mathbb{T})$, are known at the inception of the variance swap.

Yield-Based Yield Volatility

A second measure of yield volatility re-scales the percentage index by the model-free yield $y_\mathcal{B}(t, T, \mathbb{T})$ defined in Eq. (4.35), as follows:

$$\text{GB-VI}_{\text{Y}}^{\text{bp}}(t, T, \mathbb{T}) = 100 \times y_\mathcal{B}(t, T, \mathbb{T}) \times \text{GB-VI}(t, T, \mathbb{T}),$$

or more succinctly,

$$\text{GB-VI}_{\text{Y}}^{\text{bp}}(t, T, \mathbb{T}) = 100 \times \hat{P}^{-1}\left[\frac{\text{GB-VI}^{\text{bp}}(t, T, \mathbb{T})}{\text{GB-VI}(t, T, \mathbb{T})}\right] \times \text{GB-VI}(t, T, \mathbb{T}) \quad (4.40)$$

The index $\text{GB-VI}_{\text{Y}}^{\text{bp}}(t, T, \mathbb{T})$ in Eq. (4.40) is model-free, and can be spanned and hedged through percentage variance swaps, for the same reasons outlined regarding $\text{GB-VI}_{\text{Yd}}^{\text{bp}}(t, T, \mathbb{T})$ in Eq. (4.39). Once again, in Appendix C.6, Eq. (C.20), we extend this index to the "post-issuance case," where T is higher than the date of issuance of the bond.

4.2.9 Certainty Equivalent Bond Prices as Expectations of Forward Prices

4.2.9.1 Theory

What is the link between the certainty equivalent $\mathcal{B}(t, T, \mathbb{T})$ and the forward price process $F_\tau(T, \mathbb{T})$? In Appendix C.6, we show that there exists a weighting function ω_τ, known at time t and integrating to one over $[t, T]$, such that $\mathcal{B}^2(t, T, \mathbb{T})$ can be expressed as a time-average of the entire path of the conditional second order moment of $F_\tau^2(T, \mathbb{T})$ under a new probability Q_{v^τ}:

$$\mathcal{B}(t, T, \mathbb{T}) = \sqrt{\int_t^T \omega_\tau \mathbb{E}_t^{Q_{v^\tau}} \left[F_\tau^2(T, \mathbb{T}) \right] d\tau}, \quad \text{with} \int_t^T \omega_\tau d\tau = 1, \qquad (4.41)$$

where $\mathbb{E}_t^{Q_{v^\tau}}[\cdot]$ is the conditional expectation taken under the probability Q_{v^τ} defined through the Radon–Nikodym derivative

$$\rho(\tau; T) = \left. \frac{dQ_{v^\tau}}{dQ_{FT}} \right|_{\mathbb{F}_\tau} = \frac{\|v_\tau(T, \mathbb{T})\|^2}{\mathbb{E}_t^{Q_{FT}}[\|v_\tau(T, \mathbb{T})\|^2]}. \qquad (4.42)$$

We refer to Q_{v^τ} as the *realized variance probability*. This probability distorts the forward probability so as to give more weight to the paths of $F_\tau(T, \mathbb{T})$ that have higher chances of experiencing episodes of high volatility—the states of nature with higher volatility receive higher weight.

4.2.9.2 One Example

Consider the following extension of the Ho and Lee (1986) model, in which the short-term rate r_τ is a process with stochastic volatility,

$$\begin{cases} dr_\tau = \theta_\tau d\tau + v_\tau dW_{1,\tau} \\ dv_\tau^2 = \xi v_\tau dW_{2,\tau} \end{cases} \qquad (4.43)$$

where ξ is a "volatility of variance" parameter, θ_τ is an "infinite-dimensional" parameter that allows us to fit the initial yield curve at $\tau = t$ without error, and W_i are Brownian motions under the risk-neutral probability.[6]

---

[6]The model in Eq. (4.43) can be extended to one in which the interest rate basis point variance v_τ^2 is linear mean-reverting (see Appendix C.7), such that v_τ^2 is a stationary square-root process (Cox et al. 1985), similar to the stochastic volatility model of Heston (1993) in the equity case. These models are "affine," in that their conditional characteristic function is exponential-affine in the initial state (r_t, v_t^2) (see, also, Mele et al. 2015b).

In Appendix C.7, we show that (i) the infinite-dimensional parameter θ_τ in the drift of the short-term rate in (4.43) is

$$\theta_\tau = \frac{\partial f_\$(t,\tau)}{\partial \tau} + \frac{\partial^2 C_\tau(t)}{\partial \tau^2} v_t^2, \tag{4.44}$$

where $f_\$(t,\tau)$ denotes an hypothetically *observed* instantaneous forward rate at time t for maturity τ; and (ii) the price of a zero coupon bond at time $\tau \geq t$ when the state is (r_τ, v_τ), is:

$$P_\tau\left(r_\tau, v_\tau^2, T\right) \equiv e^{\int_\tau^T (s-T)\theta_s ds - (T-\tau)r_\tau + C_T(\tau)v_\tau^2}, \tag{4.45}$$

where θ_s is as in Eq. (4.44), and $C_T(\tau)$ is the solution to the following Riccati equation,

$$\dot{C}_T(\tau) = -\frac{1}{2}(T-\tau)^2 - \frac{1}{2}\xi^2 C_T^2(\tau), \quad C_T(T) = 0, \tag{4.46}$$

and the dot indicates differentiation with respect to time τ.

It can be verified (see Appendix C.7) that this model is able to match the yield curve initially observed (i.e. at t), in that

$$P_t\left(r_t, v_t^2, T\right) = e^{-\int_t^T f_\$(t,\tau)d\tau} \equiv P_t(T), \quad \text{for all } T, \tag{4.47}$$

where $P_t(T)$ denotes the market price of a zero-coupon bond with usual notation. The motivation underlying this matching is the need to exploit the model's predictions about bond price volatility, while making sure that we fit the initial yield curve without error. Variables of interest that have no closed form solutions could be obtained by simulating the interest rate process in Eq. (4.43) with the infinite dimensional parameter θ_τ taking the values in Eq. (4.44), which are known at the time of evaluation t; the variance process v_t^2 has to be filtered though.

As Eq. (4.45) makes clear, we make predictions regarding the future yield curve and, *potentially*, expected volatility on fixed income instruments, while feeding the model with all the bond prices observed at time t and not only the short-term rate. It is a standard procedure (see Chaps. 11 and 12 in Mele 2014) where we have added a stochastic volatility component v_τ. Let us emphasize a potential and interesting property of the model, which is to allow the current yield curve to feed expected developments in volatility. This "price feedback" property only applies to coupon bearing bonds, though, as further discussed below.

Consider a forward on a zero-coupon bond for simplicity. In Appendix C.7, we show that the forward price in Eq. (4.14) is the solution to

$$\begin{cases} \dfrac{dF_\tau(T,\mathbb{T})}{F_\tau(T,\mathbb{T})} = v_\tau\left(-(\mathbb{T}-T)dW_{1,\tau}^{F^T} + \xi\left(C_\mathbb{T}(\tau) - C_T(\tau)\right)dW_{2,\tau}^{F^T}\right) \\ dv_\tau^2 = \xi^2 v_\tau^2 C_T(\tau)d\tau + \xi v_\tau dW_{2,\tau}^{F^T} \end{cases} \tag{4.48}$$

where $W_{1,\tau}^{F^T}$ and $W_{2,\tau}^{F^T}$ are two independent Brownian motions under the forward probability Q_{FT}, and $C_T(\tau)$ is the bond price variance exposure in Eq. (4.45), the solution to Eq. (4.46).

The variance process of the forward price is

$$\left\| v_\tau(T, \mathbb{T}) \right\|^2 = \phi_\tau(\mathbb{T} - T) \cdot v_\tau^2, \tag{4.49}$$

where $\phi_\tau(T, \mathbb{T})$ is a deterministic process given by

$$\phi_\tau(T, \mathbb{T}) \equiv (\mathbb{T} - T)^2 + \xi^2 \big(C_{\mathbb{T}}(\tau) - C_T(\tau)\big)^2.$$

Moreover, in Appendix C.7, we show that the weighting function in Eq. (4.41) is

$$\omega_\tau = \frac{\bar{\phi}_\tau(t, T, \mathbb{T})}{\int_t^T \bar{\phi}_\tau(t, T, \mathbb{T}) d\tau}, \quad \bar{\phi}_\tau(t, T, \mathbb{T}) \equiv \phi_\tau(T, \mathbb{T}) e^{\xi^2 \int_t^\tau C_T(s) ds}, \tag{4.50}$$

and that the variance of the forward price process in Eq. (4.48), v_s^2, is the solution to

$$dv_s^2 = \xi^2 \big(1 + v_s^2 C_T(s)\big) ds + \xi v_s dW_{2,s}^{v^\tau}, \quad s \in [t, \tau], \tag{4.51}$$

where $W_{2,s}^{v^\tau}$ is a Brownian motion under Q_{v^τ}. Equation (4.51) says that under the realized variance probability, Q_{v^τ}, the average path of the forward price volatility is higher than that under the original forward probability, Q_{FT}, by the additional factor, $\xi^2 ds$. That is, certainty equivalence has to be scaled up for volatility.

4.2.9.3 Pricing Government Bond Volatility Products: An Introductory Example

Note that the expected variance under the forward probability Q_{FT} is (see Eq. (C.27) in Appendix C.7),

$$\mathbb{E}_t^{Q_{FT}}\left(\left\| v_\tau(T, \mathbb{T}) \right\|^2\right) = \bar{\phi}_\tau(t, T, \mathbb{T}) \cdot v_t^2, \tag{4.52}$$

so that the percentage volatility index, and hence the (square of the) fair value of a government bond variance swap predicted by the model in this section, is

$$\text{GB-VI}(v_t; t, T, \mathbb{T}) \equiv \sqrt{\frac{1}{T - t} \int_t^T \bar{\phi}_\tau(t, S, \mathbb{T}) d\tau \cdot v_t}. \tag{4.53}$$

Note that the model we are studying does not lead to "price feedbacks" in the context of zero-coupon bond volatility. While the entire yield curve at t affects the very same yield curve at τ through the parameter θ_τ (see Eqs. (4.44) and (4.45)), it does not provide information about developments in expected volatility. Indeed, Eq. (4.53) reveals that this expected volatility depends only on the current short-term rate volatility (in basis points), v_t, times a "multiplier" involving the function

$\bar{\phi}_\tau(t, T, \mathbb{T})$, which is independent of the yield curve at t. Mele et al. (2015b) show that price feedbacks arise while modeling the volatility of forwards on coupon bearing bonds.

We turn to another potential issue: misspecification of the variance swap pricing probability. Consider a naïve pricing procedure, whereby one prices a government bond variance swap by taking expectation under the risk-neutral probability, Q. The result would be:

$$\mathbb{E}_t\left(\left\|v_\tau(T, \mathbb{T})\right\|^2\right) = \phi_\tau(T, \mathbb{T}) \cdot v_t^2 \qquad (4.54)$$

and the following biased estimate of GB-VI$(v_t; t, T, \mathbb{T})$,

$$\widehat{\text{GB-VI}(v_t; t, T, \mathbb{T})} \equiv \sqrt{\frac{1}{T-t} \int_t^T \phi_\tau(T, \mathbb{T}) d\tau \cdot v_t}. \qquad (4.55)$$

Pricing a variance swap based on Eq. (4.53) rather than Eq. (4.55) gives rise to an arbitrage opportunity according to the model of this section. Alternatively, consider an example of a future on the squared index in Eq. (4.53). Theoretically, this future has the same price as the squared index, GB-VI$^2(v_t; t, T, \mathbb{T})$, due to the assumption in (4.43) that v_τ^2 is a martingale under Q. But according to an evaluation based on Eq. (4.55), one might mistakenly conclude that the market is in (or is pricing a) contango or backwardation according to parameter values. One plausible assumption leading to contango or backwardation under the correctly specified market is that v_τ^2 in (4.43) is a mean-reverting process. We present such a model in Appendix C.7. In Mele et al. (2015b), we develop a three factor model with mean-reversion in the short-term rate, volatility, and a long-term volatility factor, which is able to generate contango and backwardation.

4.2.10 Early Exercise and Futures Corrections

While European-style options are available to feed the equity VIX or the interest rate swap volatility indexes in Chap. 3, typically available for trading government bond volatility are American-style option prices on government bond *futures*. This section develops corrections to turn the typically available American-style option prices on Treasury futures into theoretically fair prices of European-style options. Because the volatility indexes in this book rely on OTM European options, the corrections we would need are likely to be small: it has been well known since at least Flesaker (1993) that OTM American options typically carry negligible early exercise premiums. However, corrections might matter for less far OTM options.

Note that the corrections in this section make the typically available data fully consistent with the theoretical framework underlying the government bond volatility indexes in this book. We map the price of American *future* (not *forward*) options into fair prices of European-style *forward* (not *future*) options, i.e., the ingredients we need to calculate the index. This additional step is implemented as a by-product of the calibration procedure that maps the price of American options into that of European options.[7]

To obtain the price of European options from the market price of American ones, we need a benchmark model where absence of arbitrage would allow us to map American prices into European. In a comprehensive analysis of the S&P 500 future options returns and volatilities, Broadie et al. (2007) estimate a number of benchmark models by transforming the available American prices into hypothetical, model-based European prices—statistical inference is easier to perform using European prices than American. Bikbov and Chernov (2011) perform a similar task in their analysis of Eurodollar future options.

These approaches rely on calculating an implied volatility for American options, which is then used to calculate the price of European options. We describe an approach that represents the government bond counterpart to this procedure. Appendix C.10 provides a numerical assessment of this approach. Note that a procedure like the following explicitly takes into account the inevitable feature that bond return volatility is time-dependent, and shrinking to zero. In the equity case, a more straightforward procedure would rely on the standard Barone-Adesi and Whaley (1987) approximation.

We begin by formulating a benchmark model for the development of the short-term rate. As a non-limitative example of a benchmark, consider the Vasicek (1977) model, in which the short-term rate, r_τ, evolves over time as follows:

$$dr_\tau = \kappa(\mu - r_\tau)d\tau + \sigma dW_\tau^p, \tag{4.56}$$

where κ, μ, and σ are parameters, and W^p is a Brownian motion under the physical probability. Under the risk-neutral probability, Q, the short-term rate is a solution to:

$$dr_\tau = \kappa(\bar{r} - r_\tau)d\tau + \sigma dW_\tau, \quad \bar{r} \equiv \mu - \frac{\lambda\sigma}{\kappa}, \tag{4.57}$$

where λ is a risk-premium, the Sharpe Ratio on any zero coupon bond in this one-factor market, supposed to be constant and, as usual, W_τ is a Brownian motion under Q. Note that while this model predicts that bond return volatility is deterministic, the size of the corrections we document in Appendix C.10 is not likely to change when, say, volatility is a random mean-reverting process and is sufficiently close to its historical average.

Denote the coupon-bearing bond price predicted by the Vasicek model by $B_t(r_t; \mathbb{T}) \equiv B_t(\mathbb{T})$, and the future by

$$\tilde{F}_t(r_t; T, \mathbb{T}) \equiv \mathbb{E}_t\big[B_T(r_T, \mathbb{T})\big] = \sum_{i=1}^N \bar{C}_i \cdot e^{a_t^F(T,T_i) - b_t^F(T,T_i)r_t}, \tag{4.58}$$

---

[7]Note that the algorithms in this section could equally be applied to turn American forward option prices into European forward option prices, or European future option prices into European forward option prices.

where $\bar{C}_i \equiv \frac{C_i}{n}$, for $i = 1, \ldots, N-1$, $\bar{C}_N = \frac{C_N}{n} + 1$, $(C_i)_{i=1}^N$ denotes the series of coupons in Eqs. (4.10)–(4.11), and

$$a_t^F(T, T_i) \equiv a_T(T_i) - \left(1 - e^{-\kappa(T-t)}\right)\bar{r}b_T(T_i) + \frac{\sigma^2(1 - e^{-2\kappa(T-t)})b_T^2(T_i)}{4\kappa};$$

$$b_t^F(T, T_i) \equiv e^{-\kappa(T-t)}b_T(T_i),$$

$$(4.59)$$

and the expressions for $a_t(T)$ and $b_t(T)$ are given in Eqs. (4.64) below. Appendix C.8 provides the derivation of Eqs. (4.58)–(4.59).

Note that the expectation in Eq. (4.58) is taken under the risk-neutral probability, not under the forward. Intuitively, a future position entails a continuous marking-to-market, so that the expected instantaneous changes in the future position are zero under the risk-neutral probability, $\mathbb{E}_t[d\tilde{F}_t(r_t; T, \mathbb{T})] = 0$, whence Eq. (4.58), by the boundary condition $\tilde{F}_T(r_T; T, \mathbb{T}) = B_T(r_T, \mathbb{T})$.

The mapping procedure is implemented in three steps.

Step 1 Estimate the parameters κ, μ and σ in Eq. (4.56), using data related to variables approximating the short-term rate.

Step 2 Calibrate the price of risk, λ in Eq. (4.57), so that the market prices of the out-of-the-money American options are matched by the prices predicted by the model. Denote this value of λ by $\hat{\lambda}$.

Step 3 Use the previously obtained value of $\hat{\lambda}$ and calculate the no-arbitrage prices of the out-of-the-money European options written on the forward coupon-bearing bond. The prices of these options are obtained through the Jamshidian (1989) formula, which is provided below for completeness (see Eq. (4.66)). Finally, use the same calibrated $\hat{\lambda}$ to calculate the value of the forward.

Step 2 can be implemented by alternative means. We describe two non-limitative examples.

Step 2.a Let $O^\$(K)$ denote the market price of an American option on the coupon-bearing bond (be it a put or a call) with strike K, and entering into the volatility indexes of Eq. (4.21) and Eq. (4.28). Let $O(K; \lambda)$ be the price of the American option predicted by the model when the price of risk is set equal to λ, which could be obtained by a number of methods combining time discretizations and Monte Carlo simulations (see, e.g., Longstaff and Schwartz 2001). As an example of a discretization, let $C_\tau(r_\tau; K)$ denote the price at time τ of an American future option with strike K when the short-term rate is equal to r_τ. This price satisfies

$$C_\tau(r_\tau; K) = \max\left\{\psi\left(\tilde{F}_\tau(r_\tau; T, \mathbb{T})\right), e^{-r_\tau \Delta\tau}\mathbb{E}_t\left[C_{\tau+\Delta\tau}(r_{\tau+\Delta\tau}; K)\right]\right\}, \quad (4.60)$$

where $\tilde{F}_\tau(r_\tau; T, \mathbb{T})$ is the model-based future price, as defined in Eq. (4.58), the initial value of the short-term rate, r_t, is calculated by inverting the future pricing equation (4.58) for r_t, i.e. $r_t : \tilde{F}_t(r_t; T, \mathbb{T}) = \tilde{F}_t^\$(T, \mathbb{T})$, where $\tilde{F}_t^\$(T, \mathbb{T})$ denotes the current future price; finally, the payoff, $\psi(\cdot)$, is

$$\psi(\tilde{F}_\tau) = \tilde{F}_\tau - K, \quad \text{for a call;} \quad \text{and} \quad \psi(\tilde{F}_\tau) = K - \tilde{F}_\tau, \quad \text{for a put.}$$

Then, one solves recursively for $C_\tau(r_\tau; K)$ and sets $O(K; \lambda) \equiv C_\tau(r_\tau; K)|_{\tau=t}$. The fundamental element of this step is to estimate λ so as to minimize the distance of the model predictions from market data, as in the following example:

$$\hat{\lambda} = \arg\min_{\lambda \in \Lambda} \int_0^\infty \left(O(K; \lambda) - O^\$(K) \right)^2 \omega(K) dK, \tag{4.61}$$

where ω is a given weighting function, and Λ is some compact set. The previous minimization problem can be implemented using a finite number of options, as follows:

$$\hat{\lambda} = \arg\min_{\lambda \in \Lambda} \sum_{j=1}^M \left(O(K_j; \lambda) - O^\$(K_j) \right)^2 \omega(K_j), \tag{4.62}$$

where M denotes the number of available American OTM options. The price of OTM European options in Step 3 can now be obtained by plugging $\hat{\lambda}$ into Jamshidian's formula, obtaining $\mathcal{O}(K_i; \hat{\lambda})$, which we define as the European counterpart to $O(K_i; \hat{\lambda})$. The forward price of the coupon bearing bond predicted by the Vasicek model, $F_t(r_t; T, \mathbb{T})$ say, is obtained by using Eq. (4.68) below.

Alternatively, consider the following approach.

Step 2.b For each strike K, calibrate λ so that the price of the American option predicted by the model exactly matches the corresponding market price. This calibration leads to a "risk-premium skew," defined as the function $\hat{\lambda}(K)$ such that $O(K; \hat{\lambda}(K)) = O^\$(K)$ for each K. We can now calculate the price of OTM European options on forwards in Step 3, by plugging $\hat{\lambda}(K)$ into Jamshidian's formula, obtaining the European option on the forward predicted by Jamshidian's formula, $\mathcal{O}(K_i^*; \hat{\lambda}(K_i^*))$, where $K_i^* \equiv K_i \frac{F_t(T, \mathbb{T})}{\tilde{F}_t^\$(T, \mathbb{T})}$ and $\tilde{F}_t^\$(T, \mathbb{T})$ and $F_t(r_t; T, \mathbb{T})$ denote, as usual, the market future price and the model-based forward price, obtained through Eq. (4.68) below. We tilt the risk-premium skew from K_i to K_i^*, because the calibration of $\hat{\lambda}(\cdot)$, which relates to options on futures prices, has to be plugged into a formula for options on forward prices.

As for the Jamshidian formula, assume first that the maturity of the forward, T, is less than the issuance date, T_0, $T < T_0$. Define the price of the coupon-bearing bond predicted by the Vasicek (1977) model as

$$B_t(r_t, \mathbb{T}) \equiv \sum_{i=1}^N \bar{C}_i P_t(r_t, T_i), \tag{4.63}$$

where $P_t(r, T)$ is the price of a zero

$$P_t(r, T) = e^{a_t(T) - b_t(T) \cdot r},$$

$$a_t(T) = \left(\frac{1 - e^{-\kappa(T-t)}}{\kappa} - (T - t) \right) \left(\bar{r} - \frac{1}{2} \left(\frac{\sigma}{\kappa} \right)^2 \right) - \frac{\sigma^2}{4\kappa^3} \left(1 - e^{-\kappa(T-t)} \right)^2,$$

$$b_t(T) = \frac{1}{\kappa} \left(1 - e^{-\kappa(T-t)} \right). \tag{4.64}$$

One finds the current value of the short-term rate, r_t, by inverting the future pricing equation (4.58) for r_t, i.e. $r_t : \tilde{F}_t(r_t; T, \mathbb{T}) = \tilde{F}_t^{\$}(T, \mathbb{T})$. Next, define the value of the short-term rate, say $r^*(K)$, which solves

$$B_T\left(r^*(K), \mathbb{T} \right) = K. \tag{4.65}$$

The Jamshidian formula for the price of a call option on the coupon bearing bond forward is

$$\text{Call}_t(K, T, \mathbb{T}) = \sum_{i=1}^{n} \bar{C}_i \cdot \overline{\text{Call}}_t \left(T; P_t(T_i), \mathcal{K}_i^*(K), v_i \right), \tag{4.66}$$

where $\mathcal{K}_i^*(K) \equiv P_T(r^*(K), T_i)$, r^* solves Eq. (4.65), and

$$\overline{\text{Call}}_t(T; P_i, \mathcal{K}_i^*(K), v_i) = P_i \Phi(d_{1,i}) - \mathcal{K}_i^*(K) P_t(T) \Phi(d_{1,i} - v_i),$$

$$d_{1,i} = \frac{\ln \frac{P_i}{\mathcal{K}_i^*(K) P_t(T)} + \frac{1}{2} v_i^2}{v_i}, \qquad v_i = \sigma \sqrt{\frac{1 - e^{-2\kappa(T-t)}}{2\kappa}} b_T(T_i),$$

where Φ denotes the cumulative standard Normal distribution. The value of a put is found through the put-call parity,

$$\text{Put}_t(K, T, \mathbb{T}) \equiv \text{Call}_t(K, T, \mathbb{T}) + P_t(r_t; T)\left(K - F_t(r_t; T, \mathbb{T}) \right), \tag{4.67}$$

where $F_t(r_t; T, \mathbb{T})$, the forward price of the coupon-bearing bond, is determined as

$$F_t(r_t; T, \mathbb{T}) = \frac{B_t(r_t; \mathbb{T})}{P_t(r_t; T)}, \tag{4.68}$$

using either the constant $\hat{\lambda}$ obtained in Step 2.a or the value $\hat{\lambda}(K_{\text{atm}}^*)$ obtained in Step 2.b, where the value $K_{\text{atm}}^* \equiv F_t(r_t; T, \mathbb{T})$. Note that the procedure in Step 2.b leads to a fixed point problem. One starts with a guess for $\hat{\lambda}(K_{\text{atm}}^*)$, say $\hat{\lambda}^{(0)}$, which is used to calculate the forward price, say $F_t^{(1)} \equiv F_t(r_t; T, \mathbb{T}; \hat{\lambda}^{(0)})$, where $F_t(r_t; T, \mathbb{T}; \lambda)$ denotes the forward price in Eq. (4.68) when the risk-premium is fixed at λ. The guess, $\hat{\lambda}^{(0)}$, could be obtained from the initial futures risk-premium skew, as $\hat{\lambda}^{(0)} = \hat{\lambda}(K_{\text{atm}})$, where $K_{\text{atm}} = \tilde{F}_t^{\$}(T, \mathbb{T})$. The iteration

$$\hat{\lambda}^{(i)} = \hat{\lambda}\left(F_t^{(i)} \right), \qquad F_t^{(i+1)} = F_t\left(r_t; T, \mathbb{T}; \hat{\lambda}^{(i)} \right),$$

is repeated until some convergence criterion is achieved.

The case where $T \geq T_0$ is dealt with a change in notation. In this case, the critical value of the short-term rate $r^*(K)$ in Eq. (4.65) is replaced with $r^*(K)$, which solves

$$\sum_{i=i_T}^{N} \bar{C}_i P_T\left(r^*(K), T_i\right) = K,$$

where i_T is defined as in Eq. (4.11).

4.2.11 Implementation Example

We provide an example of the main steps involved in the calculation of the government bond volatility indexes T-VI$(t, T, \mathbb{T})$ and T-VI$^{\mathrm{bp}}(t, T, \mathbb{T})$ in Eqs. (4.21) and (4.28). We use data representing hypothetical market conditions on April 27, 2012, relating to 1-month options written on a 1-month forward on 10-year US Treasury Notes. We do not make early exercise corrections of the type discussed in the previous section; we are assuming these data are for European options on forwards.

We approximate the theoretical values of the indexes while relying on the same discretization schemes as in Chap. 3 (see Sect. 3.3.3). The first column of Table 4.3 reports strikes K. The ATM strike is $K = 132$. The third column provides option prices. For reference, the second column also reports percentage implied volatilities for each strike ("Percentage Implied Vol").

Table 4.4 contains the calculations leading to the determination of the two indexes: the second column displays the type of OTM options; the third column has the premiums; the fourth and fifth columns report the weights the prices bear in the computation of the index before the rescaling of $\frac{2}{T-t}$; finally, the sixth and seventh columns report out-of-the-money option prices corrected by the appropriate weight: each price in the third column multiplied by the corresponding weight in the fourth column, for the "Basis Point Contribution," and each price in the third column multiplied by the corresponding weight in the fifth column, for the "Percentage Contribution."

The two indexes are calculated by evaluating Eq. (4.21) and Eq. (4.28),

$$\text{GB-VI} = 100 \times \sqrt{\frac{1}{0.9980} \frac{2}{(1/12)} \times 1.0268 \cdot 10^{-4}} = 4.9692$$

$$\text{GB-VI}^{\mathrm{bp}} = 100^2 \times \sqrt{\frac{1}{0.9980} \frac{2}{(1/12)} \times 1.7757 \cdot 10^{-4}} = 653.4751$$

where the rescaling factor inside the square roots, $\frac{1}{0.9980}$, is the inverse of a zero coupon bond expiring in 1 month. In comparison, ATM implied volatilities are 4.53 % and 4.53 × 132 = 597.96 basis points.

Table 4.3 Implied volatilities and option premiums

Strike price (%)	Percentage implied Vol	Premiums	
		Put option	Call option
125.00	9.10	$0.2343 \cdot 10^{-3}$	$7.0234 \cdot 10^{-2}$
125.50	8.53	$0.2346 \cdot 10^{-3}$	$6.5234 \cdot 10^{-2}$
126.00	7.32	$0.1326 \cdot 10^{-3}$	$6.0132 \cdot 10^{-2}$
126.50	6.78	$0.1328 \cdot 10^{-3}$	$5.5132 \cdot 10^{-2}$
127.00	7.24	$0.3423 \cdot 10^{-3}$	$5.0342 \cdot 10^{-2}$
127.50	6.64	$0.3465 \cdot 10^{-3}$	$4.5346 \cdot 10^{-2}$
128.00	6.33	$0.4516 \cdot 10^{-3}$	$4.0451 \cdot 10^{-2}$
128.50	6.15	$0.6567 \cdot 10^{-3}$	$3.5656 \cdot 10^{-2}$
129.00	5.81	$0.8557 \cdot 10^{-3}$	$3.0855 \cdot 10^{-2}$
129.50	5.63	$1.2506 \cdot 10^{-3}$	$2.6250 \cdot 10^{-2}$
130.00	5.35	$1.7225 \cdot 10^{-3}$	$2.1722 \cdot 10^{-2}$
130.50	5.05	$2.3656 \cdot 10^{-3}$	$1.7365 \cdot 10^{-2}$
131.00	4.82	$3.3632 \cdot 10^{-3}$	$1.3363 \cdot 10^{-2}$
131.50	4.71	$4.9229 \cdot 10^{-3}$	$9.9229 \cdot 10^{-3}$
132.00 (ATM)	4.53	$6.8864 \cdot 10^{-3}$	$6.8864 \cdot 10^{-3}$
132.50	4.43	$9.5398 \cdot 10^{-3}$	$4.5398 \cdot 10^{-3}$
133.00	4.40	$1.2865 \cdot 10^{-2}$	$2.8655 \cdot 10^{-3}$
133.50	4.38	$1.6705 \cdot 10^{-2}$	$1.7053 \cdot 10^{-3}$
134.00	4.40	$2.0979 \cdot 10^{-2}$	$0.9793 \cdot 10^{-3}$
134.50	4.58	$2.5619 \cdot 10^{-2}$	$0.6192 \cdot 10^{-3}$
135.00	4.78	$3.0400 \cdot 10^{-2}$	$0.4000 \cdot 10^{-3}$
135.50	4.93	$3.5246 \cdot 10^{-2}$	$0.2462 \cdot 10^{-3}$
136.00	5.17	$4.0169 \cdot 10^{-2}$	$0.1696 \cdot 10^{-3}$
136.50	5.21	$4.5090 \cdot 10^{-2}$	$9.0837 \cdot 10^{-5}$

Finally, we calculate the basis point yield volatility indexes in Sect. 4.2.8. First, we calculate the certainty equivalent for the price of the coupon-bearing bond in a model-free fashion using Eq. (4.34):

$$\mathcal{B} = \frac{\text{GB-VI}^{\text{bp}}}{\text{GB-VI}} = \frac{653.4751}{4.9692} = 131.5121.$$

Second, we assume that in Eq. (4.35), $n = 1$, $N = 10$ and the annual coupon $C_i = 4$, and invert for the yield corresponding to $\mathcal{B}$, obtaining $y_\mathcal{B} = \hat{P}^{-1}(\mathcal{B}) = 7.2226 \cdot 10^{-3}$, leading to a modified duration equal to $D_\mathcal{B} = 8.6048$. The model-free index of basis point volatility based on duration, GB-VI$^{\text{bp}}_{\text{Yd}}$ in Eq. (4.39), is

$$\text{GB-VI}^{\text{bp}}_{\text{Yd}} = 100 \times \frac{\text{GB-VI}}{D_\mathcal{B}} = 100 \times \frac{4.9692}{8.6048} = 57.749,$$

Table 4.4 Calculation of volatility indexes

Strike price (%)	Option type	Premiums	Weights		Contributions to strikes	
			Basis point ΔK_i	Percentage $\Delta K_i / K_i^2$	Basis point contribution	Percentage contribution
125.00	Put	$0.2343 \cdot 10^{-3}$	0.005	$3.2000 \cdot 10^{-3}$	$1.1715 \cdot 10^{-6}$	$7.4976 \cdot 10^{-7}$
125.50	Put	$0.2346 \cdot 10^{-3}$	0.005	$3.1745 \cdot 10^{-3}$	$1.1733 \cdot 10^{-6}$	$7.4494 \cdot 10^{-7}$
126.00	Put	$0.1326 \cdot 10^{-3}$	0.005	$3.1494 \cdot 10^{-3}$	$6.6302 \cdot 10^{-7}$	$4.1762 \cdot 10^{-7}$
126.50	Put	$0.1328 \cdot 10^{-3}$	0.005	$3.1245 \cdot 10^{-3}$	$6.6429 \cdot 10^{-7}$	$4.1512 \cdot 10^{-7}$
127.00	Put	$0.3423 \cdot 10^{-3}$	0.005	$3.1000 \cdot 10^{-3}$	$1.7118 \cdot 10^{-6}$	$1.0613 \cdot 10^{-6}$
127.50	Put	$0.3465 \cdot 10^{-3}$	0.005	$3.0757 \cdot 10^{-3}$	$1.7326 \cdot 10^{-6}$	$1.0658 \cdot 10^{-6}$
128.00	Put	$0.4516 \cdot 10^{-3}$	0.005	$3.0517 \cdot 10^{-3}$	$2.2580 \cdot 10^{-6}$	$1.3781 \cdot 10^{-6}$
128.50	Put	$0.6567 \cdot 10^{-3}$	0.005	$3.0280 \cdot 10^{-3}$	$3.2838 \cdot 10^{-6}$	$1.9887 \cdot 10^{-6}$
129.00	Put	$0.8557 \cdot 10^{-3}$	0.005	$3.0046 \cdot 10^{-3}$	$4.2785 \cdot 10^{-6}$	$2.5710 \cdot 10^{-6}$
129.50	Put	$1.2506 \cdot 10^{-3}$	0.005	$2.9814 \cdot 10^{-3}$	$6.2534 \cdot 10^{-6}$	$3.7289 \cdot 10^{-6}$
130.00	Put	$1.7225 \cdot 10^{-3}$	0.005	$2.9585 \cdot 10^{-3}$	$8.6128 \cdot 10^{-6}$	$5.0963 \cdot 10^{-6}$
130.50	Put	$2.3656 \cdot 10^{-3}$	0.005	$2.9359 \cdot 10^{-3}$	$1.1828 \cdot 10^{-5}$	$6.9454 \cdot 10^{-6}$
131.00	Put	$3.3632 \cdot 10^{-3}$	0.005	$2.9135 \cdot 10^{-3}$	$1.6816 \cdot 10^{-5}$	$9.7990 \cdot 10^{-6}$
131.50	Put	$4.9229 \cdot 10^{-3}$	0.005	$2.8914 \cdot 10^{-3}$	$2.4614 \cdot 10^{-5}$	$1.4234 \cdot 10^{-5}$
132.00	ATM	$6.8864 \cdot 10^{-3}$	0.005	$2.8696 \cdot 10^{-3}$	$3.4431 \cdot 10^{-5}$	$1.9761 \cdot 10^{-5}$
132.50	Call	$4.5398 \cdot 10^{-3}$	0.005	$2.8479 \cdot 10^{-3}$	$2.2699 \cdot 10^{-5}$	$1.2929 \cdot 10^{-5}$
133.00	Call	$2.8655 \cdot 10^{-3}$	0.005	$2.8266 \cdot 10^{-3}$	$1.4327 \cdot 10^{-5}$	$8.0999 \cdot 10^{-6}$
133.50	Call	$1.7053 \cdot 10^{-3}$	0.005	$2.8054 \cdot 10^{-3}$	$8.5265 \cdot 10^{-6}$	$4.7842 \cdot 10^{-6}$
134.00	Call	$0.9793 \cdot 10^{-3}$	0.005	$2.7845 \cdot 10^{-3}$	$4.8969 \cdot 10^{-6}$	$2.7271 \cdot 10^{-6}$
134.50	Call	$0.6192 \cdot 10^{-3}$	0.005	$2.7632 \cdot 10^{-3}$	$3.0963 \cdot 10^{-6}$	$1.7116 \cdot 10^{-6}$
135.00	Call	$0.4000 \cdot 10^{-3}$	0.005	$2.7434 \cdot 10^{-3}$	$2.0002 \cdot 10^{-6}$	$1.0975 \cdot 10^{-6}$
135.50	Call	$0.2462 \cdot 10^{-3}$	0.005	$2.7232 \cdot 10^{-3}$	$1.2312 \cdot 10^{-6}$	$6.7062 \cdot 10^{-7}$
136.00	Call	$0.1696 \cdot 10^{-3}$	0.005	$2.7032 \cdot 10^{-3}$	$8.4830 \cdot 10^{-7}$	$4.5864 \cdot 10^{-7}$
136.50	Call	$9.0837 \cdot 10^{-5}$	0.005	$2.6835 \cdot 10^{-3}$	$4.5418 \cdot 10^{-7}$	$2.4376 \cdot 10^{-7}$
SUMS					$1.7757 \cdot 10^{-4}$	$1.0268 \cdot 10^{-4}$

and the model-free index of basis point yield volatility based on yield, GB-VI$_Y^{bp}$ in Eq. (4.40), is

$$\text{GB-VI}_Y^{bp} = 100 \times y_B \times \text{GB-VI} = 100 \times 7.2226 \cdot 10^{-3} \times 4.9692 = 3.5891.$$

Option-implied expected volatility might seem to be relatively small. To put this figure in context, we calculate the realized volatility of the 1 month-10 year US Treasury Note future returns over a sample period covering daily data from January 3, 2000 to April 27, 2012. Figure 4.4 depicts the time series behavior of this realized volatility.

Figure 4.4 shows government future price volatility oscillating significantly, ranging from levels as low as 2 % to record highs of more than 18 %. The last

Fig. 4.4 Estimates of 20-day realized volatility of the 1-month 10-year US Treasury Note future returns, annualized, percent,
$$100\sqrt{12\sum_{t=1}^{21}\ln^2\frac{F_{t-i+1}}{F_{t-i}}},$$
where F_t denotes the future price. The sample includes daily data from January 3, 2000 to April 27, 2012, for a total of 3093 observations

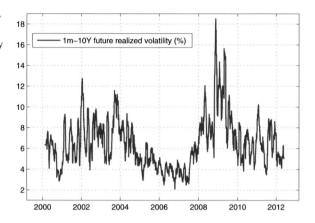

Fig. 4.5 The CBOE/CBOT TYVIX index of volatility on the 10-Year US Treasury Note

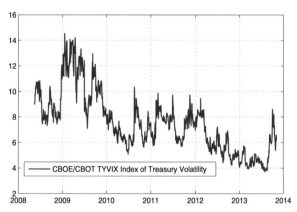

day of our sample is April 27, 2012, which is the same date the index calculations in Tables 4.3 and 4.4 refer to. On this day, realized volatility equals 5.0158 %, and its average since the beginning of year 2012 is 5.1332 %. These figures compare with the value of the percentage index, 4.9642 %, suggesting the absence of variance risk-premiums over the first four months of 2012. This finding parallels research conducted on the equity VIX (Corradi et al. 2013), which shows that equity variance risk premiums are relatively low when realized equity volatility is low.[8]

Figure 4.5 plots the CBOE/CBOT TYVIX volatility index of the 10-Year US Treasury Note, a percentage index calculated as proposed in this section.[9] The reader is referred to the CBOE website (see Chicago Board Options Exchange 2013) for additional details regarding its implementation, empirical properties, and listed products on it, as well as Mele and Obayashi (2015) and Sect. 1.2 in Chap. 1, for additional empirical properties of the TYVIX index.

---

[8] See Fig. 2.7 in Chap. 2 for an estimate of the volatility risk-premium regarding the Treasury space.

[9] The index was launched by CBOE in May 2013 under the ticker name "VXTYN." The ticker change to "TYVIX" occurred in May 2015.

4.2.12 Jumps

The framework in the previous section can be extended to one in which the forward price follows a jump-diffusion process. Relying on the general framework of Chap. 2 (Sect. 2.7), one can show that *basis point* government bond volatility can still be priced in a model-free fashion, whereas *percentage* volatility cannot, as the latter depends on parametric assumptions regarding the jumps generating process.

4.3 Time Deposits

This section develops security designs for variance swaps leading to volatility indexes for time deposits such as the Eurodollar. The risks underlying this asset class relate to changes in forward interest rates, which are martingales under the forward probability, just as in the forward prices of Sect. 4.2. Therefore, much of the theoretical framework that underlies the contracts in Sect. 4.2 applies to time deposit volatility, with some differences arising while implementing the early exercise corrections (see Sect. 4.3.4).

4.3.1 The Underlying Risks

Let $l_t(\Delta) \equiv L(t, t + \Delta)$ be the simply compounded interest rate on a deposit one party makes with another for the time period from t to $t + \Delta$. As an example, the two counterparties could be financial institutions, and $l_t(\Delta)$ could be an interest rate referenced to the LIBOR rate. For simplicity, we refer to $l_t(\Delta)$ as the LIBOR, although this interest rate may relate to any deposit any institution can make with another.

Define a forward contract whereby at time t one party agrees to pay a counterparty a payoff equal to $100 \times (1 - l_T(\Delta)) - Z_t(T, T + \Delta)$ at time T. The forward LIBOR price, $Z_t(T, T + \Delta)$, is agreed at time t, so that in the absence of arbitrage,

$$Z_t(T, T + \Delta) = 100 \times \left(1 - f_t(T, T + \Delta)\right), \qquad (4.69)$$

where $f_t(T, T + \Delta)$ is the forward LIBOR, which satisfies $f_t(T, T + \Delta) = \mathbb{E}_t^{Q_{FT}}(l_T(\Delta))$. Because $l_T(\Delta) = f_T(T, T + \Delta)$, $f_t(T, T + \Delta)$ is a martingale under Q_{FT}. Therefore, assuming that the information in this market is driven by Brownian motions, the forward price, $Z_t(T, T + \Delta)$, satisfies:

$$\frac{dZ_\tau(T, T + \Delta)}{Z_\tau(T, T + \Delta)} = v_\tau^z(T, \Delta) dW_\tau^{FT}, \qquad \tau \in [t, T], \qquad (4.70)$$

where, as usual, $W_\tau^{F^T}$ is a multidimensional Brownian motion under Q_{FT}, and $v_\tau^z(T, \Delta)$ is the vector of instantaneous volatilities, adapted to $W_\tau^{F^T}$.[10] Similarly as in the government bond case, we can generalize this setting to one with jump-diffusion processes through a mere change in notation.

We now provide details regarding variance contracts and volatility indexes for time deposits, based on both percentage and basis point volatility.

4.3.2 Variance Contracts and Volatility Indexes

Consider the following payoff of a time-deposit variance swap:

$$\pi_z(T, \Delta) \equiv V^z(t, T, \Delta) - \mathbb{P}_z(t, T, \Delta), \quad V^z(t, T, \Delta) \equiv \int_t^T \left\| v_\tau^z(T, \Delta) \right\|^2 d\tau,$$

where $\mathbb{P}_z(t, T, \Delta)$, the fair value of the strike, is

$$\mathbb{P}_z(t, T, \Delta) = \frac{1}{P_t(T)} \mathbb{E}_t \left(e^{-\int_t^T r_\tau d\tau} V^z(t, T, \Delta) \right) = \mathbb{E}_t^{Q_{FT}} \left(V^z(t, T, \Delta) \right),$$

and the second equality follows by the usual change of probability. Because the forward LIBOR price $Z_\tau(T, T + \Delta)$ in Eq. (4.70) is a martingale under Q_{FT} like the government bond forward price $F_\tau(T, \mathbb{T})$ in Eq. (4.14), we can apply the same spanning arguments leading to pricing government bond volatility in Sect. 4.2. For the log-contract, we have:

$$\mathbb{P}_z(t, T, \Delta) = \frac{2}{P_t(T)} \left(\int_0^{Z_t(T,T+\Delta)} \text{Put}_t^z(K, T, \Delta) \frac{1}{K^2} dK \right.$$
$$\left. + \int_{Z_t(T,T+\Delta)}^\infty \text{Call}_t^z(K, T, \Delta) \frac{1}{K^2} dK \right),$$

where $\text{Put}_t^z(K, T, \Delta)$ and $\text{Call}_t^z(K, T, \Delta)$ denote the prices of OTM European puts and calls struck at K, expiring at time T, and written at t on the forward LIBOR price, $Z_t(T, T + \Delta)$. An index of *percentage* time deposit volatility is

$$\boxed{\text{TD-VI}(t, T, \Delta) \equiv 100 \times \sqrt{\frac{\mathbb{P}_z(t, T, \Delta)}{T - t}}} \tag{4.71}$$

---

[10]The timing convention we follow in this section differs from that underlying a standard forward rate agreement (see, e.g., Chap. 3 (Sect. 3.2)). In the standard case, the settlement of the payoff, $\Delta \times (l_T(\Delta) - K)$, occurs at time $T + \Delta$, so that the clearing rate K at time t is $f_t^\Delta(T, T + \Delta)$ (using the same notation of Sect. 3.2 in Chap. 3), where $f_\tau^\Delta(T, T + \Delta) \equiv \mathbb{E}_\tau^{Q_{FT+\Delta}}(l_T(\Delta))$ is a martingale under $Q_{FT+\Delta}$, not Q_{FT}. The forward LIBOR in this section, $f_t(T, T + \Delta)$, is a martingale under Q_{FT}, not under $Q_{FT+\Delta}$.

For the basis point version, consider the following time deposit variance contract

$$\pi_z^{bp}(T, \Delta) \equiv V^{z,bp}(t, T, \Delta) - \mathbb{P}_z^{bp}(t, T, \Delta),$$

$$V^{z,bp}(t, T, \Delta) \equiv \int_t^T Z_\tau^2(T, T+\Delta) \| v_\tau^z(T, \Delta) \|^2 d\tau,$$

where the fair value of the strike can be found by spanning a quadratic contract through OTM options, just as in Sect. 4.2,

$$\mathbb{P}_z^{bp}(t, T, \Delta)$$

$$= \frac{1}{P_t(T)} \mathbb{E}_t \left(e^{-\int_t^T r_\tau d\tau} V^{z,bp}(t, T, \Delta) \right)$$

$$= \mathbb{E}_t^{Q_{FT}} \left(V^{z,bp}(t, T, \Delta) \right)$$

$$= \frac{2}{P_t(T)} \left(\int_0^{Z_t(T,T+\Delta)} \text{Put}_t^z(K, T, \Delta) dK + \int_{Z_t(T,T+\Delta)}^\infty \text{Call}_t^z(K, T, \Delta) dK \right).$$

The basis point time deposit volatility index is

$$\boxed{\text{TD-VI}^{bp}(t, T, \Delta) \equiv 100 \times \sqrt{\frac{\mathbb{P}_z^{bp}(t, T, \Delta)}{T - t}}} \tag{4.72}$$

Note that unlike the basis point volatility indexes in the other chapters, we rescale (the square root of) the variance swap fair value by 100, rather than 100^2, due to the fact that the forward LIBOR price in Eq. (4.69) is expressed in percentage terms.

Moreover, the marking-to-market value of the percentage deposit variance swap is the same as that for government bonds in Eq. (4.29):

$$\text{M-Var}_\tau^z(T, \Delta) \equiv P_\tau(T) \left[V^z(t, \tau, \Delta) + \mathbb{P}_z(\tau, T, \Delta) - \mathbb{P}_z(t, T, \Delta) \right],$$

and that of the Basis Point is the same as M-Var$_\tau^z(T, \Delta)$, but with $V^{z,bp}(t, T, \Delta)$ replacing $V^z(t, T, \Delta)$.

Finally, in the absence of price quotes for ATM options, we could adopt the same approach as in Sect. 4.2.7 regarding the government bond volatility case (see Eq. (4.30) and Eq. (4.31)). That is, let $K_{0,z}$ denote the first strike below the current forward price $Z_t(T, T + \Delta)$, where in case $Z_t(T, T + \Delta)$ is not observed, it is approximated by the strike price at which the absolute difference between the call and put prices is smallest; set $K_{0,z} = Z_t$ only when an option exists that is struck at Z_t. Then, the percentage index and the basis point index are approximated by the following expressions:

$$\boxed{\begin{array}{l} \text{TD-VI}_o(t, T, \Delta) \equiv 100 \times \sqrt{\dfrac{\mathbb{P}_{z,o}(t, T, \Delta)}{T - t}}, \\[2em] \text{TD-VI}_o^{bp}(t, T, \Delta) \equiv 100 \times \sqrt{\dfrac{\mathbb{P}_{z,o}^{bp}(t, T, \Delta)}{T - t}} \end{array}} \tag{4.73}$$

where

$$\mathbb{P}_{z,o}(t, T, \Delta)$$

$$\equiv \frac{2}{P_t(T)} \left(\int_0^{K_{0,z}} \text{Put}_t^z(K, T, \Delta) \frac{1}{K^2} dK + \int_{K_{0,z}}^{\infty} \text{Call}_t^z(K, T, \Delta) \frac{1}{K^2} dK \right)$$

$$- \left(\frac{Z_t(T, T + \Delta) - K_{0,z}}{K_{0,z}} \right)^2, \tag{4.74}$$

and

$$\mathbb{P}_{z,o}^{\text{bp}}(t, T, \Delta)$$

$$\equiv \frac{2}{P_t(T)} \left(\int_0^{K_{0,z}} \text{Put}_t^z(K, T, \Delta) dK + \int_{K_{0,z}}^{\infty} \text{Call}_t^z(K, T, \Delta) dK \right)$$

$$- \left(Z_t(T, T + \Delta) - K_{0,z} \right)^2. \tag{4.75}$$

4.3.3 Yield Volatility

We develop the model-free time deposit yield counterpart to the government bond yield volatility in Sect. 4.2.8. First, consider the following representation of basis point expected price volatility:

$$\text{TD-VI}^{\text{bp}}(t, T, \Delta)$$

$$= 100 \times \sqrt{\frac{1}{P_t(T)} \mathbb{E}_t \left(e^{-\int_t^T r_\tau d\tau} \int_t^T Z_\tau^2(T, T + \Delta) \left\| v_\tau^z(T, \Delta) \right\|^2 d\tau \right)}$$

$$= \mathcal{Z}(t, T, \Delta) \times \text{TD-VI}(t, T, \Delta), \tag{4.76}$$

where $\mathcal{Z}(t, T, \Delta)$ is the guaranteed price of the LIBOR contract at time T, so that the basis point volatility on the LIBOR forward, $\text{TD-VI}^{\text{bp}}(t, T, \Delta)$, is the same as the basis point volatility in a certainty equivalent market. Similarly as with the steps made to show that $y_B(t, T, \mathbb{T})$ in Eq. (4.35) is well-defined, we can show that $\mathcal{Z}(t, T, \Delta) \leq 100$.

As for the time-deposit counterpart to the duration-based yield volatility index in Eq. (4.39), consider the modified duration of $\mathcal{Z}(t, T, \Delta)$,

$$D_{\mathcal{Z}}(t, T, \Delta) = \frac{\Delta}{1 + y_{\mathcal{Z}}(t, T, \Delta)}, \tag{4.77}$$

where $y_{\mathcal{Z}}(t, T, \Delta)$ is the solution to the following equation:

$$y_{\mathcal{Z}}(t, T, \Delta) : \mathcal{Z}(t, T, \Delta) = \frac{\text{TD-VI}^{\text{bp}}(t, T, \Delta)}{\text{TD-VI}(t, T, \Delta)} = \hat{Z}\big(y_{\mathcal{Z}}(t, T, \Delta)\big)$$

$$\equiv 100 \times \big(1 + y_{\mathcal{Z}}(t, T, \Delta)\big)^{-\Delta}, \tag{4.78}$$

so that the model-free duration-based yield volatility index for time-deposit is

$$\text{TD-VI}_{\text{Yd}}^{\text{bp}}(t, T, \Delta) = 100 \times \frac{\text{TD-VI}(t, T, \Delta)}{D_{\mathcal{Z}}(t, T, \Delta)}.$$

We can use Eqs. (4.77)–(4.78) to express this index in terms of the basis point and percentage indexes $\text{TD-VI}^{\text{bp}}(t, T, \Delta)$ and $\text{TD-VI}(t, T, \Delta)$, as follows:

$$\text{TD-VI}_{\text{Yd}}^{\text{bp}}(t, T, \Delta) = \frac{100 \times (1 + \hat{Z}^{-1}[\frac{\text{TD-VI}^{\text{bp}}(t,T,\Delta)}{\text{TD-VI}(t,T,\Delta)}]) \times \text{TD-VI}(t, T, \Delta)}{\Delta} \quad (4.79)$$

where $\hat{Z}^{-1}$ is the inverse function of $\hat{Z}$ in Eq. (4.78).

Finally, we introduce a model-free measure of yield volatility by mapping the new representation of basis point expected price volatility in Eq. (4.76) into its yield equivalent as follows:

$$\text{TD-VI}_{\text{Y}}^{\text{bp}}(t, T, \Delta) = 100 \times y_{\mathcal{Z}}(T) \times \text{TD-VI}(t, T, \Delta),$$

so that

$$\text{TD-VI}_{\text{Y}}^{\text{bp}}(t, T, \Delta) = 100 \times \hat{Z}^{-1}\left[\frac{\text{TD-VI}^{\text{bp}}(t, T, \Delta)}{\text{TD-VI}(t, T, \Delta)}\right] \times \text{TD-VI}(t, T, \Delta) \quad (4.80)$$

Equations (4.79) and (4.80) are model-free indexes of time-deposit yield volatility, as they relate to the certainty equivalent price for the LIBOR price, $\mathcal{Z}(t, T, \Delta)$, which is model-free, as it is recovered through Eq. (4.76) based on the market price of OTM options. Moreover, yield volatility can be spanned and hedged through dedicated percentage variance swaps since $y_{\mathcal{Z}}(t, T, \Delta)$ is known at the time of the inception of the variance swap, t.

4.3.4 American Future Corrections

We describe two algorithms to calculate our volatility indexes on time deposits. Similar to the government bond case (see Sect. 4.2.10), listed products on time deposit contracts are typically American options on futures. For example, typically traded are American-style option prices on Eurodollar *futures* rather than European-style options on forwards. We map these American-style option prices on futures into theoretically fair prices of European-style options on forwards. The mapping algorithm is the same as that in Sect. 4.2.10; for example, we could still use the Vasicek (1977) model as the benchmark (see Eqs. (4.56)–(4.57)) to ensure the mapping takes place while preserving absence of arbitrage. However, we need different evaluation formulae for the different asset class dealt with in this section.

To anticipate, the difference between the pricing formulae in Sect. 4.2.10 and those in this section relates to Eq. (4.58) and Eqs. (4.66)–(4.67) (referring to the future on a coupon bearing bond and its associated European options), and Eq. (4.81) and Eqs. (4.85)–(4.87) (referring to the future LIBOR and its associated European options).

Note that Vasicek's model is a special case of that considered by Bikbov and Chernov (2011). We focus on a simpler model than Bikbov's and Chernov's because our use of a benchmark model is different from theirs. The mapping we create in this section applies to the entire skew, which is used as inputs for the indexes, and a parsimonious model is likely to perform this task in a relatively fast and efficient way. For example, we provide closed-form expressions for the future LIBOR and the price of European options on the forward LIBOR. Our focus on the Vasicek model serves as a non-limitative illustration.[11]

Denote the LIBOR predicted by the Vasicek model by $l_t(r_t; \Delta) \equiv l_t(\Delta)$, and the future LIBOR by $\tilde{z}_t(r_t; T, \Delta)$. We have

$$\tilde{z}_t(r_t; T, \Delta) = \mathbb{E}_t\left[l_T(r_T; \Delta)\right] = \frac{1}{\Delta}\left(e^{a_t^Z(T,\Delta)+b_t^Z(T,\Delta)r_t} - 1\right), \qquad (4.81)$$

where

$$a_t^Z(T, \Delta) \equiv -a_T(T + \Delta) + \left(1 - e^{-\kappa(T-t)}\right)\bar{r}b_T(T + \Delta)$$
$$+ \frac{\sigma^2(1 - e^{-2\kappa(T-t)})b_T^2(T + \Delta)}{4\kappa} \qquad (4.82)$$
$$b_t^Z(T, \Delta) \equiv e^{-\kappa(T-t)}b_T(T + \Delta),$$

and the expressions for $a_t(T)$ and $b_t(T)$ are given in Eqs. (4.64). We have that the future price is

$$\tilde{Z}_t(r_t; T, \Delta) = 100 \times \left(1 - \tilde{z}_t(r_t; T, \Delta)\right). \qquad (4.83)$$

Appendix C.9 provides the derivation of Eqs. (4.81)–(4.82). Naturally, the expectation in Eq. (4.81) is taken under the risk-neutral probability, not under the forward, as we are pricing LIBOR futures and not forwards.

The mapping procedure is implemented through the three steps of Sect. 4.2.8, which we repeat here for clarity of exposition.

Steps 1–2 Estimate κ, μ and σ in Eq. (4.56) using data on the short-term rate, and calibrate the price of risk, λ in Eq. (4.57), so that the model-based prices of the OTM American options match the corresponding market prices. As in Sect. 4.2.10, we can either (i) calibrate a single λ or (ii) calibrate λ on the skew:

(i) We choose $\hat{\lambda}$ similarly as we did in Eq. (4.61), as follows:

$$\hat{\lambda} = \arg\min_{\lambda \in \Lambda} \int_0^\infty \left(O_z(K; \lambda) - O_z^\$(K)\right)^2 \omega(K)dK,$$

---

[11] Note that while our analytical formulae provided below are new, they still link to a special case of the multifactor affine model considered by Bikbov and Chernov (2011), for which no closed-form solution is available in general.

where $O_z^\$(K)$ is the market price of an American option on the future (be it a put or a call) with strike K entering the volatility indexes of Eq. (4.71) and Eq. (4.72), and $O_z(K; \lambda)$ is its model-counterpart. The price, $O_z(K; \lambda)$, can be obtained through the same time discretizations and Monte Carlo simulations mentioned in Sect. 4.2.10. As a non-limitative example of a discretization, let $C_\tau^z(r_\tau; K)$ denote the price at time τ of an American future option with strike K when the short-term rate is equal to r_τ. This price satisfies

$$C_\tau^z(r_\tau; K) = \max\left\{\psi\left(\tilde{Z}_\tau(r_\tau; T, \Delta)\right), e^{-r_\tau \Delta\tau}\mathbb{E}_t\left[C_{\tau+\Delta\tau}^z(r_{\tau+\Delta\tau}; K)\right]\right\},$$
(4.84)

where $\tilde{Z}_\tau(r_\tau; T, \Delta)$ is as in Eq. (4.83), the initial value of the short-term rate, r_t, is calculated by inverting the future pricing equation (4.83) for r_t, i.e. $r_t : \tilde{Z}_t(r_t; T, \Delta) = \tilde{Z}_t^\$(T, \Delta)$, where $\tilde{Z}_t^\$(T, \Delta)$ denotes the current future price; and finally, the payoff, $\psi(\cdot)$, is

$$\psi(\tilde{Z}_\tau) = \tilde{Z}_\tau - K \quad \text{for a call;} \quad \text{and} \quad \psi(\tilde{Z}_\tau) = K - \tilde{Z}_\tau, \quad \text{for a put.}$$

Then, one solves recursively for $C_\tau^z(r_\tau; K)$ and set $O_z(K; \lambda) \equiv C_\tau^z(r_\tau; K)|_{\tau=t}$. Note that the pricing equation (4.84) is formally the same as Eq. (4.60) in Sect. 4.2.10, although the asset underlying the option is the future time deposit price, $\tilde{Z}_\tau$, and not the future coupon-bearing bond price, $\tilde{F}_\tau$.

(ii) Alternatively, for each strike K, calibrate λ so that the price of the American option predicted by the model exactly matches the corresponding market price. This calibration leads to a "risk-premium skew," defined as the function $\hat{\lambda}(K)$ such that $O_z(K; \hat{\lambda}(K)) = O_z^\$(K)$, for each K.

Step 3 Use the calibrated values of λ and calculate the prices of the OTM European options written on the forward LIBOR, $\text{Call}_t^z(K, T, \Delta)$ and $\text{Put}_t^z(K, T, \Delta)$. In case (i) of the previous steps, use a single $\hat{\lambda}$ in Eq. (4.85) below. In case (ii), use the "tilted" risk-premium skew, $\hat{\lambda}(K_i^*)$, where $K_i^* = K_i \frac{Z_t(r_t; T, T+\Delta)}{\tilde{Z}_t^\$(T, T+\Delta)}$, $\tilde{Z}_t^\$(T, T+\Delta)$ is the future market price, and $Z_t(r_t; T, T+\Delta) \equiv Z_t(T, T+\Delta)$ denotes the forward price predicted by the Vasicek model. Similarly as in the government bond case of Sect. 4.2.10, we tilt the risk-premium skew from K_i to K_i^* because the calibration of $\hat{\lambda}(\cdot)$, which originally relates to options on futures, has to be plugged into a formula for options on forwards. In Appendix C.9, we show that the price of the European call is given by

$$\text{Call}_t^z(K, T, \Delta)$$

$$= \mathbb{E}_t\left(e^{-\int_t^T r_\tau d\tau}\left(Z_T(r_T; T, T+\Delta) - K\right)^+\right)$$

$$= 100 \times P_t(T)\frac{1}{\Delta}\left[K^o(K) \cdot \Phi(\delta_t) - K^o(Z_t) \cdot \Phi\left(\delta_t - b_\Delta\sqrt{\mathbb{V}_t(r_T)}\right)\right],$$
(4.85)

where

$$a_\Delta = \left(\frac{1-e^{-\kappa\Delta}}{\kappa} - \Delta\right)\left(\bar{r} - \frac{1}{2}\left(\frac{\sigma}{\kappa}\right)^2\right) - \frac{\sigma^2}{4\kappa^3}\left(1-e^{-\kappa\Delta}\right)^2,$$

$$b_\Delta = \frac{1-e^{-\kappa\Delta}}{\kappa},$$

and

$$K^o(x) \equiv 1 + \left(1 - \frac{x}{100}\right)\Delta, \qquad \delta_t \equiv \frac{\ln\frac{K^o(K)}{K^o(Z_t)} + \frac{1}{2}b_\Delta^2\,\mathbb{V}_t(r_T)}{b_\Delta\sqrt{\mathbb{V}_t(r_T)}},$$

and, finally,

$$\mathbb{E}_t^{Q_{FT}}(r_T) = \bar{r} + e^{-\kappa(T-t)}(r_t - \bar{r}) - \frac{\sigma^2}{2\kappa^2}\left(1 - e^{-\kappa(T-t)}\right)^2,$$

$$\mathbb{V}_t(r_T) = \frac{\sigma^2}{2\kappa}\left(1 - e^{-2\kappa(T-t)}\right),$$

(4.86)

denote the expectation under Q_{FT} and the variance of the short-term rate, taken conditionally upon the information set of time t.[12] Moreover, by the put-call parity, the price of the put is

$$\mathrm{Put}_t^z(K,T,\Delta) = \mathrm{Call}_t^z(K,T,\Delta) + P_t(r_t;T)\big(K - Z_t(r_t;T,T+\Delta)\big). \quad (4.87)$$

In Appendix C.9, we show that the expression for the forward price predicted by the Vasicek model is

$$Z_t(r_t;T,T+\Delta) = 100 \times \left(1 - \frac{1}{\Delta}\left(e^{-a_\Delta + b_\Delta\cdot\mathbb{E}_t^{Q_{FT}}(r_T) + \frac{1}{2}b_\Delta^2\cdot\mathbb{V}_t(r_T)} - 1\right)\right), \quad (4.88)$$

where $\mathbb{E}_t^{Q_{FT}}(r_T)$ and $\mathbb{V}_t(r_T)$ are as in Eqs. (4.86). Note the second difference with the evaluation formulae applying to the government bond case: in Sect. 4.2.10, we rely on the Jamshidian formula in Eq. (4.66), whereas in the time-deposit case of this section, we rely on the closed-form expression in Eq. (4.85).

The expression for $Z_t(r_t;T,T+\Delta)$ in Eq. (4.88) can be calculated using either the constant $\hat{\lambda}$ obtained in Steps 1–2(i) or the value $\hat{\lambda}(K_{\mathrm{atm}}^*)$ obtained in Steps 1–2(ii), where the value $K_{\mathrm{atm}}^* \equiv Z_t(r_t;T,T+\Delta)$. In the latter case, we are led to a fixed point problem similar to that arising in Sect. 4.2.10. We proceed similarly as follows. We start with a guess for $\hat{\lambda}(K_{\mathrm{atm}}^*)$, say $\hat{\lambda}^{(0)}$, which is used to calculate the forward price, say $Z_t^{(1)} \equiv Z_t(r_t;T,T+\Delta;\hat{\lambda}^{(0)})$, where $Z_t(r_t;T,T+\Delta;\lambda)$ denotes

[12] We utilize the expression for $\mathbb{E}_t^{Q_{FT}}(r_T)$ below (see Eq. (4.88)).

Table 4.5 Implied volatilities and option premiums

Strike price (%)	Implied volatility		Premiums	
	Black's implied Vol (%)	Basis point implied Vol	Put option	Call option
98.750	0.91	89.75	$9.0197 \cdot 10^{-5}$	$7.5871 \cdot 10^{-3}$
98.875	0.79	78.18	$9.2048 \cdot 10^{-5}$	$6.3395 \cdot 10^{-5}$
99.000	0.67	66.27	$9.6557 \cdot 10^{-5}$	$5.0945 \cdot 10^{-3}$
99.125	0.59	58.79	$1.3910 \cdot 10^{-4}$	$3.8876 \cdot 10^{-3}$
99.250	0.49	48.66	$1.9261 \cdot 10^{-4}$	$2.6916 \cdot 10^{-3}$
99.375	0.38	38.20	$2.8781 \cdot 10^{-4}$	$1.5373 \cdot 10^{-3}$
99.500 (ATM)	0.27	27.22	$5.3566 \cdot 10^{-4}$	$5.3566 \cdot 10^{-4}$
99.625	0.24	23.66	$1.3405 \cdot 10^{-3}$	$9.1043 \cdot 10^{-5}$
99.750	0.24	24.25	$2.5069 \cdot 10^{-3}$	$7.9202 \cdot 10^{-6}$
99.875	0.31	27.50	$3.7628 \cdot 10^{-3}$	$3.8729 \cdot 10^{-6}$

the forward price in Eq. (4.88) when the risk-premium is λ. The guess, $\hat{\lambda}^{(0)}$, could be obtained from the initial futures risk-premium skew, as $\hat{\lambda}^{(0)} = \hat{\lambda}(K_{\text{atm}})$, where $K_{\text{atm}} = \tilde{Z}_t^{\$}(T, T + \Delta)$. The iteration is

$$\hat{\lambda}^{(i)} = \hat{\lambda}\big(Z_t^{(i)}\big), \quad Z_t^{(i+1)} = Z_t\big(r_t; T, T + \Delta; \hat{\lambda}^{(i)}\big),$$

and goes through until achievement of some convergence criterion.

4.3.5 Implementation Example

We provide a numerical example of the main steps involved in the calculation of the time deposit volatility indexes TD-VI(t, T, Δ) and TD-VI$^{\text{bp}}(t, T, \Delta)$ in Eqs. (4.71) and (4.72). We use data representing hypothetical market conditions on June 26, 2012, relating to 3-month options on 3-month Eurodollar forwards. We do not make early exercise corrections as we are assuming these hypothetical data relate to European options on forwards.

We approximate the theoretical values of the indexes through the same discretization schemes adopted in Chap. 3 (see Sect. 3.3.3). The first column of Table 4.5 reports strikes, K. The ATM strike is $K = 99.50$. The fourth and fifth columns provide option premiums used to calculate the index. For reference, the second and third ("Basis Point Implied Vol") columns also report both the log-normal ("Percentage Implied Vol") and normal ("Basis Point Implied Vol") skews implied by these premiums, special cases of definitions provided in Chap. 2 (see Sect. 2.4.2).

Table 4.6 provides the calculations leading to the determination of the two indexes: the second column displays the type of OTM option entering into the calculation; the third column has the premiums of the used option; the fourth and fifth

Table 4.6 Calculation of volatility indexes

Strike price (%)	Option type	Premiums	Weights		Contributions to strikes	
			Basis point ΔK_i	Percentage $\Delta K_i / K_i^2$	Basis point contribution	Percentage contribution
98.750	Put	$9.0197 \cdot 10^{-5}$	$0.1250 \cdot 10^{-2}$	$0.1281 \cdot 10^{-2}$	$1.1274 \cdot 10^{-7}$	$1.1561 \cdot 10^{-7}$
98.875	Put	$9.2048 \cdot 10^{-5}$	$0.1250 \cdot 10^{-2}$	$0.1278 \cdot 10^{-2}$	$1.1506 \cdot 10^{-7}$	$1.1769 \cdot 10^{-7}$
99.000	Put	$9.6557 \cdot 10^{-5}$	$0.1250 \cdot 10^{-2}$	$0.1275 \cdot 10^{-2}$	$1.2069 \cdot 10^{-7}$	$1.2314 \cdot 10^{-7}$
99.125	Put	$1.3910 \cdot 10^{-4}$	$0.1250 \cdot 10^{-2}$	$0.1272 \cdot 10^{-2}$	$1.7388 \cdot 10^{-7}$	$1.7696 \cdot 10^{-7}$
99.250	Put	$1.9261 \cdot 10^{-4}$	$0.1250 \cdot 10^{-2}$	$0.1268 \cdot 10^{-2}$	$2.4077 \cdot 10^{-7}$	$2.4442 \cdot 10^{-7}$
99.375	Put	$2.8781 \cdot 10^{-4}$	$0.1250 \cdot 10^{-2}$	$0.1265 \cdot 10^{-2}$	$3.5976 \cdot 10^{-7}$	$3.6431 \cdot 10^{-7}$
99.500 (ATM)	ATM	$5.3566 \cdot 10^{-4}$	$0.1250 \cdot 10^{-2}$	$0.1262 \cdot 10^{-2}$	$6.6958 \cdot 10^{-7}$	$6.7632 \cdot 10^{-7}$
99.625	Call	$9.1043 \cdot 10^{-5}$	$0.1250 \cdot 10^{-2}$	$0.1259 \cdot 10^{-2}$	$1.1381 \cdot 10^{-7}$	$1.1466 \cdot 10^{-7}$
99.750	Call	$7.9202 \cdot 10^{-6}$	$0.1250 \cdot 10^{-2}$	$0.1256 \cdot 10^{-2}$	$9.9002 \cdot 10^{-9}$	$9.9499 \cdot 10^{-9}$
99.875	Call	$3.8729 \cdot 10^{-6}$	$0.1250 \cdot 10^{-2}$	$0.1253 \cdot 10^{-2}$	$4.8412 \cdot 10^{-9}$	$4.8527 \cdot 10^{-9}$
SUMS					$1.9210 \cdot 10^{-6}$	$1.9479 \cdot 10^{-6}$

columns report the weights the prices bear in the final computation of the index, before the final rescaling, $\frac{2}{T-t}$; finally, the sixth and seventh columns report OTM weighted option prices, that is, each price in the third column multiplied by the weight in the fourth column ("Basis Point Contribution"), and each price in the third column multiplied by the weight in the fifth column ("Percentage Contribution").

The two indexes are calculated by evaluating Eq. (4.71) and Eq. (4.72),

$$\text{TD-VI} = 100 \times \sqrt{\frac{1}{0.9996} \frac{2}{(3/12)} \times 1.9479 \cdot 10^{-6}} = 0.39483$$

$$\text{TD-VI}^{bp} = 100^2 \times \sqrt{\frac{1}{0.9996} \frac{2}{(3/12)} \times 1.9210 \cdot 10^{-6}} = 39.2100$$

where $\frac{1}{0.9996}$ is the inverse of a zero-coupon bond expiring in 3 months. Note that we calculate the basis point index, TD-VI^{bp}, by rescaling the outcome in Table 4.6 by 100^2, rather than 100 as suggested by Eq. (4.72), because the strikes in Tables 4.5 and 4.6 are expressed in decimals, not in percentages as in Eq. (4.69).

In comparison, the ATM log-normal and the ATM normal implied volatilities are 0.27 % and 27.22 basis points, and the arithmetic average log-normal and normal skews are 0.49 % and 48.24 basis points.

To calculate the yield volatility indexes in Eqs. (4.79) and (4.80), we evaluate the certainty equivalent $\mathcal{Z}$ in Eq. (4.78) for the LIBOR price as the ratio of the basis point to the percentage index:

$$\frac{\text{TD-VI}^{bp}}{\text{TD-VI}} = \frac{39.2100}{0.39483} = 99.6460,$$

and solve for $y_{\mathcal{Z}}(T)$ in Eq. (4.78), with $\Delta = \frac{1}{4}$:

$$\frac{99.3063}{100} = (1 + y_{\mathcal{Z}})^{-\frac{1}{4}} \implies y_{\mathcal{Z}} = 1.4286 \times 10^{-2},$$

leading to,

$$\text{TD-VI}^{bp}_{Yd} = \frac{100 \times (1 + 1.4286 \times 10^{-2}) \times 0.39483}{(1/4)} = 160.19,$$

for the duration-based yield volatility index in Eq. (4.79), and

$$\text{TD-VI}^{bp}_{Y} = 1.4286 \times 0.39483 = 0.5641,$$

for the yield-based yield volatility index in Eq. (4.80). Note that if we replace the certainty equivalent $\mathcal{Z} = 99.3063$ with the ATM price from Table 4.6 (i.e., 99.50), we would obtain $y_{\mathcal{Z}} = 2.0253 \times 10^{-2}$, then $\text{TD-VI}^{bp}_{Yd} = 161.68$ and $\text{TD-VI}^{bp}_{Y} = 0.80238$.

4.3.6 LIBOR Variance Contracts and Volatility Indexes

This section develops variance swap security designs based on *rates*, rather than *prices*. Accordingly, we consider the forward LIBOR equivalent to Eq. (4.70):

$$\frac{df_\tau(T, T + \Delta)}{f_\tau(T, T + \Delta)} = v_\tau^f(T, \Delta) dW_\tau^{F^T}, \quad \tau \in [t, T], \tag{4.89}$$

where, by Eq. (4.69) and Itô's lemma, $v_\tau^f(T, \Delta) \equiv (1 - f_\tau^{-1}(T, T + \Delta))v_\tau^z(T, \Delta)$.

We define the basis point and percentage LIBOR integrated variances as

$$V^{f,\mathrm{bp}}(t, T, \Delta) \equiv \int_t^T f_\tau^2(T, T + \Delta) \left\| v_\tau^f(T, \Delta) \right\|^2 d\tau,$$

$$V^f(t, T, \Delta) \equiv \int_t^T \left\| v_\tau^f(T, \Delta) \right\|^2 d\tau,$$

so that, by the usual arguments, the fair values of the two time deposit *rate*-variance swaps at time t, $\mathbb{P}_f^{\mathrm{bp}}(t, T, \Delta)$ and $\mathbb{P}_f(t, T, \Delta)$ with payoffs equal to

$$V^{f,\mathrm{bp}}(t, T, \Delta) - \mathbb{P}_f^{\mathrm{bp}}(t, T, \Delta) \quad \text{and} \quad V^f(t, T, \Delta) - \mathbb{P}_f(t, T, \Delta),$$

are

$$\mathbb{P}_f^{\mathrm{bp}}(t, T, \Delta) = \frac{2}{P_t(T)} \left(\int_0^{f_t(T, T+\Delta)} \mathrm{Put}_t^f(K_f, T, \Delta) dK_f \right.$$

$$\left. + \int_{f_t(T, T+\Delta)}^\infty \mathrm{Call}_t^f(K_f, T, \Delta) dK_f \right),$$

and

$$\mathbb{P}_f(t, T, \Delta) = \frac{2}{P_t(T)} \left(\int_0^{f_t(T, T+\Delta)} \mathrm{Put}_t^f(K_f, T, \Delta) \frac{1}{K_f^2} dK_f \right.$$

$$\left. + \int_{f_t(T, T+\Delta)}^\infty \mathrm{Call}_t^f(K_f, T, \Delta) \frac{1}{K_f^2} dK_f \right),$$

where

$$\mathrm{Put}_t^f(K_f, T, \Delta) = \frac{\mathrm{Call}_t^z(100(1 - K_f), T, \Delta)}{100},$$

$$\mathrm{Call}_t^f(K_f, T, \Delta) = \frac{\mathrm{Put}_t^z(100(1 - K_f), T, \Delta)}{100}, \tag{4.90}$$

and $\mathrm{Call}_t^z(\cdot, T, \Delta)$ and $\mathrm{Put}_t^z(\cdot, T, \Delta)$ are the same OTM and at-the-money (ATM) European options defined in Sect. 4.3.2. Note that $\mathrm{Put}_t^f(K_f, T, \Delta)$ and $\mathrm{Call}_t^f(K_f, T, \Delta)$ are OTM European options on the forward LIBOR.

Accordingly, an index of *basis point* time deposit rate-volatility is

$$
\text{TD-VI}_f^{\text{bp}}(t, T, \Delta) \equiv 100^2 \times \sqrt{\frac{\mathbb{P}_f^{\text{bp}}(t, T, \Delta)}{T - t}}
\tag{4.91}
$$

and an index of *percentage* time deposit rate-volatility is

$$
\text{TD-VI}_f(t, T, \Delta) \equiv 100 \times \sqrt{\frac{\mathbb{P}_f(t, T, \Delta)}{T - t}}
\tag{4.92}
$$

The reason we now rescale by 100^2 in Eq. (4.91), rather than by 100 as in Eq. (4.72), is simply that unlike in the forward LIBOR price case (see Eq. (4.69)), we now follow the convention that the forward LIBOR rate in Eq. (4.89) is expressed in decimals, not in percentage terms.

In the absence of price quotes for ATM options, we proceed as usual (see Sect. 4.3.2), through the following approximations of the basis point and the percentage index, respectively:

$$
\begin{aligned}
\text{TD-VI}_{f,o}^{\text{bp}}(t, T, \Delta) &\equiv 100^2 \times \sqrt{\frac{\mathbb{P}_{f,o}^{\text{bp}}(t, T, \Delta)}{T - t}}, \\
\text{TD-VI}_{f,o}(t, T, \Delta) &\equiv 100 \times \sqrt{\frac{\mathbb{P}_{f,o}(t, T, \Delta)}{T - t}}
\end{aligned}
\tag{4.93}
$$

where

$$
\begin{aligned}
&\mathbb{P}_{f,o}^{\text{bp}}(t, T, \Delta) \\
&\equiv \frac{2}{P_t(T)} \left(\int_0^{K_{0,f}} \text{Put}_t^f(K_f, T, \Delta) dK_f + \int_{K_{0,f}}^{\infty} \text{Call}_t^{\dot{z}}(K_f, T, \Delta) dK_f \right) \\
&\quad - \left(f_t(T, T + \Delta) - K_{0,f} \right)^2,
\end{aligned}
$$

and

$$
\begin{aligned}
&\mathbb{P}_{f,o}(t, T, \Delta) \\
&\equiv \frac{2}{P_t(T)} \left(\int_0^{K_{0,f}} \text{Put}_t^f(K_f, T, \Delta) \frac{1}{K_f^2} dK_f + \int_{K_{0,f}}^{\infty} \text{Call}_t^f(K_f, T, \Delta) \frac{1}{K_f^2} dK_f \right) \\
&\quad - \left(\frac{f_t(T, T + \Delta) - K_{0,f}}{K_{0,f}} \right)^2,
\end{aligned}
$$

and $K_{0,f}$ denotes the first strike below the current forward LIBOR $f_t(T, T + \Delta)$, where in case $f_t(T, T + \Delta)$ is not observed, it is approximated by the strike price at which the absolute difference between the call and put prices is smallest, and setting $K_{0,f} = f_t$ whenever an option exists, which is struck at f_t.

Finally, the American future corrections in Sect. 4.3.4 can be utilized to deal with the indexes of this section. Given the market price of American options on future prices, we obtain the price of European options on forward prices, as described in Sect. 4.3.4, which we then turn into the price of European options on forward LIBOR by interchanging puts and calls as indicated by Eq. (4.90).

4.4 Maturity Mismatch

The working assumption adopted until now in this chapter is that the maturity of the forward we want to price the volatility of is the same as the maturity of the options spanning the variance contracts. This assumption may be invalid in practice. For example, we could conceive of situations in which the option maturity is, say, 1 month, and the underlying is a forward expiring in 5 years. Furthermore, we can still conceive of post-issuance cases similar to those explained at the end of Sect. 4.2.8.2 and dealt with in Appendix C.6, where the date of issuance of the coupon-bearing bond occurs before the forward maturity.

This section shows that, to cope with these issues, we need to apply some correction terms to the fair values of variance swaps developed thus far. The correction terms are model-dependent for the following reasons. Let T be the maturity of the option and S the maturity of the forward, with $T \leq S$, such that "option spanning" operates under the T-forward probability, whereas the forward risk and its volatility are defined under the S-forward probability. Unless $T = S$, we would price the forward risk volatility with the "wrong" probability, a situation reminiscent of convexity problems arising in fixed income security evaluation (see, e.g., Veronesi 2010, Chap. 21). We illustrate these facts as applied to the two asset classes considered in this chapter. In Appendix C.10, we implement an experiment that reveals that the impact of this mismatch is quite small when $S - T$ is just a couple of months.

The plan of this section is the following. In Sects. 4.4.1 and 4.4.2, we provide corrections taking into account these maturity mismatches for both government bond and time deposit volatility. Section 4.4.3 provides alternative corrections that use different representations of the same issue, and Sect. 4.4.4 explains in more detail the origin of the issues, notably in connection with the theory of forward contract multipliers of Chap. 2.

4.4.1 Government Bonds

We utilize the same notation as in Sect. 4.2, and let $F_t(S, \mathbb{T})$ be the forward price at t for delivery at S of the coupon bearing bond expiring at $\mathbb{T}$, as defined in Eq. (4.12), with S naturally replacing T. Accordingly, $F_t(S, \mathbb{T})$ satisfies

$$\frac{dF_\tau(S, \mathbb{T})}{F_\tau(S, \mathbb{T})} = v_\tau(S, \mathbb{T}) \cdot dW_\tau^{FS}, \quad \tau \in [t, S], \tag{4.94}$$

where W_τ^{FS} is a multidimensional Brownian motion under Q_{FS}, and $v_\tau(S, \mathbb{T})$ is defined as in Eq. (4.14) with an obvious change in notation. Note that while the martingale dynamics of $F_t(T, \mathbb{T})$ in Eq. (4.14) are defined under the forward probability Q_{FT}, the martingale dynamics of $F_t(S, \mathbb{T})$ in Eq. (4.94) are defined under Q_{FS}. It is trivial, but the entire point of this section concerns the following question: How do we price the volatility of a forward expiring at S with OTM options expiring at $T < S$?

4.4.1.1 Percentage Volatility

We define the percentage integrated variance of the forward price in Eq. (4.94) as

$$V(t, T, S, \mathbb{T}) \equiv \int_t^T \left\| v_\tau(S, \mathbb{T}) \right\|^2 d\tau.$$

The fair value of the variance swap referenced to $V(t, T, S, \mathbb{T})$ is

$$\mathbb{P}(t, T, S, \mathbb{T}) = \mathbb{E}_t^{Q_{FT}}\left(V(t, T, S, \mathbb{T}) \right). \tag{4.95}$$

Define the change of probability,

$$\left. \frac{dQ_{FS}}{dQ_{FT}} \right|_{\mathbb{F}_S} \equiv \xi(t, T, S) \equiv \frac{P_t(T)}{P_t(S)} e^{-\int_T^S r_\tau d\tau}, \tag{4.96}$$

and

$$C_{ot}(T, S, \mathbb{T}) \equiv \mathrm{Cov}_t^{Q_{FT}}\left(V(t, T, S, \mathbb{T}), \xi(t, T, S) \right). \tag{4.97}$$

We have

$$-2\mathbb{E}_t^{Q_{FS}}\left(\ln \frac{F_T(S, \mathbb{T})}{F_t(S, \mathbb{T})} \right) = \mathbb{E}_t^{Q_{FS}}\left(V(t, T, S, \mathbb{T}) \right)$$

$$= \mathbb{E}_t^{Q_{FT}}\left[V(t, T, S, \mathbb{T}) \cdot \xi(t, T, S) \right]$$

$$= \mathbb{E}_t^{Q_{FT}}\left(V(t, T, S, \mathbb{T}) \right) + C_{ot}(T, S, \mathbb{T})$$

$$= \mathbb{P}(t, T, S, \mathbb{T}) + C_{ot}(T, S, \mathbb{T}), \tag{4.98}$$

where the first equality follows by Itô's lemma, the second by a change of probability, the third by the definition of $C_{ot}(\cdot)$ in Eq. (4.97) and the fourth by Eq. (4.95).

As for the log-contract, note that by the usual spanning arguments,

$$-\mathbb{E}_t^{Q_{FS}}\left(\ln\frac{F_T(S,\mathbb{T})}{F_t(S,\mathbb{T})}\right)$$

$$=\int_0^{F_t(S,\mathbb{T})}\mathbb{E}_t^{Q_{FS}}\left(K-F_T(S,\mathbb{T})\right)^+\frac{1}{K^2}dK$$

$$+\int_{F_t(S,\mathbb{T})}^{\infty}\mathbb{E}_t^{Q_{FS}}\left(F_T(S,\mathbb{T})-K\right)^+\frac{1}{K^2}dK. \qquad (4.99)$$

All in all, the fair value, $\mathbb{P}(t,T,S,\mathbb{T})$, cannot be expressed in a model-free format unless $S=T$. For example, the expectation $\mathbb{E}_t^{Q_{FS}}(F_T(S,\mathbb{T})-K)^+$ is not the price of a European option with underlying the forward at T, $F_T(S,\mathbb{T})$. Moreover, $\mathrm{Cot}(\cdot)$ in Eq. (4.97) is non-zero, except of course when $S=T$, as $\xi(t,T,T)=1$ in this case.

Then let us define

$$C_{1t}(K,T,S,\mathbb{T})\equiv\mathrm{Cov}_t^{Q_{FT}}\left((K-F_T(S,\mathbb{T}))^+,\xi(t,T,S)\right),\quad K\le F_t(S,\mathbb{T}),$$

$$C_{2t}(K,T,S,\mathbb{T})\equiv\mathrm{Cov}_t^{Q_{FT}}\left((F_T(S,\mathbb{T})-K)^+,\xi(t,T,S)\right),\quad K\ge F_t(S,\mathbb{T}),$$

so that by a change in probability similar to that in Eq. (4.98),

$$\mathbb{E}_t^{Q_{FS}}\left(K-F_T(S,\mathbb{T})\right)^+=\mathbb{E}_t^{Q_{FT}}\left[(K-F_T(S,\mathbb{T}))^+\cdot\xi(t,T,S)\right]$$

$$=\mathbb{E}_t^{Q_{FT}}\left[(K-F_T(S,\mathbb{T}))^+\right]+C_{1t}(K,T,S,\mathbb{T})$$

$$=\frac{1}{P_t(T)}\mathrm{Put}_t(K,T,S,\mathbb{T})+C_{1t}(K,T,S,\mathbb{T}), \qquad (4.100)$$

and, similarly,

$$\mathbb{E}_t^{Q_{FS}}\left(F_T(S,\mathbb{T})-K\right)^+=\frac{1}{P_t(T)}\mathrm{Call}_t(K,T,S,\mathbb{T})+C_{2t}(K,T,S,\mathbb{T}), \qquad (4.101)$$

where $\mathrm{Put}_t(K,T,S,\mathbb{T})$ and $\mathrm{Call}_t(K,T,S,\mathbb{T})$ denote the price of OTM European puts and calls, struck at K, expiring at T, and referencing a forward price expiring at S, $F_T(S,\mathbb{T})$.

We can now match Eq. (4.98) to Eq. (4.99), use the expressions for the expectations in Eqs. (4.100) and (4.101) taken under the "wrong probability," Q_{FS}, and obtain the fair value of the percentage variance swap:

$$\mathbb{P}(t,T,S,\mathbb{T})=\mathbb{P}_{\mathrm{vix}}(t,T,S,\mathbb{T})+\mathbb{C}(t,T,S,\mathbb{T}), \qquad (4.102)$$

where

$$\mathbb{P}_{\text{vix}}(t, T, S, \mathbb{T}) \equiv \frac{2}{P_t(T)} \left(\int_0^{F_t(S,\mathbb{T})} \text{Put}_t(K, T, S, \mathbb{T}) \frac{1}{K^2} dK \right.$$
$$\left. + \int_{F_t(S,\mathbb{T})}^{\infty} \text{Call}_t(K, T, S, \mathbb{T}) \frac{1}{K^2} dK \right),$$

$$\mathbb{C}(t, T, S, \mathbb{T}) \equiv 2 \left(\int_0^{F_t(S,\mathbb{T})} C_{1t}(K, T, S, \mathbb{T}) \frac{1}{K^2} dK \right.$$
$$\left. + \int_{F_t(S,\mathbb{T})}^{\infty} C_{2t}(K, T, S, \mathbb{T}) \frac{1}{K^2} dK \right) - C_{ot}(T, S, \mathbb{T}).$$

Therefore, we can generalize the index of percentage government bond volatility in Eq. (4.21) through the following measurement:

$$\boxed{\text{GB-VI}(t, T, S, \mathbb{T}) \equiv 100 \times \sqrt{\frac{\mathbb{P}(t, T, S, \mathbb{T})}{T - t}}} \tag{4.103}$$

This index is obviously not model-free because $\mathbb{P}(t, T, S, \mathbb{T})$ depends on the correction term $\mathbb{C}(t, T, S, \mathbb{T})$.

4.4.1.2 Basis Point

The steps leading to an index of basis point government bond volatility are similar. First, define the basis point integrated variance of the forward price in Eq. (4.94) as

$$V^{\text{bp}}(t, T, S, \mathbb{T}) \equiv \int_t^T F_\tau^2(S, \mathbb{T}) \| v_\tau(S, \mathbb{T}) \|^2 d\tau.$$

We have, by arguments similar to those leading to Eq. (4.98), that

$$\mathbb{E}_t^{Q_{FS}} \left(F_T^2(S, \mathbb{T}) \right) - F_t^2(S, \mathbb{T}) = \mathbb{P}^{\text{bp}}(t, T, S, \mathbb{T}) + C_{ot}^{\text{bp}}(T, S, \mathbb{T}), \tag{4.104}$$

where

$$\mathbb{P}^{\text{bp}}(t, T, S, \mathbb{T}) = \mathbb{E}_t^{Q_{FT}} \left(V^{\text{bp}}(t, T, S, \mathbb{T}) \right) \tag{4.105}$$

is the fair value of a basis point variance swap, and

$$C_{ot}^{\text{bp}}(T, S, \mathbb{T}) \equiv \text{Cov}_t^{Q_{FT}} \left(V^{\text{bp}}(t, T, S, \mathbb{T}), \xi(t, T, S) \right).$$

Furthermore, using again Eqs. (4.100)–(4.101),

$$\mathbb{E}_t^{Q_{FS}} \left(F_T^2(S, \mathbb{T}) \right) - F_t^2(S, \mathbb{T}) = \mathbb{P}_{\text{vix}}^{\text{bp}}(t, T, S, \mathbb{T}) + \mathbb{C}^{\text{bp}}(t, T, S, \mathbb{T}) + C_{ot}^{\text{bp}}(T, S, \mathbb{T}), \tag{4.106}$$

where

$$\mathbb{P}^{\mathrm{bp}}_{\mathrm{vix}}(t, T, S, \mathbb{T})$$

$$\equiv \frac{2}{P_t(T)} \left(\int_0^{F_t(S,\mathbb{T})} \mathrm{Put}_t(K, T, S, \mathbb{T}) dK + \int_{F_t(S,\mathbb{T})}^{\infty} \mathrm{Call}_t(K, T, S, \mathbb{T}) dK \right),$$

$$\mathbb{C}^{\mathrm{bp}}(t, T, S, \mathbb{T})$$

$$\equiv 2 \left(\int_0^{F_t(S,\mathbb{T})} C_{1t}(K, T, S, \mathbb{T}) dK + \int_{F_t(S,\mathbb{T})}^{\infty} C_{2t}(K, T, S, \mathbb{T}) dK \right)$$

$$- C^{\mathrm{bp}}_{ot}(T, S, \mathbb{T}).$$

Therefore, matching Eq. (4.104) and Eq. (4.106) leaves the following expression for the fair value of the basis point government bond variance swap:

$$\mathbb{P}^{\mathrm{bp}}(t, T, S, \mathbb{T}) = \mathbb{P}^{\mathrm{bp}}_{\mathrm{vix}}(t, T, S, \mathbb{T}) + \mathbb{C}^{\mathrm{bp}}(t, T, S, \mathbb{T}). \qquad (4.107)$$

An index of basis point government bond volatility generalizing that in Eq. (4.28) is

$$\boxed{\mathrm{GB\text{-}VI}^{\mathrm{bp}}(t, T, S, \mathbb{T}) \equiv 100^2 \times \sqrt{\frac{\mathbb{P}^{\mathrm{bp}}(t, T, S, \mathbb{T})}{T - t}}} \qquad (4.108)$$

It is not model-free due to the term $\mathbb{C}^{\mathrm{bp}}(t, T, S, \mathbb{T})$, which collapses to zero when $S = T$.

4.4.1.3 Basis Point Yield Volatility

We generalize the basis point yield volatility in Sect. 4.2.8 by introducing the notion of a certainty equivalent price, $\mathcal{B}(t, T, S, \mathbb{T})$, as the guaranteed price of a *hypothetical* coupon-bearing bond price, deliverable at T, time to maturity and coupon payments still running over a fixed period equal to $\mathbb{T} - S$, and such that the basis point volatility in Eq. (4.108) is the same as the basis point volatility in this certainty equivalent market.

By following steps similar to those in Sect. 4.2.8, we find that this certainty equivalent price is the ratio of the basis point expected volatility in Eq. (4.103) to the percentage in Eq. (4.108), viz

$$\mathrm{GB\text{-}VI}^{\mathrm{bp}}(t, T, S, \mathbb{T}) = \mathcal{B}(t, T, S, \mathbb{T}) \times \mathrm{GB\text{-}VI}(t, T, S, \mathbb{T}). \qquad (4.109)$$

Accordingly, let $\hat{P}$ denote the mapping in Eq. (4.36) and $\hat{P}^{-1}$ its functional inverse. Further, let N denote the number of coupon payments over the interval $[S, \mathbb{T}]$, and n the frequency of coupon payments.

A model-free measure of duration-based yield expected volatility generalizing that in Eq. (4.39) is

$$
\begin{aligned}
&\text{GB-VI}_{\text{Yd}}^{\text{bp}}(t, T, S, \mathbb{T}) \\
&= \frac{100 \times (1 + \frac{1}{n}\hat{P}^{-1}[\frac{\text{GB-VI}^{\text{bp}}(t,T,S,\mathbb{T})}{\text{GB-VI}(t,T,S,\mathbb{T})}]) \times \text{GB-VI}^{\text{bp}}(t, T, S, \mathbb{T})}{\sum_{i=1}^{N} \frac{C_i}{n}(1 + \frac{1}{n}\hat{P}^{-1}[\frac{\text{GB-VI}^{\text{bp}}(t,T,S,\mathbb{T})}{\text{GB-VI}(t,T,S,\mathbb{T})}])^{-i}\frac{i}{n} + 100(1 + \frac{1}{n}\hat{P}^{-1}[\frac{\text{GB-VI}^{\text{bp}}(t,T,S,\mathbb{T})}{\text{GB-VI}(t,T,S,\mathbb{T})}])^{-N}\frac{N}{n}}
\end{aligned}
$$

$$(4.110)$$

Likewise, a model-free measure of yield-based yield expected volatility generalizing that in Eq. (4.40) is

$$
\text{GB-VI}_{\text{Y}}^{\text{bp}}(t, T, S, \mathbb{T}) = 100 \times \hat{P}^{-1}\left[\frac{\text{GB-VI}^{\text{bp}}(t, T, S, \mathbb{T})}{\text{GB-VI}(t, T, S, \mathbb{T})}\right] \times \text{GB-VI}(t, T, S, \mathbb{T})
$$

$$(4.111)$$

Note that the two indexes in Eqs. (4.110) and (4.111) are not model-free, as their constituents, $\text{GB-VI}(t, T, S, \mathbb{T})$ in Eq. (4.103) and $\text{GB-VI}^{\text{bp}}(t, T, S, \mathbb{T})$ in Eq. (4.108), are not.

In Appendix C.6, we extend the definition of the indexes in Eq. (4.110) and Eq. (4.111) to cover the "post-issuance case," where the maturity S of the forward can be higher than the date of issuance of the bond (see Eq. (C.21) and Eq. (C.22)).

4.4.1.4 Forward Adjustments

In the absence of quotes for ATM options, the value of the indexes in Eq. (4.103) and Eq. (4.108) can be approximated similarly as in Sect. 4.2.7, as follows:

$$
\begin{aligned}
&\text{GB-VI}_o(t, T, S, \mathbb{T}) \equiv 100 \times \sqrt{\frac{\mathbb{P}_o(t, T, S, \mathbb{T})}{T - t}}, \\
&\text{GB-VI}_o^{\text{bp}}(t, T, S, \mathbb{T}) \equiv 100^2 \times \sqrt{\frac{\mathbb{P}_o^{\text{bp}}(t, T, S, \mathbb{T})}{T - t}}
\end{aligned}
$$

$$(4.112)$$

where

$$
\mathbb{P}_o(t, T, S, \mathbb{T}) \equiv \mathbb{P}_{K_0}(t, T, S, \mathbb{T}) - \left(\frac{F_t(S, \mathbb{T}) - K_0}{K_0}\right)^2,
$$

$$
\mathbb{P}_o^{\text{bp}}(t, T, S, \mathbb{T}) \equiv \mathbb{P}_{K_0}^{\text{bp}}(t, T, S, \mathbb{T}) - \left(F_t(S, \mathbb{T}) - K_0\right)^2,
$$

and K_0 denotes the first strike below $F_t(S, \mathbb{T})$, and $F_t(S, \mathbb{T})$ is approximated by the strike price at which the absolute difference between the call and put prices is smallest (with $K_0 = F_t(S, \mathbb{T})$, should an option exist with strike equal to $F_t(S, \mathbb{T})$) and, finally, $\mathbb{P}_{K_0}(t, T, S, \mathbb{T})$ and $\mathbb{P}_{K_0}^{\text{bp}}(t, T, S, \mathbb{T})$ are defined as in Eq. (4.102) and

Eq. (4.107), respectively, but with K_0 replacing $F_t(S, \mathbb{T})$ in the integration limits of
$\mathbb{P}(t, T, S, \mathbb{T})$ and $\mathbb{P}^{\text{bp}}(t, T, S, \mathbb{T})$.

Once again, the two volatility indexes in Eqs. (4.112) are not model-free as the
fair values, $\mathbb{P}_o(t, T, S, \mathbb{T})$ and $\mathbb{P}_o^{\text{bp}}(t, T, S, \mathbb{T})$, are not.

4.4.2 Time Deposits

Consider a forward contract expiring at time S, delivering a payoff equal to $100 \times
(1 - l_S(\Delta)) - Z_t(S, S + \Delta)$. By the usual arguments, the forward price satisfies
$Z_t(S, S+\Delta) = 100 \times (1 - f_t(S, S+\Delta))$, where $f_t(S, S+\Delta)$ is the forward LIBOR,
which now satisfies

$$\frac{dZ_\tau(S, S + \Delta)}{Z_\tau(S, S + \Delta)} = v_\tau^z(S, \Delta)dW_\tau^{FS}, \quad \tau \in [t, S], \tag{4.113}$$

where W_τ^{FS} is a multidimensional Brownian motion under Q_{FS}, and $v_\tau^z(S, \Delta)$ is as
in Eq. (4.70) with a change in notation. Similar to the government bond case (see
Sect. 4.4.1), the dynamics of $Z_\tau(T, T + \Delta)$ in Eq. (4.70) are defined under Q_{FT},
whereas the dynamics of $Z_\tau(S, S + \Delta)$ in Eq. (4.113) are defined under Q_{FS}.

Parallel to the dynamics of $f_\tau(T, T + \Delta)$ under Q_{FT} in Eq. (4.89), we define the
dynamics of the forward LIBOR $f_\tau(S, S + \Delta)$ under Q_{FS},

$$\frac{df_\tau(S, S + \Delta)}{f_\tau(S, S + \Delta)} = v_\tau^f(S, \Delta)dW_\tau^{FS}, \quad \tau \in [t, S], \tag{4.114}$$

where $v_\tau^f(S, \Delta) \equiv (1 - f_\tau^{-1}(S, S + \Delta))v_\tau^z(S, \Delta)$.

4.4.2.1 Prices

Define the basis point and the percentage integrated variance of the forward price in
Eq. (4.113) as

$$V^{z,\text{bp}}(t, T, S, \Delta) \equiv \int_t^T Z_\tau^2(S, S + \Delta) \left\| v_\tau^z(S, \Delta) \right\|^2 d\tau,$$

$$V^z(t, T, S, \Delta) \equiv \int_t^T \left\| v_\tau^z(S, \Delta) \right\|^2 d\tau. \tag{4.115}$$

We utilize the arguments made in the government bond case (see Sect. 4.4.1),
and deal with the fair value of variance swaps for time deposits. The fair value in
the basis point case is

$$\mathbb{P}_z^{\text{bp}}(t, T, S, \Delta) \equiv \mathbb{E}_t^{Q_{FS}}\left(Z_T^2(S, S + \Delta)\right) - Z_t^2(S, S + \Delta) - C_{ot}^{z,\text{bp}}(T, S, \Delta),$$

where

$$C_{ot}^{z,\text{bp}}(T, S, \Delta) \equiv \text{Cov}_t^{Q_{FT}} \left(V^{z,\text{bp}}(t, T, S, \Delta), \xi(t, T, S) \right),$$

and $\xi(t, T, S)$ is the Radon–Nikodym derivative defined in Eq. (4.96). We have, as usual,

$$\mathbb{E}_t^{Q_{FS}} \left(Z_T^2(S, S + \Delta) \right) - Z_t^2(S, S + \Delta)$$
$$= 2 \left(\int_0^{Z_t(S,S+\Delta)} \mathbb{E}_t^{Q_{FS}} \left(K - Z_T(S, S + \Delta) \right)^+ dK \right.$$
$$\left. + \int_{Z_t(S,S+\Delta)}^{\infty} \mathbb{E}_t^{Q_{FS}} \left(Z_T(S, S + \Delta) - K \right)^+ dK \right),$$

and, by arguments similar to those leading to Eqs. (4.100) and (4.101),

$$\mathbb{E}_t^{Q_{FS}} \left(K - Z_T(S, S + \Delta) \right)^+ = \frac{1}{P_t(T)} \text{Put}_t^z(K, T, S, \Delta) + C_{1t}^z(K, T, S, \Delta),$$

$$\mathbb{E}_t^{Q_{FS}} \left(Z_T(S, S + \Delta) - K \right)^+ = \frac{1}{P_t(T)} \text{Call}_t^z(K, T, S, \Delta) + C_{2t}^z(K, T, S, \Delta),$$

where $\text{Put}_t^z(K, T, S, \Delta)$ and $\text{Call}_t^z(K, T, S, \Delta)$ denote the price of out-of-the-money European puts and calls, struck at K, expiring at T, and referenced to forward LI-BOR prices for time S, and, finally,

$$C_{1t}^z(K, T, S, \Delta) \equiv \text{Cov}_t^{Q_{FT}} \left(\left(K - Z_T(S, S + \Delta) \right)^+, \xi(t, T, S) \right),$$
$$K \leq Z_t(S, S + \Delta), \tag{4.116}$$

$$C_{2t}^z(K, T, S, \Delta) \equiv \text{Cov}_t^{Q_{FT}} \left(\left(Z_T(S, S + \Delta) - K \right)^+, \xi(t, T, S) \right),$$
$$K \geq Z_t(S, S + \Delta). \tag{4.117}$$

Let us gather all these findings. We have:

$$\mathbb{P}_z^{\text{bp}}(t, T, S, \Delta) = \mathbb{P}_{z,\text{vix}}^{\text{bp}}(t, T, S, \Delta) + \mathbb{C}_z^{\text{bp}}(t, T, S, \Delta), \tag{4.118}$$

where

$$\mathbb{P}_{z,\text{vix}}^{\text{bp}}(t, T, S, \Delta)$$
$$\equiv \frac{2}{P_t(T)} \left(\int_0^{Z_t(S,S+\Delta)} \text{Put}_t^z(K, T, S, \Delta) dK \right.$$
$$\left. + \int_{Z_t(S,S+\Delta)}^{\infty} \text{Call}_t^z(K, T, S, \Delta) dK \right),$$

$$\mathbb{C}_z^{bp}(t, T, S, \Delta)$$

$$\equiv 2\left(\int_0^{Z_t(S,S+\Delta)} C_{1t}^z(K, T, S, \Delta)dK + \int_{Z_t(S,S+\Delta)}^{\infty} C_{2t}^z(K, T, S, \Delta)dK\right)$$

$$- C_{ot}^{z,bp}(T, S, \Delta).$$

Therefore, an index of basis point time deposit volatility generalizing that in Eq. (4.72) is

$$\boxed{\text{TD-VI}^{bp}(t, T, S, \Delta) \equiv 100 \times \sqrt{\frac{\mathbb{P}_z^{bp}(t, T, S, \Delta)}{T - t}}} \qquad (4.119)$$

Likewise, an index of percentage time deposit volatility generalizing that in Eq. (4.71) is

$$\boxed{\text{TD-VI}(t, T, S, \Delta) \equiv 100 \times \sqrt{\frac{\mathbb{P}_z(t, T, S, \Delta)}{T - t}}} \qquad (4.120)$$

where $\mathbb{P}_z(t, T, S, \Delta) \equiv \mathbb{E}_t^{Q_{FT}}(V^z(t, T, S, \Delta))$ is the fair value of a percentage variance contract, and $V^z(t, T, S, \Delta)$ is defined as in the second of Eqs. (4.115), so that

$$\mathbb{P}_z(t, T, S, \Delta) = \mathbb{P}_{z,vix}(t, T, S, \Delta) + \mathbb{C}_z(t, T, S, \Delta), \qquad (4.121)$$

with

$$\mathbb{P}_{z,vix}(t, T, S, \Delta)$$

$$\equiv \frac{2}{P_t(T)}\left(\int_0^{Z_t(S,S+\Delta)} \text{Put}_t^z(K, T, S, \Delta)\frac{1}{K^2}dK\right.$$

$$\left. + \int_{Z_t(S,S+\Delta)}^{\infty} \text{Call}_t^z(K, T, S, \Delta)\frac{1}{K^2}dK\right),$$

$$\mathbb{C}_z(t, T, S, \Delta)$$

$$\equiv 2\left(\int_0^{Z_t(S,S+\Delta)} C_{1t}^z(K, T, S, \Delta)\frac{1}{K^2}dK + \int_{Z_t(S,S+\Delta)}^{\infty} C_{2t}^z(K, T, S, \Delta)\frac{1}{K^2}dK\right)$$

$$- C_{ot}^z(T, S, \Delta),$$

and

$$C_{ot}^z(T, S, \Delta) \equiv \text{Cov}_t^{Q_{FT}}(V^z(t, T, S, \Delta), \xi(t, T, S)).$$

Note, as usual, that the Radon–Nikodym derivative in Eq. (4.96) collapses to 1 when $S = T$, $\xi(t, T, T) = 1$, in which case $C_{ot}^z(T, S, \Delta) = 0$ and the two terms in Eqs. (4.116)–(4.117) collapse to $C_{1t}^z(K, T, T, \Delta) = C_{2t}^z(K, T, T, \Delta) = 0$, so that

$\mathbb{C}_z^{bp}(t, T, T, \Delta) = \mathbb{C}_z(t, T, T, \Delta) = 0$. That is, for $S = T$, we recover the variance contracts and indexes in Sect. 4.3.

In the general case where $S \geq T$, the two indexes in Eq. (4.119) and Eq. (4.120) are not model-free, because the adjustment terms $\mathbb{C}_z^{bp}(t, T, S, \Delta)$ and $\mathbb{C}_z(t, T, S, \Delta)$ are not.

4.4.2.2 Yields

We generalize the time deposit basis point yield volatility indexes in Eq. (4.79) and Eq. (4.80), by introducing the duration-based index,

$$\text{TD-VI}_{\text{Yd}}^{bp}(t, T, S, \Delta) = \frac{100 \times (1 + \hat{Z}^{-1}[\frac{\text{TD-VI}^{bp}(t,T,S,\Delta)}{\text{TD-VI}(t,T,S,\Delta)}]) \times \text{TD-VI}(t, T, S, \Delta)}{\Delta}$$

(4.122)

and the yield-based index,

$$\text{TD-VI}_Y^{bp}(t, T, S, \Delta) = 100 \times \hat{Z}^{-1}\left[\frac{\text{TD-VI}^{bp}(t, T, S, \Delta)}{\text{TD-VI}(t, T, S, \Delta)}\right] \times \text{TD-VI}(t, T, S, \Delta)$$

(4.123)

where $\hat{Z}^{-1}$ is the inverse function of $\hat{Z}$ in Eq. (4.78), and $\text{TD-VI}^{bp}(t, T, S, \Delta)$ and $\text{TD-VI}(t, T, S, \Delta)$ are as in Eq. (4.119) and Eq. (4.120), respectively.

The interpretation of the ratio $\frac{\text{TD-VI}^{bp}}{\text{TD-VI}}$ is similar to that of $\mathcal{B}(t, T, S, \mathbb{T})$ in Eq. (4.109), given in the government bond case. It is the guaranteed price of a *hypothetical* LIBOR contract at time T, so that the basis point volatility index, $\text{TD-VI}^{bp}(t, T, S, \Delta)$, is the same as the basis point volatility in this certainty equivalent market.

Once again, the indexes in Eq. (4.122) and Eq. (4.123) are not model-free, because the two indexes, $\text{TD-VI}^{bp}(t, T, S, \Delta)$ and $\text{TD-VI}(t, T, S, \Delta)$, are not.

4.4.2.3 Rates

Basis point and percentage integrated variance of the forward LIBOR in Eq. (4.114) are defined as

$$V^{f,bp}(t, T, S, \Delta) \equiv \int_t^T f_\tau^2(S, S + \Delta) \|v_\tau^f(S, \Delta)\|^2 d\tau,$$

$$V^f(t, T, S, \Delta) \equiv \int_t^T \|v_\tau^f(T, S, \Delta)\|^2 d\tau.$$

We can follow the usual steps in this section and show that an index of basis point time deposit *rate*-volatility generalizing that in Eq. (4.91) is

$$\text{TD-VI}_f^{bp}(t, T, S, \Delta) \equiv 100^2 \times \sqrt{\frac{\mathbb{P}_f^{bp}(t, T, S, \Delta)}{T - t}}$$

(4.124)

where $\mathbb{P}_f^{bp}(t, T, S, \Delta)$ is the fair value of a basis point variance swap on $V^{f,bp}(t, T, S, \Delta)$, $\mathbb{P}_f^{bp}(t, T, S, \Delta) = \mathbb{E}_t^{Q_{FT}}(V^{f,bp}(t, T, S, \Delta))$. It equals

$$\mathbb{P}_f^{bp}(t, T, S, \Delta) = \mathbb{P}_{f,\text{vix}}^{bp}(t, T, S, \Delta) + \mathbb{C}_f^{bp}(t, T, S, \Delta), \tag{4.125}$$

where

$$\mathbb{P}_{f,\text{vix}}^{bp}(t, T, S, \Delta) \equiv \frac{2}{P_t(T)} \left(\int_0^{f_t(S,S+\Delta)} \text{Put}_t^f(K_f, T, \Delta) dK_f \right.$$
$$\left. + \int_{f_t(S,S+\Delta)}^{\infty} \text{Call}_t^f(K_f, T, \Delta) dK_f \right),$$

and

$$\text{Put}_t^f(K_f, T, S, \Delta) = \frac{\text{Call}_t^z(100(1 - K_f), T, S, \Delta)}{100},$$

$$\text{Call}_t^f(K_f, T, S, \Delta) = \frac{\text{Put}_t^z(100(1 - K_f), S, T, \Delta)}{100},$$

$$\mathbb{C}_f^{bp}(t, T, S, \Delta)$$
$$\equiv 2 \left(\int_0^{f_t(S,S+\Delta)} C_{1t}^f(K_f, T, S, \Delta) dK_f + \int_{f_t(S,S+\Delta)}^{\infty} C_{2t}^f(K_f, T, S, \Delta) dK_f \right)$$
$$- C_{ot}^{f,bp}(T, S, \Delta),$$

and

$$C_{ot}^{f,bp}(T, S, \Delta) \equiv \text{Cov}_t^{Q_{FT}} \left(V^{f,bp}(t, T, S, \Delta), \xi(t, T, S) \right)$$
$$C_{1t}^f(K, T, S, \Delta) \equiv \text{Cov}_t^{Q_{FT}} \left((K_f - f_T(S, S + \Delta))^+, \xi(t, T, S) \right),$$
$$K_f \leq f_t(S, S + \Delta),$$
$$C_{2t}^f(K, T, S, \Delta) \equiv \text{Cov}_t^{Q_{FT}} \left((f_T(S, S + \Delta) - K_f)^+, \xi(t, T, S) \right),$$
$$K_f \geq f_t(S, S + \Delta),$$

and, finally, $\xi(t, T, S)$ is the Radon–Nikodym derivative in Eq. (4.96).

Likewise, an index of percentage time deposit *rate*-volatility generalizing that in Eq. (4.92) is

$$\boxed{\text{TD-VI}_f(t, T, S, \Delta) \equiv 100 \times \sqrt{\frac{\mathbb{P}_f(t, T, S, \Delta)}{T - t}}} \tag{4.126}$$

where $\mathbb{P}_f(t, T, S, \Delta) = \mathbb{E}_t^{Q_{FT}}(V^f(t, T, S, \Delta))$ is the fair value of a percentage variance swap on $V^f(t, T, S, \Delta)$. It equals

$$\mathbb{P}_f(t, T, S, \Delta) = \mathbb{P}_{f,\text{vix}}(t, T, S, \Delta) + \mathbb{C}_f(t, T, S, \Delta), \qquad (4.127)$$

where

$$\mathbb{P}_{f,\text{vix}}(t, T, S, \Delta)$$

$$\equiv \frac{2}{P_t(T)} \left(\int_0^{f_t(S, S+\Delta)} \text{Put}_t^f(K_f, T, S, \Delta) \frac{1}{K_f^2} dK_f \right.$$

$$\left. + \int_{f_t(S, S+\Delta)}^{\infty} \text{Call}_t^f(K_f, T, S, \Delta) \frac{1}{K_f^2} dK_f \right),$$

and

$$\mathbb{C}_f(t, T, S, \Delta)$$

$$\equiv 2 \left(\int_0^{f_t(S, S+\Delta)} C_{1t}^f(K_f, T, S, \Delta) dK_f + \int_{f_t(S, S+\Delta)}^{\infty} C_{2t}^f(K_f, T, S, \Delta) dK_f \right)$$

$$- C_{ot}^f(T, S, \Delta),$$

where

$$C_{ot}^f(T, S, \Delta) \equiv \text{Cov}_t^{Q_{FT}}\left(V^f(t, T, S, \Delta), \xi(t, T, S)\right).$$

As with the two *price*-volatility indexes in Eq. (4.119) and Eq. (4.120), the two *rate*-volatility indexes in Eq. (4.124) and Eq. (4.126) are not model-free, as the adjustment terms $\mathbb{C}_f^{\text{bp}}(t, T, S, \Delta)$ and $\mathbb{C}_f(t, T, S, \Delta)$ are not.

4.4.2.4 Forward Adjustments

Similarly as in the government bond volatility case (see Sect. 4.4.1), we can extend the indexes in Eqs. (4.119), (4.120), (4.124) (4.126), to deal with cases in which no ATM quotes are available.

First, consider price volatility, and let $K_{0,z}$ be defined just as it is in Eqs. (4.73). We generalize the indexes in Eqs. (4.73) as follows:

$$\boxed{\begin{aligned} \text{TD-VI}_o(t, T, S, \Delta) &\equiv 100 \times \sqrt{\frac{\mathbb{P}_{z,o}(t, T, S, \Delta)}{T - t}}, \\ \\ \text{TD-VI}_o^{\text{bp}}(t, T, S, \Delta) &\equiv 100 \times \sqrt{\frac{\mathbb{P}_{z,o}^{\text{bp}}(t, T, S, \Delta)}{T - t}} \end{aligned}} \qquad (4.128)$$

where

$$\mathbb{P}_{z,o}(t, T, S, \Delta) \equiv \mathbb{P}_{K_{0,z},o}(t, T, S, \Delta) - \left(\frac{Z_t(S, S+\Delta) - K_{0,z}}{K_{0,z}}\right)^2,$$

$$\mathbb{P}_{z,o}^{bp}(t, T, S, \Delta) \equiv \mathbb{P}_{K_{0,z},o}^{bp}(t, T, S, \Delta) - \left(Z_t(S, S+\Delta) - K_{0,z}\right)^2,$$

and, finally, $\mathbb{P}_{K_{0,z},o}(t, T, S, \Delta)$ and $\mathbb{P}_{K_{0,z},o}^{bp}(t, T, S, \Delta)$ are defined as in Eq. (4.121) and Eq. (4.118), respectively, but with $K_{0,z}$ replacing $Z_t(S, S+\Delta)$ in the integration limits of $\mathbb{P}_z(t, T, S, \Delta)$ and $\mathbb{P}_z^{bp}(t, T, S, \Delta)$.

Accordingly, indexes of basis point yield volatility can be implemented by replacing TD-VI$_o(t, T, S, \Delta)$ with TD-VI$_o^{bp}(t, T, S, \Delta)$, and TD-VI$_o(t, T, S, \Delta)$ with TD-VI$_o(t, T, S, \Delta)$, in Eqs. (4.122)–(4.123).

Finally, consider rates volatility, and let $K_{0,f}$ be defined as it is in Eqs. (4.93). The indexes in Eqs. (4.93) can be generalized as follows:

$$\text{TD-VI}_{f,o}^{bp}(t, T, S, \Delta) \equiv 100^2 \times \sqrt{\frac{\mathbb{P}_{f,o}^{bp}(t, T, S, \Delta)}{T - t}},$$

$$\text{TD-VI}_{f,o}(t, T, S, \Delta) \equiv 100 \times \sqrt{\frac{\mathbb{P}_{f,o}(t, T, S, \Delta)}{T - t}} \tag{4.129}$$

where

$$\mathbb{P}_{f,o}^{bp}(t, T, S, \Delta) \equiv \mathbb{P}_{K_{0,f},o}^{bp}(t, T, S, \Delta) - \left(\frac{f_t(S, S+\Delta) - K_{0,f}}{K_{0,f}}\right)^2,$$

$$\mathbb{P}_{f,o}(t, T, S, \Delta) \equiv \mathbb{P}_{K_{0,f},o}(t, T, S, \Delta) - \left(f_t(S, S+\Delta) - K_{0,f}\right)^2,$$

and, finally, $\mathbb{P}_{K_{0,f},o}^{bp}(t, T, S, \Delta)$ and $\mathbb{P}_{K_{0,f},o}(t, T, S, \Delta)$ are defined as in Eq. (4.125) and Eq. (4.127), respectively, but with $K_{0,f}$ replacing $f_t(S, S+\Delta)$ in the integration limits of $\mathbb{P}_f^{bp}(t, T, S, \Delta)$ and $\mathbb{P}_f(t, T, S, \Delta)$.

Once again, the four indexes in Eqs. (4.128) and (4.129) are not model-free, as the fair values they are built upon are not.

4.4.3 Alternative Characterizations of Variance Contracts and Indexes

4.4.3.1 Percentage

We provide alternative characterizations of the fair values of the variance swaps, $\mathbb{P}(t, T, S, \mathbb{T})$ in Eq. (4.102) and $\mathbb{P}^{bp}(t, T, S, \mathbb{T})$ in Eq. (4.107). Similar characterizations apply regarding the time-deposit case, which obtain with a mere change of notation.

We begin with the percentage case. Recall the definition of the forward volatility, $v_\tau(x, \mathbb{T}) \equiv \sigma_\tau^B(\mathbb{T}) - \sigma_\tau(x)$, where $\sigma_\tau^B(\mathbb{T})$ denotes the vector of instantaneous volatilities of the coupon bearing bond expiring at $\mathbb{T}$, and $\sigma_\tau(T)$ is the vector of volatilities of a zero-coupon bond expiring at T. By Girsanov's theorem, the forward price, $F_t(S, \mathbb{T})$ in Eq. (4.94), satisfies

$$\frac{dF_\tau(S, \mathbb{T})}{F_\tau(S, \mathbb{T})} = v_\tau(S, \mathbb{T})\big(v_\tau(S, \mathbb{T}) - v_\tau(T, \mathbb{T})\big)d\tau + v_\tau(S, \mathbb{T}) \cdot dW_\tau^{F^T}, \quad \tau \in [t, T],$$

(4.130)

where

$$dW_\tau^{F^T} = dW_\tau^{F^S} - \big(\sigma_\tau(T) - \sigma_\tau(S)\big)d\tau$$
$$= dW_\tau^{F^S} - \big(v_\tau(S, \mathbb{T}) - v_\tau(T, \mathbb{T})\big)$$

is a multidimensional Brownian motion under Q_{FT}.

Therefore, by Itô's lemma and the definition of the fair value $\mathbb{P}(t, T, S, \mathbb{T})$ in Eq. (4.95),

$$-\mathbb{E}_t^{Q_{FT}}\left(\ln\frac{F_T(S, \mathbb{T})}{F_t(S, \mathbb{T})}\right) = -\mathbb{E}_t^{Q_{FT}}\big(\tilde{\ell}(t, T, S, \mathbb{T})\big) + \frac{1}{2}\mathbb{P}(t, T, S, \mathbb{T}),$$

(4.131)

where

$$\tilde{\ell}(t, T, S, \mathbb{T}) \equiv \int_t^T v_\tau(S, \mathbb{T})\big(v_\tau(S, \mathbb{T}) - v_\tau(T, \mathbb{T})\big)d\tau.$$

(4.132)

Moreover, the value of a log-contract satisfies

$$-\mathbb{E}_t^{Q_{FT}}\left(\ln\frac{F_T(S, \mathbb{T})}{F_t(S, \mathbb{T})}\right) = 1 - \frac{\mathbb{E}_t^{Q_{FT}}(F_T(S, \mathbb{T}))}{F_t(S, \mathbb{T})} + \frac{1}{2}\mathbb{P}_{\text{vix}}(t, T, S, \mathbb{T})$$
$$= 1 - \mathbb{E}_t^{Q_{FT}}\big(e^{\tilde{\ell}(t,T,S,\mathbb{T})}\big) + \frac{1}{2}\mathbb{P}_{\text{vix}}(t, T, S, \mathbb{T}),$$

(4.133)

where $\mathbb{P}_{\text{vix}}(t, T, S, \mathbb{T})$ is as in Eq. (4.102), and the second line follows by Eq. (4.130). Matching Eq. (4.131) to Eq. (4.133) leaves

$$\mathbb{P}(t, T, S, \mathbb{T}) = \mathbb{P}_{\text{vix}}(t, T, S, \mathbb{T}) + 2\big[1 - \mathbb{E}_t^{Q_{FT}}\big(e^{\tilde{\ell}(t,T,S,\mathbb{T})} - \tilde{\ell}(t, T, S, \mathbb{T})\big)\big].$$ (4.134)

This expression can be substituted into Eq. (4.103) to provide an alternative representation of the percentage government bond volatility index GB-VI$(t, T, S, \mathbb{T})$.

4.4.3.2 Basis Point

As for the basis point index, note that by Itô's lemma and the definition of the fair value $\mathbb{P}^{\text{bp}}(t, T, S, \mathbb{T})$ in Eq. (4.105),

$$\mathbb{E}_t^{Q_{FT}}\big(F_T^2(S, \mathbb{T})\big) - F_t^2(S, \mathbb{T}) = 2\mathbb{E}_t^{Q_{FT}}\big(\tilde{\ell}^{\text{bp}}(t, T, S, \mathbb{T})\big) + \mathbb{P}^{\text{bp}}(t, T, S, \mathbb{T}),$$ (4.135)

where

$$\tilde{\ell}^{bp}(t, T, S, \mathbb{T}) \equiv \int_t^T F_\tau^2(S, \mathbb{T}) v_\tau(S, \mathbb{T}) \big(v_\tau(S, \mathbb{T}) - v_\tau(T, \mathbb{T})\big) d\tau.$$

The value of a quadratic contract satisfies

$$\mathbb{E}_t^{Q_{FT}} \big(F_T^2(S, \mathbb{T})\big) - F_t^2(S, \mathbb{T})$$

$$= 2F_t(S, \mathbb{T}) \big(\mathbb{E}_t^{Q_{FT}} \big(F_T(S, \mathbb{T})\big) - F_t(S, \mathbb{T})\big) + \mathbb{P}_{vix}^{bp}(t, T, S, \mathbb{T})$$

$$= 2F_t^2(S, \mathbb{T}) \big(\mathbb{E}_t^{Q_{FT}} \big(e^{\tilde{\ell}(t, T, S, \mathbb{T})}\big) - 1\big) + \mathbb{P}_{vix}^{bp}(t, T, S, \mathbb{T}), \qquad (4.136)$$

where $\mathbb{P}_{vix}^{bp}(t, T, S, \mathbb{T})$ is defined as in Eq. (4.107), and the second line follows, again, by Eq. (4.130). Matching Eq. (4.135) to Eq. (4.136) yields

$$\mathbb{P}^{bp}(t, T, S, \mathbb{T}) = \mathbb{P}_{vix}^{bp}(t, T, S, \mathbb{T})$$

$$+ 2\big(F_t^2(S, \mathbb{T})\big(\mathbb{E}_t^{Q_{FT}} \big(e^{\tilde{\ell}(t, T, S, \mathbb{T})}\big) - 1\big) - \mathbb{E}_t^{Q_{FT}} \big(\tilde{\ell}^{bp}(t, T, S, \mathbb{T})\big)\big).$$

Substituting this expression into Eq. (4.108) leads to an alternative representation of the basis point government bond volatility index GB-VI$^{bp}(t, T, S, \mathbb{T})$.

4.4.4 Tilting the Variance Payoff

Appendix C.10 demonstrates that the maturity mismatch leads to small index biases when $S - T$ is as small as two months. We now show how to mitigate these biases for any mismatch by designing a variance swap that accounts for the market numéraire. This section also elucidates theoretical reasons, based on Chap. 2, that explain a lack of model-free representation of a variance swap in the presence of a maturity mismatch.

We provide explanations regarding the government bond case. Consider the following payoff, tilted by a random multiplier equal to the price of a zero coupon bond expiring at S:

$$\pi^*(t, T, S, \mathbb{T}) \equiv P_T(S) \cdot \big(V(t, T, S, \mathbb{T}) - \mathbb{P}^*(t, T, S, \mathbb{T})\big), \qquad (4.137)$$

where the fair value of the payoff is

$$\mathbb{P}^*(t, T, S, \mathbb{T}) = \mathbb{E}_t^{Q_{FS}} \big(V(t, T, S, \mathbb{T})\big).$$

That is, the payoff in Eq. (4.137) is designed so that the fair value of the variance swap is the expectation of realized variance under Q_{FS}, not under Q_{FT} as in

Eq. (4.95). Therefore, by Itô's lemma,

$$-\mathbb{E}_t^{Q_{FS}}\left(\ln\frac{F_T(S,\mathbb{T})}{F_t(S,\mathbb{T})}\right)=\frac{1}{2}\mathbb{P}^*(t,T,S,\mathbb{T}).$$

We now have a neat link between the fair value of a variance swap and that of a log-contract. However, the value of this log-contract cannot be expressed in a model-free fashion, as explained in Sect. 4.4.1.[13] An index stemming from the security design in Eq. (4.137) is slightly simpler than that in Eq. (4.103):

$$\boxed{\text{GB-VI}^*(t,T,S,\mathbb{T})\equiv 100\times\sqrt{\frac{\mathbb{P}^*(t,T,S,\mathbb{T})}{T-t}}}$$

where

$$\mathbb{P}^*(t,T,S,\mathbb{T})=\mathbb{P}_{\text{vix}}(t,T,S,\mathbb{T})+\mathbb{C}(t,T,S,\mathbb{T})+C_{ot}(T,S,\mathbb{T}),$$

and the remaining notation is as in Sect. 4.4.1.

What is the connection between these findings and the theoretical results in Chap. 2? The security design in Chap. 2 prescribes tilting of the variance swap with the price of the numéraire of each market of interest, just as we do in Eq. (4.137). However, the issue in this section is that the potentially spanning option prices cannot be cast as expectations under the price of the market numéraire, $P_t(S)$, as required by Definition 2.2 (Sect. 2.3 in Chap. 2), but only under $P_t(T)$.

In other words, it is a lack of existence of a "model-free" pricing benchmark that we are facing in this section. Alternatively, consider the approach we adopted previously in this section, by which one relies on the forward multiplier, $P_t(T)$, i.e., one which is trivially 1\$ at T. In this case, the reason we still cannot achieve a model-free representation of an interest rate variance contract is that the forward risk is not a martingale under Q_{FT}, as required by Definition 2.2.

4.5 Index Design with Heterogeneous Market Data

In practice, government bond and time deposit futures and future options may trade in listing cycles based on standardized roll dates. For example, the CME Group's Chicago Board of Trade (CBOT) currently lists the first five 10-Year Treasury futures contracts in the March, June, September and December quarterly cycle, and lists options on them for the first three consecutive contract months (two serial expirations and one quarterly expiration)[14] plus the next four months in the quarterly cycle. In these cases, the time to expiry of traded options and underlying futures

---

[13] See the explanations provided after Eq. (4.99).

[14] Options in a serial expiration then expire strictly before the underlying: it is a "maturity mismatch" (see Sect. 4.4).

change every day and complicates the task of calculating a government bond or a time deposit volatility index reflecting a constant maturity.

This section takes quarterly cycles as an example and develops two non-limitative estimation methodologies of expected volatility based on the model-free expressions in the previous sections. The first estimator, which we refer to as a "sandwich combination" weighs options with different maturities (e.g. near- and next-quarter) so as to ensure that at each point in time, the resulting weighted volatility index has a constant maturity, e.g. three months. The second estimator, which we refer to as a "rolling index," uses options in the same cycle to estimate the expected volatility left to the expiration of this cycle.

4.5.1 Sandwich Combinations

We cast this discussion in terms of abstract indexes, and initially assume that the target horizon of these indexes is 3 months. Let T_i denote the quarterly "option reset," so that $T_i - T_{i-1} = T_{i+1} - T_i = 3d$, and $T_{i+1} - T_{i-1} = 6d$, where d is the number of days within a month. For any point in time $t \in [T_{i-1}, T_i]$, define

$$\mathcal{I}_t \equiv \sqrt{\frac{1}{(3/12)}\left(x_t V_t(T_i) + (1 - x_t)V_t(T_{i+1})\right)}, \quad t \in [T_{i-1}, T_i], \qquad (4.138)$$

where x_t is a positive weight defined below (see Eq. (4.140)) and $V_t(T_i)$ is one of the model-free indexes of variance derived in this chapter, and built upon options expiring at T_i (say in June) and, accordingly, $V_t(T_{i+1})$ is an index of variance built upon options expiring at T_{i+1} (September).

Concretely, $V_t(\cdot)$ can be $\mathbb{P}$ in the percentage government bond volatility index of Eq. (4.21) or $\mathbb{P}_z$ in the percentage time deposit volatility index of Eq. (4.71), or it can be $\mathbb{P}^{bp}$ in the basis point government bond volatility index of Eq. (4.28) or $\mathbb{P}_z^{bp}$ in the basis point time deposit volatility index of Eq. (4.72). Alternatively, $V_t(\cdot)$ can be built upon the forward approximations in Eq. (4.30) and Eq. (4.31) (for the government bond volatility indexes), or the forward approximations in Eq. (4.74) and Eq. (4.75) (for the time deposit volatility indexes). Furthermore, $V_t(\cdot)$ can be built upon any of the expected volatility index in the maturity mismatch case of Sect. 4.4.

Finally, one can create a *yield* volatility index, such as those in Sects. 4.2.8 or 4.3.3, by using as inputs the sandwich combinations of percentage and basis point volatility indexes. For example, the sandwich combination counterpart to the government bond duration based yield volatility in Eq. (4.39) is

$$\mathcal{I}_{Yd}^{bp} = \frac{100 \times (1 + \frac{1}{n}\hat{P}^{-1}[\frac{\mathcal{I}_t^{bp}}{\mathcal{I}_t^{perc}}]) \times \mathcal{I}_t^{bp}}{\sum_{i=1}^{N} \frac{c_i}{n}(1 + \frac{1}{n}\hat{P}^{-1}[\frac{\mathcal{I}_t^{bp}}{\mathcal{I}_t^{perc}}])^{-i}\frac{i}{n} + 100(1 + \frac{1}{n}\hat{P}^{-1}[\frac{\mathcal{I}_t^{bp}}{\mathcal{I}_t^{perc}}])^{-N}\frac{N}{n}}, \qquad (4.139)$$

Fig. 4.6 Percentage weight carried by options expiring in September from the beginning of March (day no. 61) to the beginning of September (day no. 241)

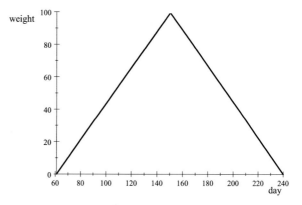

where $\hat{P}^{-1}$ denotes the inverse function of $\hat{P}$ in Eq. (4.36) and, for $j \in \{\text{perc}, \text{bp}\}$,

$$\mathcal{I}_t^j \equiv \sqrt{\frac{1}{(3/12)}\left(x_t V_t^j(T_i) + (1 - x_t)V_t^j(T_{i+1})\right)}, \quad t \in [T_{i-1}, T_i],$$

where $V_t^{\text{perc}}(T_i)$ and $V_t^{\text{bp}}(T_i)$ are the (square of the) government bond volatility indexes in Eq. (4.21) and Eq. (4.28), rescaled by maturity, with straightforward notation. In Appendix C.6, we show that this index is indeed well-defined.

Let us focus on the percentage case as the basis point case is a mere change in notation. We determine x_t in Eq. (4.138) by requiring that the average time-to-maturity of the options expiring at T_i and T_{i+1} equals 3 months, as follows:

$$x_t \frac{T_i - t}{12 \cdot d} + (1 - x_t)\frac{T_{i+1} - t}{12 \cdot d} = \frac{3}{12}, \quad t \in \{T_{i-1}, T_{i-1} + 1, \ldots, T_i\}. \quad (4.140)$$

Solving for x_t yields,

$$x_t = \frac{T_{i+1} - t}{3d} - 1, \quad t \in \{T_{i-1}, T_{i-1} + 1, \ldots, T_i\},$$

which substituted into Eq. (4.138) leaves,

$$\boxed{\begin{aligned}\mathcal{I}_t &\equiv \sqrt{\frac{1}{(3/12)}\left[\left(\frac{T_{i+1} - t}{3d} - 1\right)V_t(T_i) + \left(2 - \frac{T_{i+1} - t}{3d}\right)V_t(T_{i+1})\right]},\\ &t \in \{T_{i-1}, T_{i-1} + 1, \ldots, T_i\}\end{aligned}}$$
$$(4.141)$$

According to this design, the volatility index $\mathcal{I}_t$ in Eq. (4.141) is fed by option prices that have a constant average maturity equal to three months, and builds upon options of the same cycle which are utilized for 6 months.

As an example, set the number of days within a month, $d = 30$. Figure 4.6 depicts the pattern of percentage weight carried by options expiring in 6 months, say at the beginning of September (day number 241), from the beginning of March (day

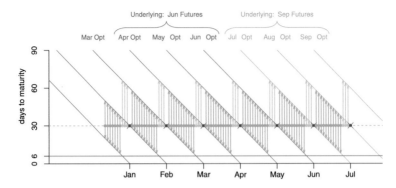

Fig. 4.7 Time-averaging of options of different cycles leading to a constant maturity expected volatility index

number 61): the weight is $1 - x_t$ up to day number 151 and x_t after it. During the first days in March, $\mathcal{I}_t$ mostly reflects options expiring in June, so that the weight carried by options expiring in September is relatively small. As time passes, the options expiring in June carry less weight, with the options expiring in September counting more. At the beginning of June (day number 151), the options expiring in September receive a 100 % weight, with their relative importance decaying until their expiration in September.

The previous examples can be generalized to deal with more elaborate time-weighted averages. For example, the CBOT lists options on 10-Year T-Note futures in a quarterly March, June, September and December cycle plus one or two closest serial months, and each option expires on the last trading day of the month prior to its expiry month. One can construct an index of expected Treasury volatility (30 days, say) with options that have at least a certain expiration in order to reduce the impact of any idiosyncratic behavior of super short-dated options.

For example, when the first month options have 1 week or longer until expiry, the first and second contract months are used. Otherwise, the second and third contract months are used. Except on days when one series of options has exactly 30 days to expiry, a time-weighted average of the expected variances implied by the two option maturities is used to infer the 30-day expected volatility.

Figure 4.7 illustrates how the option maturities used for the index calculation evolve over time. In any given one-month period, the time-averaging goes through a three-part cycle: (i) the green arrows show the first and second months being used to *interpolate* the 30-day point; (ii) the yellow arrows show the second and third months being used to *extrapolate* the 30-day point; and, finally, (iii) the red points show the one day of the month when only one option maturity is required to calculate the index.

4.5.2 Rolling Indexes

A rolling index runs until expiration of the listing cycle to which its constituent options belong. One rationale for such a design may be to provide standardization for volatility trading of each maturity cycle. Accordingly, the index captures the notion of expected government bond or time deposit volatility, *remaining* up to the completion of a given cycle.

As an example, consider a 3-month rolling volatility index (be it government bond or time deposit). At inception, the index tracks the average expected volatility for the next 3 months. It builds upon options expiring in 3 months, and it is the same as the 3 month sandwich index due to the weighting design the latter relies on, as explained in the previous section. However, after 1 month, the rolling index tracks expected volatility to arise within the following 2 months; and after 2 months, the index tracks volatility expected to arise in the following month.

Note that under regularity conditions, the rolling index does not shrink to zero. Consider, for example, the basis point time-deposit volatility index in Eq. (4.72). Under regularity conditions,

$$\lim_{t \to T} \text{TD-VI}^{\text{bp}}(t, T, \Delta) = Z_T(T, T + \Delta)\sqrt{\left\| v_T^z(T, \Delta) \right\|^2}.$$

That is, the rolling index approaches the *instantaneous* basis point volatility. The closer the maturity of the options is within a cycle, the closer the index is to the current realized volatility. The farther the option maturity, the higher the information content the index brings concerning the expectation of future developments in realized volatility.

Appendix C: Appendix on Government Bonds and Time Deposit Markets

C.1 The Equity VIX with Stochastic Interest Rates

The VIX for equity market volatility maintained by the Chicago Board Options Exchange relies on the assumption that interest rates are constant. This appendix relaxes this assumption and shows that, under certain conditions, equity volatility can be priced under the forward probability. Let S_τ be the price of a stock index at time τ. We assume that it satisfies

$$\frac{dS_\tau}{S_\tau} = r_\tau dt + \sigma_\tau \cdot dW_\tau,$$

where W_τ now denotes a vector Brownian motion under the risk-neutral probability, Q, and both the short-term rate r_τ and the instantaneous stock index volatility σ_τ are adapted to W_τ.

Next, let $F_\tau^e(T) = \frac{S_\tau}{P_\tau(T)}$ denote the forward stock index, where $P_\tau(T)$ is the price at τ of a zero coupon bond expiring at some T, and satisfies

$$\frac{dP_\tau(T)}{P_\tau(T)} = r_\tau dt + \sigma_\tau(T) \cdot dW_\tau,$$

and $\sigma_\tau(T)$ is the process of the return volatility of the zero, a vector-valued process adapted to W_τ. Naturally, the forward stock index satisfies

$$\frac{dF_\tau^e(T)}{F_\tau^e(T)} = v_\tau(T) \cdot dW_\tau^{F^T},$$

where $W_\tau^{F^T}$ is a vector of Brownian motions under the forward probability Q_{FT}, and the instantaneous forward stock volatility is $v_\tau(T) \equiv \sigma_\tau - \sigma_\tau(T)$.

By standard arguments, the fair value of the variance strike for forward volatility delivery, $\mathbb{P}^e(t,T)$ say, is

$$\mathbb{P}^e(t,T) = \frac{1}{P_t(T)} \mathbb{E}_t \left(e^{-\int_t^T r_\tau d\tau} \int_t^T \|v_\tau(T)\|^2 d\tau \right).$$

Similar to the arguments relating to the government bond variance swap contract (see Eq. (4.18) in the main text), we have that

$$-\mathbb{E}_t^{Q_{FT}} \left(\ln \frac{F_T^e(T)}{F_t^e(T)} \right) = \frac{1}{2} \mathbb{E}_t^{Q_{FT}} \left(\int_t^T \|v_\tau(T)\|^2 d\tau \right) = \frac{1}{2} \mathbb{P}^e(t,T). \qquad \text{(C.1)}$$

The term on the L.H.S. can be cast in the usual model-free format by spanning the log-contract on the forward through OTM equity options

$$\mathbb{E}_t^{Q_{FT}} \left(\ln \frac{F_T^e(T)}{F_t^e(T)} \right)$$
$$= -\frac{1}{P_t(T)} \left(\int_0^{F_t^e(T)} \text{Put}_t^e(K,T) \frac{1}{K^2} dK + \int_{F_t^e(T)}^\infty \text{Call}_t^e(K,T) \frac{1}{K^2} dK \right), \qquad \text{(C.2)}$$

with obvious notation.

Combining Eq. (C.1) and Eq. (C.2) delivers

$$\mathbb{P}^e(t,T) = \frac{2}{P_t(T)} \left(\int_0^{F_t^e(T)} \text{Put}_t^e(K,T) \frac{1}{K^2} dK + \int_{F_t^e(T)}^\infty \text{Call}_t^e(K,T) \frac{1}{K^2} dK \right),$$

leading to the following index:

$$\boxed{\text{VIX}_r(t,T) \equiv \sqrt{\frac{1}{T-t} \mathbb{P}^e(t,T)}}$$

This expression collapses to that underlying the CBOE calculations when rates are constant. Clearly, VIX indexes calculated at different maturities necessitate different discount rates as the yield curve is not necessarily flat when rates are random, implying that the entire yield curve, $P_t(T)$, needs to be used as an input.

We study the bias in an experiment. We assume the short-term rate is generated by the Vasicek (1977) model, where the short-term rate is a solution

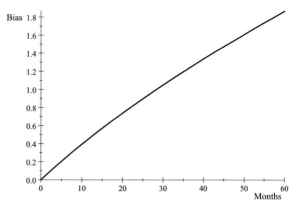

Fig. 4.8 Percentage bias arising while pricing equity volatility under the incorrect assumption that interest rate are constant. Biases are obtained assuming interest rates are generated by the Vasicek (1977) model

to Eq. (4.57) under the risk-neutral probability. Define the yield-to-maturity $y_t(T - t) \equiv -\frac{1}{T-t} \ln P_t^v(T)$, where $P_t^v(T)$ is the price of the zero generated by the Vasicek model.

We assume the current short-term rate is $r_t = 0.01$, implying that the yield-to-maturity for 1 month is $y_t(\frac{1}{12}) = -12 \cdot \ln P_t(\frac{1}{12}) = 0.0109$. Assume we use the yield-to-maturity for 1 month to determine the VIX for maturities larger than 1 month. The percentage bias underlying these calculations is

$$\text{Bias}(t, T) = \frac{\text{VIX}(t, T) - \text{VIX}_r(t, T)}{\text{VIX}_r(t, T)} = \sqrt{e^{-(y_t(T-t) - y_t(\frac{1}{12}))\cdot(T-t)}} - 1.$$

Figure 4.8 depicts this bias calculated with the same parameter values used in the experiments of Appendix B of Chap. 3: $\kappa = 0.3807$, $\bar{r} = 0.072$, and $\sigma = 3.3107 \times 10^{-2}$. In this experiment, we observe mild biases arising while assuming interest rates are constant when in fact they are not.

C.2 Naïve Model-Free Methodology and Bias in Vasicek's Market

THE BIAS. We calculate the bias in Eq. (4.8) assuming that the short-term rate is generated by the Vasicek (1977) model where the short-term rate is the solution to Eq. (4.57). Below, we show that

$$\mathbb{P}(t, T, \mathbb{T}) = \frac{\sigma^2}{\kappa^2} \left((T - t) + \frac{e^{-2\kappa(\mathbb{T}-T)} - e^{-2\kappa(\mathbb{T}-t)} - 4e^{-\kappa(\mathbb{T}-T)} + 4e^{-\kappa(\mathbb{T}-t)}}{2\kappa} \right),$$
$$\tag{C.3}$$

and

$$\text{Bias}(t, T) = -\frac{\sigma^2}{\kappa^2} \left[T - t - \frac{1}{2\kappa} \left((2 - e^{-\kappa(T-t)})^2 - 1 \right) \right]. \tag{C.4}$$

Figure 4.9 depicts $\sqrt{\frac{\mathbb{P}(t,T,\mathbb{T})}{T-t}}$ and $\sqrt{\frac{\mathbb{P}_{\text{VIX}}(t,T,\mathbb{T})}{T-t}}$ in Eq. (4.8), where $\mathbb{P}(t, T, \mathbb{T})$ is obtained through Eq. (C.3) and $\mathbb{P}_{\text{VIX}}(t, T, \mathbb{T})$ through Eq. (4.8) and Eq. (C.4). We

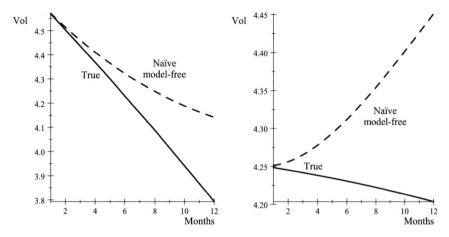

Fig. 4.9 Biases arising while estimating expected volatility with standard model-free methodology. This figure depicts the true expected volatility on a 2-year (*left panel*) and 5-year (*right panel*) zero-coupon bond log-return arising in the Vasicek (1977) market (*solid line*, in percent), along with that estimated by applying a naïve model-free methodology (*dashed line*, in percent), over a number of months. The *right panel* depicts the expected volatility arising while keeping the instantaneous basis point volatility of the short-term rate equal to that in the *left panel*, but with a halved persistence of interest rate shocks

consider a 2-year zero-coupon bond, $\mathbb{T} = 2$, a 5-year zero-coupon bond, $\mathbb{T} = 5$, maturities T ranging from one to 12-months. For the 2-year zero, we use the same parameter values reported in Appendix B of Chap. 3, i.e. $\sigma = 3.3107 \times 10^{-2}$ and $\kappa = 0.3807$. For the 5-year zero, we use the same values but halved persistence, $\kappa = 2 \times 0.3807$.

PROOF OF EQS. (C.3) AND (C.4). Some of the derivations below are taken as given, and will be made available in Appendix C.9 below. In particular, it is easy to show (see Eq. (C.32) below) that the expression for the instantaneous bond return volatility predicted by Vasicek's model is

$$\sigma_{\mathrm{v},\tau}(T) \equiv -\frac{\sigma(1 - e^{-\kappa(T-\tau)})}{\kappa}.$$

Equation (C.3) follows by evaluating the integral

$$\mathbb{P}(t, T, \mathbb{T}) = \int_t^T \sigma_{\mathrm{v},\tau}^2(T) d\tau.$$

Moreover, it can be shown (see Eq. (C.33) below) that

$$\mathbb{E}_t^{Q_{FT}}(r_\tau) = \bar{r} + e^{-\kappa(\tau-t)}(r_t - \bar{r})$$

$$- \frac{\sigma^2}{2\kappa^2}\left[2(1 - e^{-\kappa(\tau-t)}) + e^{-\kappa(T-t)}(e^{-\kappa(\tau-t)} - e^{\kappa(\tau-t)})\right].$$

For example, in the special case where $\tau = T \equiv \theta$ (say), we get

$$\mathbb{E}_t^{Q_{F^\theta}}(r_\theta) = \bar{r} + e^{-\kappa(\theta-t)}(r_t - \bar{r}) - \frac{\sigma^2}{2\kappa^2}\left(1 - e^{-\kappa(\theta-t)}\right)^2,$$

so that

$$\mathbb{E}_t^{Q_{F^T}}(r_\tau) - \mathbb{E}_t^{Q_{F^\tau}}(r_\tau) = -\frac{\sigma^2}{2\kappa^2}\left[\left(1 - e^{-2\kappa(\tau-t)}\right)\left(1 - e^{-\kappa(T-t)}e^{\kappa(\tau-t)}\right)\right]. \quad \text{(C.5)}$$

By substituting Eq. (C.5) into Eq. (4.9), and integrating, we are left with the expression in Eq. (C.4).   □

C.3 Marking to Market

The value as of time τ of the percentage variance swap payoff in Eq. (4.15) is

$$\mathbb{E}_\tau\left(e^{-\int_\tau^T r_u du}\left(V(t, T, \mathbb{T}) - \mathbb{P}(t, T, \mathbb{T})\right)\right)$$

$$= \mathbb{E}_\tau\left(e^{-\int_\tau^T r_u du}\left(V(t, \tau, \mathbb{T}) + V(\tau, T, \mathbb{T}) - \mathbb{P}(t, T, \mathbb{T})\right)\right)$$

$$= V(t, \tau, \mathbb{T})P_\tau(T) + \mathbb{E}_\tau\left(e^{-\int_\tau^T r_u du}V(\tau, T, \mathbb{T})\right) - \mathbb{P}(t, T, \mathbb{T})P_\tau(T).$$

By Eq. (4.17), we have that

$$\mathbb{E}_\tau\left(e^{-\int_\tau^T r_u du}V(\tau, T, \mathbb{T})\right) = \mathbb{P}(\tau, T, \mathbb{T})P_\tau(T),$$

so that, by rearranging terms, we obtain the expression for M-Var$_\tau(T, \mathbb{T})$ in Eq. (4.29). The marking to market updates regarding the basis point variance swap follow by nearly identical arguments applied to the payoff in Eq. (4.22).

C.4 Replication of Variance Swaps

PERCENTAGE. We have:

$$\ln\frac{F_T(T, \mathbb{T})}{F_t(T, \mathbb{T})}$$

$$= \frac{F_T(T, \mathbb{T}) - F_t(T, \mathbb{T})}{F_t(T, \mathbb{T})}$$

$$- \left(\int_0^{F_t(T,\mathbb{T})}\left(K - F_T(T, \mathbb{T})\right)^+ \frac{1}{K^2}dK + \int_{F_t(T,\mathbb{T})}^\infty\left(F_T(T, \mathbb{T}) - K\right)^+ \frac{1}{K^2}dK\right).$$

$$\text{(C.6)}$$

By Itô's lemma,

$$V(t, T, \mathbb{T}) = 2 \int_t^T \frac{dF_s(T, \mathbb{T})}{F_s(T, \mathbb{T})} - 2 \ln \frac{F_T(T, \mathbb{T})}{F_t(T, \mathbb{T})}, \qquad (C.7)$$

where, by Eq. (C.6), the second term on the R.H.S. of Eq. (C.7) is the time T pay-off delivered by a portfolio set up at t, which is two times: (a) short $1/F_t(T, \mathbb{T})$ units a forward struck at $F_t(T, \mathbb{T})$, and (b) long a continuum of out-of-the-money options with weights $K^{-2}dK$. By Eq. (4.20), the cost of this position at time t is $P(t, T, \mathbb{T}) P_t(T)$, as indicated by row (ii) of Table 4.1 in the main text. We borrow $\mathbb{P}(t, T, \mathbb{T}) P_t(T)$ at time t, which we repay back at time T, as in row (iii) of Table 4.1.

Finally, we derive the self-financing portfolio in row (i) of Table 4.1. The value of this portfolio must be zero at time t, and replicate the first term on the R.S.H. of Eq. (C.7). Consider a self-financed strategy investing in (a) the forward, and (b) a money market account that has value M_τ that grows over time according to: $dM_\tau = r_\tau M_\tau d\tau$. The total value of the strategy is

$$\upsilon_\tau = \theta_\tau F_\tau(T, \mathbb{T}) + \psi_\tau M_\tau,$$

where θ_τ are the units in the forward, and ψ_τ are the units in the money market account. Let

$$\hat{\theta}_\tau F_\tau(T, \mathbb{T}) = 1, \qquad \hat{\psi}_\tau M_\tau = \int_t^\tau \frac{dF_s(T, \mathbb{T})}{F_s(T, \mathbb{T})} - 1, \qquad (C.8)$$

so that $\hat{\upsilon}_\tau = \hat{\theta}_\tau F_\tau + \hat{\psi}_\tau M_\tau$ satisfies

$$\hat{\upsilon}_\tau = \int_t^\tau \frac{dF_s(T, \mathbb{T})}{F_s(T, \mathbb{T})}, \qquad (C.9)$$

with

$$\hat{\upsilon}_t = 0, \quad \text{and} \quad \hat{\upsilon}_T = \int_t^T \frac{dF_s(T, \mathbb{T})}{F_s(T, \mathbb{T})}.$$

Therefore, by going long two portfolios $(\hat{\theta}_\tau, \hat{\psi}_\tau)$, the first term on the R.H.S. and by the previous results, the whole of the R.H.S. of Eq. (C.7) can be replicated, provided $(\hat{\theta}_\tau, \hat{\psi}_\tau)$ is self-financed. To show that $(\hat{\theta}_\tau, \hat{\psi}_\tau)$ is self-financed, we have

$$\begin{aligned}
d\hat{\upsilon}_\tau &= \hat{\theta}_\tau F_\tau(T, \mathbb{T}) \frac{dF_\tau(T, \mathbb{T})}{F_\tau(T, \mathbb{T})} + \hat{\psi}_\tau M_\tau \frac{dM_\tau}{M_\tau} \\
&= \left(\frac{dF_\tau(T, \mathbb{T})}{F_\tau(T, \mathbb{T})} - r_\tau d\tau \right) + \left(\int_t^\tau \frac{dF_\tau(T, \mathbb{T})}{F_\tau(T, \mathbb{T})} \right) r_\tau d\tau \\
&= \left(\frac{dF_\tau(T, \mathbb{T})}{F_\tau(T, \mathbb{T})} - r_\tau d\tau \right) + r_\tau \hat{\upsilon}_\tau d\tau \\
&= \hat{\theta}_\tau F_\tau(T, \mathbb{T}) \left(\frac{dF_\tau(T, \mathbb{T})}{F_\tau(T, \mathbb{T})} - r_\tau d\tau \right) + r_\tau \hat{\upsilon}_\tau d\tau, \qquad (C.10)
\end{aligned}$$

where the second line follows by Eq. (C.8), the third by Eq. (C.9), and the fourth, again, by Eq. (C.8). The dynamics of $\hat{v}_\tau$ in Eq. (C.10) are those of a self-financed strategy.

BASIS POINT. We have

$$
\begin{aligned}
F_T^2(T, \mathbb{T}) &- F_t^2(T, \mathbb{T}) \\
&= 2F_t(T, \mathbb{T})\big(F_T(T, \mathbb{T}) - F_t(T, \mathbb{T})\big) \\
&\quad + 2\bigg(\int_0^{F_t(T,\mathbb{T})} \big(K - F_T(T, \mathbb{T})\big)^+ dK + \int_{F_t(T,\mathbb{T})}^\infty \big(F_T(T, \mathbb{T}) - K\big)^+ dK\bigg).
\end{aligned}
$$
(C.11)

By Itô's lemma,

$$
V^{\mathrm{bp}}(t, T, \mathbb{T}) = -2\int_t^T F_s(T, \mathbb{T})dF_s(T, \mathbb{T}) + \big[F_T^2(T, \mathbb{T}) - F_t^2(T, \mathbb{T})\big]. \quad \text{(C.12)}
$$

By Eq. (C.11), the second term on the R.H.S. of Eq. (C.12) is the payoff at T of a portfolio set up at t, which is: (a) long $2F_t(T, \mathbb{T})$ units a government bond forward struck at $F_t(T, \mathbb{T})$, and (b) long a continuum of out-of-the-money options with weights $2dK$. It is the static position in row (ii) of Table 4.2 in the main text. By Eq. (4.27), its cost is $\mathbb{P}^{\mathrm{bp}}(t, T, \mathbb{T})P_t(T)$, which we borrow at t, to repay it back at T, as in row (iii) of Table 4.2. To obtain the self-financed portfolio to be shorted in row (i) of Table 4.2, we proceed as in row (i) of Table 4.1, but with the portfolio,

$$
\hat{\psi}_\tau M_\tau = \int_t^\tau 2F_s(T, \mathbb{T})dF_s(T, \mathbb{T}) - 1,
$$

replacing that in Eq. (C.8).

C.5 Estimates Based on Forward Price Approximations

Let K_0 be the first strike below $F_t(T, \mathbb{T})$, as defined in the main text. We expand $\ln F_T(T, \mathbb{T})$ around K_0, as follows:

$$
\begin{aligned}
\ln &\frac{F_T(T, \mathbb{T})}{F_t(T, \mathbb{T})} \\
&= \ln \frac{K_0}{F_t(T, \mathbb{T})} + \frac{F_T(T, \mathbb{T}) - K_0}{K_0} \\
&\quad - \bigg(\int_0^{K_0} \big(K - F_T(T, \mathbb{T})\big)^+ \frac{1}{K^2} dK + \int_{K_0}^\infty \big(F_T(T, \mathbb{T}) - K\big)^+ \frac{1}{K^2} dK\bigg),
\end{aligned}
$$
(C.13)

so that the fair value of the government bond variance swap is

$$
\begin{aligned}
\mathbb{P}(t, T, \mathbb{T}) &= -2\mathbb{E}_t^{Q_{FT}}\left(\ln\frac{F_T(T, \mathbb{T})}{F_t(T, \mathbb{T})}\right) \\
&= -2\left(\ln\frac{K_0}{F_t(T, \mathbb{T})} + \frac{F_t(T, \mathbb{T}) - K_0}{K_0}\right) \\
&\quad + \frac{2}{P_t(T)}\left(\int_0^{K_0}\mathrm{Put}_t(K, T, \mathbb{T})\frac{1}{K^2}dK + \int_{K_0}^{\infty}\mathrm{Call}_t(K, T, \mathbb{T})\frac{1}{K^2}dK\right).
\end{aligned}
$$

(C.14)

Consider a second order expansion of the function $-\ln\frac{F_t}{K_0}$ about K_0,

$$
-\ln\frac{F_t}{K_0} \approx -\frac{1}{K_0}(F_t - K_0) + \frac{1}{2}\frac{1}{K_0^2}(F_t - K_0)^2.
$$

Substituting this approximation into the first term on the R.H.S. of Eq. (C.14) leaves the expression of $\mathbb{P}_o(t, T, \mathbb{T})$ in Eq. (4.30) of the main text.

Next, consider the correction applying to the basis point index. Similarly as for Eq. (C.13), expand $F_T^2(T, \mathbb{T})$ around K_0,

$$
\begin{aligned}
F_T^2(T, \mathbb{T}) &- F_t^2(T, \mathbb{T}) \\
&= K_0^2 - F_t^2(T, \mathbb{T}) + 2K_0\big(F_T(T, \mathbb{T}) - K_0\big) \\
&\quad + 2\left[\int_0^{K_0}\big(K - F_T(T, \mathbb{T})\big)^+ dK + \int_{K_0}^{\infty}\big(F_T(T, \mathbb{T}) - K\big)^+ dK\right],
\end{aligned}
$$

so that the fair value of the basis point government bond variance swap is

$$
\begin{aligned}
\mathbb{P}^{bp}(t, T, \mathbb{T}) &= K_0^2 - F_t^2(T, \mathbb{T}) + 2K_0\big(F_t(T, \mathbb{T}) - K_0\big) \\
&\quad + \frac{2}{P_t(T)}\left(\int_0^{K_0}\mathrm{Put}_t(K, T, \mathbb{T})dK + \int_{K_0}^{\infty}\mathrm{Call}_t(K, T, \mathbb{T})dK\right).
\end{aligned}
$$

(C.15)

A second order expansion of the function F_t^2 about K_0 yields

$$
F_t^2 - K_0^2 \approx 2K_0(F_t - K_0) + (F_t - K_0)^2.
$$

Substituting this approximation into Eq. (C.15) yields the approximation in Eq. (4.31).

C.6 Certainty Equivalence, and Existence of Basis Point Yield Volatility

We show that our model-free measures of yields exist both in the case of the regular indexes of Sect. 4.2, and the sandwich combinations of Sect. 4.5. Finally, we provide details regarding the certainty equivalent bond prices introduced in Sect. 4.2.8.

REGULAR INDEXES. We show that $y_B(t, T, \mathbb{T})$ in Eq. (4.35) exists and is positive. We have

$$\text{GB-VI}^{\text{bp}}(t, T, \mathbb{T})$$

$$\equiv \sqrt{\frac{1}{T-t} \frac{1}{P_t(T)} \mathbb{E}_t \left(e^{-\int_t^T r_\tau d\tau} \int_t^T F_\tau^2(T, \mathbb{T}) \left\| v_\tau(T, \mathbb{T}) \right\|^2 d\tau \right)}$$

$$\leq \sqrt{\frac{1}{T-t} \frac{1}{P_t(T)} \mathbb{E}_t \left(e^{-\int_t^T r_\tau d\tau} \int_t^T \left(\sup_{\tau \in (t,T)} F_\tau^2(T, \mathbb{T}) \right) \left\| v_\tau(T, \mathbb{T}) \right\|^2 d\tau \right)}$$

$$\leq \sqrt{\frac{1}{T-t} \frac{1}{P_t(T)} \mathbb{E}_t \left(e^{-\int_t^T r_\tau d\tau} \int_t^T \left\| v_\tau(T, \mathbb{T}) \right\|^2 \cdot \hat{P}^2(0) \right) d\tau}$$

$$= \hat{P}(0) \cdot \text{GB-VI}(t, T, \mathbb{T}),$$

where $\hat{P}$ is the function defined in Eq. (4.36), and the third line follows as $F_\tau(T, \mathbb{T}) = \mathbb{E}_\tau^{Q_{FT}} (B_T(\mathbb{T})) \leq \hat{P}(0)$, for all $\tau \in [t, T]$, with the last inequality holding as the price of a coupon-bearing bond is bounded by the (undiscounted) sum of the coupons plus principal, i.e., by $\hat{P}(0)$. That is, using Eq. (4.34),

$$\frac{\text{GB-VI}^{\text{bp}}(t, T, \mathbb{T})}{\text{GB-VI}(t, T, \mathbb{T})} = \mathcal{B}(t, T, \mathbb{T}) \leq \hat{P}(0).$$

Because $\hat{P}(y)$ is monotonically decreasing in y with $\lim_{y \to \infty} \hat{P}(y) = 0$, the previous equality shows that there exists a value $y_B(t, T, \mathbb{T}) \geq 0$ such that Eq. (4.35) holds true.

SANDWICH COMBINATIONS. Next, we show that the index in Eq. (4.139) is well-defined. We have

$$V_t^{\text{perc}}(T_i) = \frac{1}{T_i - t} \frac{1}{P_t(T_i)} \mathbb{E}_t \left(e^{-\int_t^{T_i} r_\tau d\tau} \int_t^{T_i} F_\tau^2(T_i, \mathbb{T}) d\tau \right),$$

$$V_t^{\text{bp}}(T_i) = \frac{1}{T_i - t} \frac{1}{P_t(T_i)} \mathbb{E}_t \left(e^{-\int_t^{T_i} r_\tau d\tau} \int_t^{T_i} F_\tau^2(T_i, \mathbb{T}) \left\| v_\tau(T_i, \mathbb{T}) \right\|^2 d\tau \right),$$

so that

$$x_t V_t^{\text{bp}}(T_i) + (1 - x_t) V_t^{\text{bp}}(T_{i+1}) \leq \left(x_t V_t^{\text{perc}}(T_i) + (1 - x_t) V_t^{\text{perc}}(T_{i+1}) \right) \hat{P}^2(0).$$

Therefore, $\frac{\mathcal{I}_t^{bp}}{\mathcal{I}_t^{perc}} \leq \hat{P}(0)$, and it follows that the index in Eq. (4.139) exists and is positive by the same arguments used to demonstrate the existence of a regular index.

YIELD VOLATILITY IN THE POST-ISSUANCE CASE. We provide details regarding the case in which the maturity of the forward exceeds that of the first date the bond pays off a coupon after it is issued. Define

$$x_B(t, T, \mathbb{T}) : \mathcal{B}(t, T, \mathbb{T}) = \frac{\text{GB-VI}^{bp}(t, T, \mathbb{T})}{\text{GB-VI}(t, T, \mathbb{T})} = \hat{P}_T\big(x_B(t, T, \mathbb{T})\big), \qquad (C.16)$$

where GB-VI$(t, T, \mathbb{T})$ and GB-VI$^{bp}(t, T, \mathbb{T})$ are as in Eq. (4.21) and Eq. (4.28), respectively, and, denoting by N_T the number of payments that still have to take place,

$$\hat{P}_T(x) = \sum_{i=1}^{N_T} \frac{C_{t_i}}{n}(1+x)^{-\frac{t_i - T}{365}} + 100(1+x)^{-\frac{t_{N_T} - T}{365}}, \qquad (C.17)$$

and the sequence of t_i tracks dates at which the coupon payments are still to take place. The modified duration of the guaranteed price is

$$\hat{D}_\mathcal{B}(t, T, \mathbb{T}) \equiv \frac{1}{1 + x_B(t, T, \mathbb{T})}\left(\sum_{i=1}^{N_T} \omega_{t_i}\frac{t_i - T}{365} + \hat{\omega}_{t_{N_T}}\frac{t_{N_T} - T}{365}\right)$$

$$\omega_{t_i} \equiv \frac{\frac{C_{t_i}}{n}/(1+x_B(t, T, \mathbb{T}))^{\frac{t_i - T}{365}}}{\mathcal{B}(t, T, \mathbb{T})}, \qquad \hat{\omega}_{t_{N_T}} \equiv \frac{100/(1+x_B(t, T, \mathbb{T}))^{\frac{t_{N_T} - T}{365}}}{\mathcal{B}(t, T, \mathbb{T})}$$

$$(C.18)$$

where $x_B(t, T, \mathbb{T})$ is defined as in Eq. (C.16). Accordingly, our model-free duration-based yield volatility is still as in Eq. (4.38), with $\hat{D}_\mathcal{B}(t, T, \mathbb{T})$ replacing $D_\mathcal{B}(t, T, \mathbb{T})$, so that, using the definitions of $\hat{P}_T$ in Eq. (C.17) and that of $\hat{D}_\mathcal{B}$ in Eqs. (C.18):

GB-VI$^{bp}_{Yd}(t, T, \mathbb{T})$

$$= \frac{100 \times (1 + \hat{P}_T^{-1}[\frac{\text{GB-VI}^{bp}(t,T,\mathbb{T})}{\text{GB-VI}(t,T,\mathbb{T})}]) \times \text{GB-VI}^{bp}(t, T, \mathbb{T})}{\sum_{i=1}^{N_T}\frac{C_{t_i}}{n}(1 + \hat{P}_T^{-1}[\frac{\text{GB-VI}^{bp}(t,T,\mathbb{T})}{\text{GB-VI}(t,T,\mathbb{T})}])^{-\frac{t_i - T}{365}}\frac{t_i - T}{365} + 100(1 + \hat{P}_T^{-1}[\frac{\text{GB-VI}^{bp}(t,T,\mathbb{T})}{\text{GB-VI}(t,T,\mathbb{T})}])^{-\frac{t_{N_T} - T}{365}}\frac{t_{N_T} - T}{365}}$$

$$(C.19)$$

where $\hat{P}_T^{-1}$ denotes the inverse function of $\hat{P}_T$ in Eq. (C.17).

Similarly, the yield-based yield volatility index in Eq. (4.40) of the main text can be replaced with

$$\text{GB-VI}^{bp}_Y(t, T, \mathbb{T}) = 100 \times \hat{P}_T^{-1}\left[\frac{\text{GB-VI}^{bp}(t, T, \mathbb{T})}{\text{GB-VI}(t, T, \mathbb{T})}\right] \times \text{GB-VI}(t, T, \mathbb{T}), \quad (C.20)$$

where $\hat{P}_T^{-1}$ is the inverse function of $\hat{P}_T$ in Eq. (C.17).

Finally, we can generalize Eq. (4.110) and Eq. (4.111) in Sect. 4.4 of the main text, as follows:

$\text{GB-VI}_{\text{Yd}}^{\text{bp}}(t, T, S, \mathbb{T})$

$$= \frac{100 \times (1 + \hat{P}_T^{-1}[\frac{\text{GB-VI}^{\text{bp}}(t,T,S,\mathbb{T})}{\text{GB-VI}(t,T,S,\mathbb{T})}])\text{GB-VI}^{\text{bp}}(t, T, S, \mathbb{T})}{\sum_{i=1}^{N_T} \frac{C_{t_i}}{n}(1 + \hat{P}_T^{-1}[\frac{\text{GB-VI}^{\text{bp}}(t,T,S,\mathbb{T})}{\text{GB-VI}(t,T,S,\mathbb{T})}])^{-\frac{t_i-T}{365}}\frac{t_i-T}{365} + 100(1 + \hat{P}_T^{-1}[\frac{\text{GB-VI}^{\text{bp}}(t,T,S,\mathbb{T})}{\text{GB-VI}(t,T,S,\mathbb{T})}])^{-\frac{t_{N_T}-T}{365}}\frac{t_{N_T}-T}{365}},$$

$$(C.21)$$

and

$$\text{GB-VI}_Y^{\text{bp}}(t, T, S, \mathbb{T}) = 100 \times \hat{P}_T^{-1}\left[\frac{\text{GB-VI}^{\text{bp}}(t, T, S, \mathbb{T})}{\text{GB-VI}(t, T, S, \mathbb{T})}\right] \times \text{GB-VI}(t, T, S, \mathbb{T}),$$

$$(C.22)$$

where $\text{GB-VI}(t, T, S, \mathbb{T})$ and $\text{GB-VI}^{\text{bp}}(t, T, S, \mathbb{T})$ are as in Eq. (4.103) and Eq. (4.108), respectively.

ADDITIONAL DETAILS REGARDING CERTAINTY EQUIVALENT PRICES. We prove that Eq. (4.41) holds true. By Eq. (4.34), we have that $\mathcal{B}(t, T, \mathbb{T})$ satisfies

$$\mathcal{B}^2(t, T, \mathbb{T}) = \frac{\int_t^T \mathbb{E}_t^{Q_{FT}}(F_\tau^2(T, \mathbb{T})\|v_\tau(T, \mathbb{T})\|^2)d\tau}{\int_t^T \mathbb{E}_t^{Q_{FT}}(\|v_\tau(T, \mathbb{T})\|^2)d\tau}$$

$$= \frac{\int_t^T \mathbb{E}_t^{Q_{FT}}(\|v_\tau(T, \mathbb{T})\|^2)\mathbb{E}_t^{Q_{v^\tau}}(F_\tau^2(T, \mathbb{T}))d\tau}{\int_t^T \mathbb{E}_t^{Q_{FT}}(\|v_\tau(T, \mathbb{T})\|^2)d\tau},$$

where $\mathbb{E}_t^{Q_{v^\tau}}$ is the conditional expectation taken under the realized variance probability defined through the Radon–Nikodym derivative in Eq. (4.42). This is Eq. (4.41) with weighting function

$$\omega_\tau \equiv \frac{\mathbb{E}_t^{Q_{FT}}(\|v_\tau(T, \mathbb{T})\|^2)}{\int_t^T \mathbb{E}_t^{Q_{FT}}(\|v_\tau(T, \mathbb{T})\|^2)d\tau}.$$

$$(C.23)$$

C.7 Illustrations with a Stochastic Volatility Model

SOLUTION OF THE STOCHASTIC VOLATILITY MODEL, EQ. (4.45). By standard arguments (see, e.g., Mele 2014, Chap. 12), the pricing function in Eq. (4.45) holds with

$$P_\tau(r_\tau, v_\tau^2, T) = e^{A_T(\tau) - (T-\tau)r_\tau + C_T(\tau)v_\tau^2},$$

$$(C.24)$$

where

$$A_T(\tau) = \int_\tau^T (s - T)\theta_s ds,$$

$$(C.25)$$

and $C_T(\cdot)$ is the solution to Eq. (4.46). Matching the instantaneous forward rate predicted by the model at t, say $f_t(r_t, v_t^2, T)$, to the hypothetically observed instantaneous forward rate at the same time t, $f_\$(t, T)$, leaves

$$f_\$(t, T) = f_t\big(r_t, v_t^2, T\big) \equiv -\frac{\partial \ln P_t(r_t, v_t^2, T)}{\partial T} = \int_t^T \theta_\tau d\tau + r_t - \frac{\partial C_T(t)}{\partial T} v_t^2,$$

(C.26)

where the last equality follows by differentiating the price function in Eq. (C.24), and the expression for $A_T(\tau)$ in Eq. (C.25). Differentiating Eq. (C.26) with respect to T leaves $\frac{\partial f_\$(t,T)}{\partial T} = \theta_T - \frac{\partial^2 C_T(t)}{\partial T^2} v_t^2$ and Eq. (4.44) of the main text. Finally, substituting Eq. (4.44) into Eq. (C.25) leaves the expression for the integral in the exponent of the price function in Eq. (4.45) of the main text.

It is easy to verify that θ_τ in Eq. (4.44) is indeed the infinite-dimensional parameter we are searching for. Substitute Eq. (4.44) into Eq. (C.26), and readily check that the result holds as an identity. Moreover, let us evaluate Eq. (4.45) when $\tau = t$,

$$P_t\big(r_t, v_t^2, T\big) = e^{\ell_T(r_t, v_t^2, (\theta_\tau)_{\tau \in [t,T]}, T)},$$

where,

$$\ell_T\big(r_t, v_t^2, (\theta_\tau)_{\tau \in [t,T]}, T\big) \equiv \int_t^T (\tau - T)\theta_\tau d\tau - (T - t)r_t + C_T(t)v_t^2$$

$$= -\int_t^T f_\$(t, \tau)d\tau,$$

and the last equality follows after integrating Eq. (C.26), establishing that Eq. (4.47) holds true.

DYNAMICS OF THE FORWARD IN THE STOCHASTIC VOLATILITY MODEL, EQ. (4.48). We prove that Eq. (4.48) holds true. By Itô's lemma, the price of a zero coupon bond expiring at S predicted by the model satisfies

$$\begin{cases} \dfrac{dP_\tau(S)}{P_\tau(S)} = r_\tau d\tau + v_\tau\big(-(S - \tau)dW_{1,\tau} + \xi C_S(\tau)dW_{2,\tau}\big) \\ dv_\tau^2 = \xi v_\tau dW_{2,\tau} \end{cases}$$

By standard arguments, we can define two Brownian motions under the forward probability Q_{FT}, $dW_{1,\tau}^{F^T} = dW_{1,\tau} + v_\tau(T - \tau)d\tau$ and $dW_{2,\tau}^{F^T} = dW_{2,\tau} - \xi C_T(\tau)v_\tau d\tau$, and use Itô's lemma to conclude that the price of the forward on a zero coupon bond expiring at time $\mathbb{T}$ is the solution to Eq. (4.48) in the main text.

EXTENSION TO MEAN-REVERTING STOCHASTIC VOLATILITY. Consider the following extension of the model in Eqs. (4.43), one for which the instantaneous basis point variance of the short-term rate is mean-reverting under the risk-neutral probability, similar to the Heston (1993) model for equity returns:

$$\begin{cases} dr_\tau = \theta_\tau d\tau + v_\tau dW_{1,\tau} \\ dv_\tau^2 = k(m - v_\tau^2)d\tau + \xi v_\tau\big(\rho dW_{1,\tau} + \sqrt{1 - \rho^2}dW_{2,\tau}\big) \end{cases}$$

where k is the speed of adjustment of v_τ^2 towards its long term mean, m, and ρ is a constant correlation between the conditional changes, dr_τ and dv_τ^2.

It is straightforward to show that the price of a zero-coupon bond is given by the following expressions generalizing Eq. (4.45) in the main text:

$$P_\tau\left(r_\tau, v_\tau^2, T\right) = e^{\int_\tau^T (s-T)\theta_s ds + km \int_\tau^T C_T(s)ds - (T-\tau)r_\tau + C_T(\tau)v_\tau^2},$$

where $C_T(\cdot)$ is the solution to the Riccati equation

$$\dot{C}_T(\tau) = \left(k + \rho\xi(T-\tau)\right)C_T(\tau) - \left(\frac{1}{2}(T-\tau)^2 + \frac{1}{2}\xi^2 C_T^2(\tau)\right), \quad C_T(T) = 0,$$

and the infinite-dimensional parameter satisfies

$$\theta_\tau = \frac{\partial f_\$(t,\tau)}{\partial\tau} + km \int_\tau^T \frac{\partial^2 C_\tau(u)}{\partial\tau^2} du + \frac{\partial^2 C_\tau(t)}{\partial\tau^2} v_t^2.$$

DETAILS REGARDING REALIZED VARIANCE PROBABILITY. We show that Eq. (4.50) holds true. We need to show that Eq. (4.41) collapses to Eq. (4.50) when the forward price solves Eqs. (4.48). By Eq. (4.49), and the second of Eqs. (4.48), we have that, for all τ,

$$\mathbb{E}_t^{Q_{FT}}\left(\|v_\tau(T,\mathbb{T})\|^2\right) = \bar{\phi}_\tau(T,\mathbb{T}) \cdot v_t^2, \tag{C.27}$$

where $\bar{\phi}_\tau(T,\mathbb{T})$ is as in the main text, so that Eq. (C.23) collapses to Eq. (4.50), as claimed in the main text.

We are left to prove that under Q_{v^τ}, the forward price variance satisfies Eq. (4.51). Define the following density process, relying on the Radon–Nikodym derivative in Eq. (4.42):

$$\begin{aligned}
\rho(s;T) &\equiv \frac{\mathbb{E}_s^{Q_{FT}}\left(\|v_\tau(T,\mathbb{T})\|^2\right)}{\mathbb{E}_t^{Q_{FT}}\left(\|v_\tau(T,\mathbb{T})\|^2\right)} \\
&= \frac{\bar{\phi}_\tau(s,T,\mathbb{T})}{\bar{\phi}_\tau(t,T,\mathbb{T})} \cdot \frac{v_s^2}{v_t^2} \\
&= e^{-\xi^2 \int_t^s C_T(u)du} \frac{v_s^2}{v_t^2} \\
&= e^{-\frac{1}{2}\int_t^s (\frac{\xi}{v_u})^2 du - \int_t^s (-\frac{\xi}{v_u})dW_{2,u}^{FT}}, \quad s \in [t,\tau],
\end{aligned}$$

so that

$$\frac{d\rho(s;T)}{\rho(s;T)} = -\left(-\frac{\xi}{v_s}\right)dW_{2,s}^{FT}, \quad s \in [t,\tau].$$

By Girsanov's theorem, we have that $dW_{2,s}^{v^\tau} = dW_{2,s}^{FT} - \frac{\xi}{v_s}ds$ is a Brownian motion under Q_{v^τ}, and Eq. (4.51) follows.

C.8 The Future Price in Vasicek's Model

We derive the second equality of Eq. (4.58). By the first equality of Eq. (4.58), and the expression of the coupon-bearing bond price predicted by Vasicek's model, Eq. (4.63), we have:

$$\tilde{F}_t(r_t; T, \mathbb{T}) = \sum_{i=i_T}^{N} \bar{C}_i \cdot \mathbb{E}_t\left(P_T(r_T, T_i)\right) \equiv \sum_{i=i_T}^{N} \bar{C}_i \cdot \tilde{f}_t(r_t; T_i),$$

where the index i_T and the coupon series $\bar{C}_i$ are defined as in the main text and, by Eq. (4.64):

$$\tilde{f}_t(r_t; T_i) \equiv \mathbb{E}_t\left(e^{a_T(T_i)-b_T(T_i)\cdot r_T}\right).$$

Note that conditionally upon the information set as of time t, the short-term rate is normally distributed, with expectation $\mathbb{E}_t(r_T)$ and variance $\mathbb{V}_t(r_T)$ given by:

$$\mathbb{E}_t(r_T) = \bar{r} + e^{-\kappa(T-t)}(r_t - \bar{r}), \qquad \mathbb{V}_t(r_T) = \frac{\sigma^2}{2\kappa}\left(1 - e^{-2\kappa(T-t)}\right), \qquad \text{(C.28)}$$

so that

$$\tilde{f}_t(r_t; T_i) = e^{a_T(T_i)-b_T(T_i)\cdot \mathbb{E}_t(r_T)+\frac{1}{2}b_T^2(T_i)\cdot \mathbb{V}_t(r_T)} = e^{a_t^F(T,T_i)-b_t^F(T,T_i)r_t},$$

where $a_t^F(T, T_i)$ and $b_t^F(T, T_i)$ are defined in Eq. (4.59), and the second equality follows by utilizing the expressions for $\mathbb{E}_t(r_T)$ and $\mathbb{V}_t(r_T)$ in Eqs. (C.28).

C.9 Future and Forward LIBOR Options in Vasicek's Model

FUTURE LIBOR. We derive the second equality of Eq. (4.81). By the first equality of Eq. (4.81), and the expression of the zero-coupon bond price predicted by Vasicek's model, Eq. (4.64), we have

$$\tilde{z}_t(r_t; T, \Delta) = \mathbb{E}_t\left(l_T(r_T, \Delta)\right) \equiv \frac{1}{\Delta}\mathbb{E}_t\left(\frac{1}{P_T(r_T; T+\Delta)} - 1\right)$$

$$= \frac{1}{\Delta}\left(\mathbb{E}_t\left(e^{-a_T(T+\Delta)+b_T(T+\Delta)\cdot r_T}\right) - 1\right).$$

We know that conditionally upon the information set as of time t, the short-term rate is normally distributed, with expectation $\mathbb{E}_t(r_T)$ and variance $\mathbb{V}_t(r_T)$ given by Eqs. (C.28), so that

$$\tilde{z}_t(r_t; T, \Delta) = \frac{1}{\Delta}\left(e^{-a_T(T+\Delta)+b_T(T+\Delta)\cdot \mathbb{E}_t(r_T)+\frac{1}{2}b_T^2(T+\Delta)\cdot \mathbb{V}_t(r_T)} - 1\right)$$

$$= \frac{1}{\Delta}\left(e^{a_t^Z(T,\Delta)+b_t^Z(T,\Delta)r_t} - 1\right),$$

where $a_t^Z(T, \Delta)$ and $b_t^Z(T, \Delta)$ are defined in Eq. (4.82), and the second equality follows by utilizing the expressions for $\mathbb{E}_t(r_T)$ and $\mathbb{V}_t(r_T)$ in Eqs. (C.28).

FORWARD LIBOR OPTIONS. We prove Eq. (4.85). At time T, the forward price predicted by Vasicek's model collapses to:

$$Z_T(r_T; T, T + \Delta) = 100 \times \left(1 - l_T(r_T; \Delta)\right)$$

$$= 100 \times \left(1 - \frac{1}{\Delta}\left(e^{a_T^Z(T,\Delta)+b_T^Z(T,\Delta)r_T} - 1\right)\right)$$

$$= 100 \times \frac{1}{\Delta}\left(1 + \Delta - \tilde{B}(r_T)\right),$$

where

$$\tilde{B}(r_T) \equiv e^{-a_\Delta + b_\Delta r_T}, \tag{C.29}$$

and $a_\Delta \equiv a_T(T + \Delta)$ and $b_\Delta \equiv b_T(T + \Delta)$, due to the expressions of $a_t^Z(T, \Delta)$ and $b_t^Z(T, \Delta)$ in Eqs. (4.82), and where $a_T(T + \Delta)$ and $b_T(T + \Delta)$ are as in Eqs. (4.64). Therefore, the price of the call on the forward can be expressed as

$$\text{Call}_t^z(K, T, \Delta) = \mathbb{E}_t\left(e^{-\int_t^T r_\tau d\tau}\left(Z_T(r_T; T, T + \Delta) - K\right)^+\right)$$

$$= 100 \times \frac{1}{\Delta}\mathbb{E}_t\left(e^{-\int_t^T r_\tau d\tau}\left(K^o(K) - \tilde{B}(r_T)\right)^+\right)$$

$$= 100 \times P_t(T)\frac{1}{\Delta}\mathbb{E}_t^{Q_{F^T}}\left(K^o(K) - \tilde{B}(r_T)\right)^+, \tag{C.30}$$

where $K^o(K) \equiv 1 + (1 - \frac{K}{100})\Delta$, the third equality follows by a change of probability and, finally, Q_{F^T} denotes the forward probability as in the main text.

The Brownian motion under the forward probability is

$$W_\tau^{F^T} = W_\tau - \int_t^\tau \sigma_u(T)du,$$

where $\sigma_\tau(T)$ is the vector of instantaneous volatilities for a zero in Eq. (4.1), and W_τ a Brownian motion under the risk-neutral probability.

Therefore, for the Vasicek model, the short-term rate is a solution to

$$dr_\tau = \left[\kappa(\bar{r} - r_\tau) + \sigma \cdot \sigma_{v,\tau}(T)\right]d\tau + \sigma dW_\tau^{F^T}, \quad \bar{r} \equiv \mu - \frac{\lambda\sigma}{\kappa}, \tag{C.31}$$

where notation is as in Eq. (4.57) of the main text, and $\sigma_{v,\tau}(T)$ is the bond return volatility predicted by Vasicek's model; by Itô's lemma,

$$\sigma_{v,\tau}(T) \equiv -\sigma b_\tau(T) = -\frac{\sigma(1 - e^{-\kappa(T-\tau)})}{\kappa}, \tag{C.32}$$

where we have used the expression for $b_t(T)$ in Eq. (4.64). By tedious but straight-forward calculations, we have that

$$\mathbb{E}_t^{Q_{FT}}(r_\tau) = \bar{r} + e^{-\kappa(\tau-t)}(r_t - \bar{r})$$

$$- \frac{\sigma^2}{2\kappa^2}\left[2\left(1 - e^{-\kappa(\tau-t)}\right) + e^{-\kappa(T-t)}\left(e^{-\kappa(\tau-t)} - e^{\kappa(\tau-t)}\right)\right]. \quad (C.33)$$

In particular, under the forward probability, and conditionally upon the information set at time t, the short-term rate at time T is normally distributed with expectation equal to

$$\mathbb{E}_t^{Q_{FT}}(r_T) = \bar{r} + e^{-\kappa(T-t)}(r_t - \bar{r}) - \frac{\sigma^2}{2\kappa^2}\left(1 - e^{-\kappa(T-t)}\right)^2,$$

and variance $\mathbb{V}_t(r_T)$ given by the expression in Eqs. (C.28).

Next, we elaborate on the expectation of Eq. (C.30). Let $\mathbb{I}_{\{\mathcal{E}\}}$ denote the indicator function of the event $\mathcal{E}$, i.e. the function taking a value equal to one if the event $\mathcal{E}$ is true and zero, otherwise. We have

$$\mathbb{E}_t^{Q_{FT}}\left(K^o(K) - \tilde{B}(r_T)\right)^+$$

$$= \mathbb{E}_t^{Q_{FT}}\left(\left(K^o(K) - \tilde{B}(r_T)\right)\mathbb{I}_{\{\tilde{B}(r_T) \leq K^o(K)\}}\right)$$

$$= K^o(K)\text{Pr}_t^{Q_{FT}}\left(\tilde{B}(r_T) - K^o(K) \leq 0\right) - \mathbb{E}_t^{Q_{FT}}\left(\tilde{B}(r_T)\mathbb{I}_{\{\tilde{B}(r_T)-K^o(K) \leq 0\}}\right),$$
$$(C.34)$$

where $\text{Pr}_t^{Q_{FT}}$ denotes the probability under the forward measure given the information set at time t. Note that

$$\tilde{B}(r_T) \equiv e^{-a_\Delta + b_\Delta r_T} = e^{-a_\Delta + b_\Delta \mathbb{E}_t^{Q_{FT}}(r_T) + \frac{1}{2}y_t^2} \cdot e^{y_t \tilde{N}_T - \frac{1}{2}y_t^2}, \quad (C.35)$$

where

$$\tilde{N}_T \equiv \frac{r_T - \mathbb{E}_t^{Q_{FT}}(r_T)}{\sqrt{\mathbb{V}_t(r_T)}}, \qquad y_t \equiv b_\Delta\sqrt{\mathbb{V}_t(r_T)}, \quad (C.36)$$

so that

$$\tilde{B}(r_T) - K^o(K) = e^{-a_\Delta + b_\Delta \mathbb{E}_t^{Q_{FT}}(r_T) + \frac{1}{2}y_t^2} \cdot \left(e^{y_t \tilde{N}_T - \frac{1}{2}y_t^2} - \hat{K}_t^o(K)\right), \quad (C.37)$$

where

$$\hat{K}_t^o(K) \equiv e^{a_\Delta - b_\Delta \mathbb{E}_t^{Q_{FT}}(r_T) - \frac{1}{2}y_t^2} \cdot K^o(K). \quad (C.38)$$

Conditionally upon the information at t, $\tilde{N}_T$ is standard normally distributed under the forward probability. Therefore, by the expression of $\tilde{B}(r_T)$ in Eq. (C.35), and by Eq. (C.37),

$$\Pr_t^{Q_{FT}}\left(\tilde{B}(r_T)-K^o(K)\le 0\right)=\Phi(\delta_t),\quad \delta_t\equiv \frac{\ln \hat{K}_t^o(K)+\tfrac{1}{2}y_t^2}{y_t}. \tag{C.39}$$

Moreover, again by Eq. (C.35) and Eq. (C.37), and setting $\tilde{n}_T\equiv -\tilde{N}_T$,

$$\mathbb{E}_t^{Q_{FT}}\left(\tilde{B}(r_T)\mathbb{I}_{\{\tilde{B}(r_T)-K^o(K)\le 0\}}\right)$$

$$= e^{-a_\Delta+b_\Delta \mathbb{E}_t^{Q_{FT}}(r_T)+\tfrac{1}{2}y_t^2}\cdot \mathbb{E}_t^{Q_{FT}}\left(e^{y_t\tilde{N}_T-\tfrac{1}{2}y_t^2}\mathbb{I}_{\{\tilde{N}_T\le \delta_t\}}\right)$$

$$\equiv e^{-a_\Delta+b_\Delta \mathbb{E}_t^{Q_{FT}}(r_T)+\tfrac{1}{2}y_t^2}\cdot \mathbb{E}_t^{Q_{FT}}\left(e^{-y_t\tilde{n}_T-\tfrac{1}{2}y_t^2}\mathbb{I}_{\{\tilde{n}_T\ge -\delta_t\}}\right)$$

$$= e^{-a_\Delta+b_\Delta \mathbb{E}_t^{Q_{FT}}(r_T)+\tfrac{1}{2}y_t^2}\cdot \Phi(\delta_t-y_t). \tag{C.40}$$

Substituting Eqs. (C.39) and (C.40) into Eq. (C.34) leaves

$$\mathbb{E}_t^{Q_{FT}}\left(K^o(K)-\tilde{B}(r_T)\right)^+ = K^o(K)\Phi(\delta_t)-e^{-a_\Delta+b_\Delta \mathbb{E}_t^{Q_{FT}}(r_T)+\tfrac{1}{2}y_t^2}\cdot \Phi(\delta_t-y_t).$$

Substituting this expression into Eq. (C.30) leaves

$$\text{Call}_t^z(K,T,\Delta)=100\times P_t(T)e^{-a_\Delta+b_\Delta \mathbb{E}_t^{Q_{FT}}(r_T)+\tfrac{1}{2}y_t^2}\frac{1}{\Delta}\left(\hat{K}^o(K)\Phi(\delta_t)-\Phi(\delta_t-y_t)\right).$$

Equation (4.85) in the main text follows by the definitions of (i) y_t in the second of Eqs. (C.36), (ii) $\hat{K}_t^o(K)$ in Eq. (C.38) and $K^o(K)$ and, finally, by the property of the Forward LIBOR that

$$K^o(Z_t)=1+\left(1-\frac{Z_t}{100}\right)\Delta=e^{-a_\Delta+b_\Delta \mathbb{E}_t^{Q_{FT}}(r_T)+\tfrac{1}{2}y_t^2},$$

a property that we show next.

FORWARD LIBOR. We determine the current forward price predicted by Vasicek's model, $Z_t(r_t;T,T+\Delta)$ in Eq. (4.88), by using Eq. (4.69):

$$Z_t(r_t;T,T+\Delta)=100\times\left(1-f_t(r_t;T,T+\Delta)\right),$$

where

$$f_t(r_t;T,T+\Delta)=\mathbb{E}_t^{Q_{FT}}\left(l_T(r_T;\Delta)\right)$$

$$=\frac{1}{\Delta}\mathbb{E}_t^{Q_{FT}}\left(\frac{1}{P_T(r_T;T+\Delta)}-1\right)$$

$$=\frac{1}{\Delta}\left(\mathbb{E}_t^{Q_{FT}}\left(e^{-a_\Delta+b_\Delta\cdot r_T}\right)-1\right)$$

$$=\frac{1}{\Delta}\left(e^{-a_\Delta+b_\Delta\cdot \mathbb{E}_t^{Q_{FT}}(r_T)+\tfrac{1}{2}b_\Delta^2\cdot \mathbb{V}_t(r_T)}-1\right),$$

and a_Δ and b_Δ are as in Eq. (C.29).

C.10 The Impact of Early Exercise Premiums and Maturity Mismatch

We implement numerical experiments to gauge the quality of the approximation of the interest rate volatility indexes in this chapter based on simplifying assumptions. We rely on the government bond case and illustrate the order of magnitude of these approximations. First, we analyze the impact of early exercise premiums and future corrections while calculating an index with American rather than European options (see Sect. 4.2.10). Second, we assess the impact of maturity mismatch arising when using European options and forwards that have different maturities (see Sect. 4.4).

EARLY EXERCISE PREMIUMS. We convert values of American options on futures into values of European options on forwards based on the algorithm of Sect. 4.2.10 Step 2.a of the main text. We consider a coupon-bearing bond that pays off an annual coupon of \$4, just as in Sect. 4.2.11, but take a maturity equal to 7 years. We assume that $\kappa = 0.3807$, $\mu = 0.01$, $\sigma = 0.033107$ and $\lambda = -0.7$, and calculate hypothetical American option prices predicted by the Vasicek model, the benchmark in this exercise, while assuming that the initial value of the short-term rate is $r = 0.01$. All options are taken to have a maturity equal to 1 month, the same as the future and the forward, and strikes that are equidistant at 50 points (10 strikes below and 10 strikes above the ATM), as in the hypothetical example of Table 4.3 in the main text, with the ATM strike set equal to the current value of the future.

As regards the weighting function ω in Eq. (4.61), we experimented with various functional forms in a Monte Carlo study described below, and found that the optimization problem in Eq. (4.62) performs accurately once we fix

$$\omega(K) = \frac{1}{O^\$(K)}.$$

In the Monte Carlo experiment, we fix (κ, μ, σ) to equal the values indicated above, and calculate the price of American options and future as explained in the main text: through a Longstaff and Schwartz (2001) approximation of $O(K; \lambda) \equiv C_\tau(r_\tau; K)|_{\tau=t}$ (price of American options), where $C_\tau(r_\tau; K)$ satisfies Eq. (4.60); and $\tilde{F}_t(r_t; T, \mathbb{T})$ in Eq. (4.58) (price of future). For reference, the price of the forward is 95.557306, obtained through Eq. (4.68). This price is much lower than that in Table 4.3 in the main text (based on hypothetical market conditions on April 27, 2012). The reason is that we are using a risk-premium parameter, λ, which is as high as 0.7 in absolute value. This value implies that the medium-long end of the yield curve is much higher than that prevailing in late April 2012, when market conditions were such that the yield curve was substantially flat at almost zero.

Note that the optimization procedure requires the value of the short-term rate to be backed up from the price of the future through Eq. (4.58), which therefore changes over different trial values of λ over the optimization of the criterion function in Eq. (4.62). (The seeds utilized to implement the Longstaff and Schwartz algorithm are the same over different trial values of λ.) Table 4.7 below reports key figures regarding the empirical distribution of the estimated risk premiums over 100 runs for an experiment in which λ is estimated searching over a fine grid on $[-1, 0]$.

Table 4.7 Minimum, first quartile, median, mean, third quartile and maximum values of the estimated values of λ over Monte Carlo experiments with 100 runs. Truth is $\lambda = -0.7$

Min	1-Q	Median	Mean	3-Q	Max
-1	-0.9	-0.7	-0.57	-0.23	0

Fig. 4.10 Government bond volatility index calculated through European options on forwards (*solid line*) and American options on futures (*dashed line*), as a function of the risk-premium parameter λ in the Vasicek (1977) model. Forward looking horizon is one month

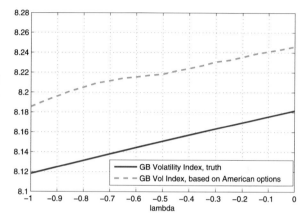

Figure 4.10 depicts the government bond volatility index as a function of the risk-premium parameter λ, calculated using both (i) European options on forwards, which represents the (square of the) true, fair value of a government bond variance swap, and (ii) American options on futures. The American-option based government bond volatility index overstates the truth, due to the presence of an early exercise premium. The American-based approximation is about one relative percentage point higher than the truth when $\lambda = -0.7$, which is quantitatively at least as important as the forward price adjustments in the equity space as documented by the VIX white paper (see Chicago Board Options Exchange 2009).

MATURITY MISMATCH. We implement an exercise with the aim of assessing the severity of the maturity mismatch issue described in Sect. 4.4. Recall that under the T-forward probability, the forward price on a coupon-bearing bond, $F_\tau(S, \mathbb{T})$, satisfies Eq. (4.130), and that by Eq. (4.134), the fair value of a government bond variance swap is

$$\mathbb{P}(t, T, S, \mathbb{T})$$

$$= 2\left(1 - \mathbb{E}_t^{Q_{FT}}\left(e^{\tilde{\ell}(t,T,S,\mathbb{T})} - \tilde{\ell}(t, T, S, \mathbb{T})\right)\right)$$

$$+ 2\left(\int_0^{F_t(S,\mathbb{T})} \frac{\mathrm{Put}_t(K, T, S, \mathbb{T})}{P_t(T)} \frac{1}{K^2}dK + \int_{F_t(S,\mathbb{T})}^\infty \frac{\mathrm{Call}_t(K, T, S, \mathbb{T})}{P_t(T)} \frac{1}{K^2}dK\right),$$

$$\tag{C.41}$$

where $\tilde{\ell}(t, T, S, \mathbb{T})$ is as in Eq. (4.132), and the remaining notation is obvious.

We assume that $T - t = \frac{1}{12}$ and $S - t = \frac{3}{12}$, and take the Vasicek (1977) model as the benchmark. By tedious but straightforward calculations, we find that the value of $\tilde{\ell}(t, T, S, \mathbb{T})$ predicted by this model is

$$\tilde{\ell}(t, T, S, \mathbb{T}) = \sigma^2 \int_t^T \sum_{i=1}^N \omega_\tau^i(S, \mathbb{T}) \Psi_\tau(T, S, T_i) d\tau,$$

where

$$\omega_\tau^i(S, \mathbb{T}) \equiv \bar{C}_i \frac{F_\tau^z(S, T_i)}{F_\tau(S, \mathbb{T})}$$

$$F_\tau^z(S, T_i) \equiv \frac{P_\tau(r_\tau, T_i)}{P_\tau(r_\tau, S)} \qquad F_\tau(S, \mathbb{T}) \equiv \sum_{i=1}^N \frac{C_i}{n} F_\tau^z(S, T_i) + F_\tau^z(S, \mathbb{T}), \qquad \mathbb{T} \equiv T_N$$

$$\Psi_\tau(T, S, T_i) \equiv \big(\varphi_{1\tau}(T_i) - \varphi_{1\tau}(S)\big)\big(\varphi_{1\tau}(T) - \varphi_{1\tau}(S)\big),$$

$$\varphi_{1\tau}(\mathcal{Y}) \equiv -\frac{1 - e^{-\kappa(\mathcal{Y}-\tau)}}{\kappa}, \qquad \text{for a generic } \mathcal{Y}$$

and $P_\tau(r_\tau, T_i)$ is the price of a zero-coupon bond predicted by Vasicek's formula (see Eq. (4.64) in the main text).

To calculate the first term of the R.H.S. of Eq. (C.41), we need to take the expectation of $e^{\tilde{\ell}(t,T,S,\mathbb{T})} - \tilde{\ell}(t, T, S, \mathbb{T})$ under the forward probability. This is accomplished through Monte Carlo simulations, in which the short-term rate is as in Eq. (C.31), viz

$$dr_\tau = \kappa \left(\bar{r} - r_\tau - \sigma^2 \frac{1 - e^{-\kappa(T-\tau)}}{\kappa}\right) d\tau + \sigma dW_\tau^{F^T}, \qquad \tau \in [t, T], \qquad \text{(C.42)}$$

where $W_\tau^{F^T}$ is a Brownian motion under the T-forward probability, and the time-varying term inside the drift captures the tilt caused by the change from the risk-neutral to the forward probability.

To calculate the second term of the R.H.S. of Eq. (C.41), we need to calculate the price of 1-month European options written on 3-month forwards. The latter could be determined by generalizing the Jamshidian (1989) formula, Eqs. (4.66) and (4.67) in the main text. However, these prices can be calculated quite rapidly as a by-product of the previous Monte Carlo simulations by approximating the following expectations:

$$\frac{\text{Put}_t(K, T, S, \mathbb{T})}{P_t(T)} = \mathbb{E}_t^{Q_{F^T}}\big((K - F_T(S, \mathbb{T}))^+\big),$$

$$\frac{\text{Call}_t(K, T, S, \mathbb{T})}{P_t(T)} = \mathbb{E}_t^{Q_{F^T}}\big((F_T(S, \mathbb{T}) - K)^+\big). \qquad \text{(C.43)}$$

We proceed as follows:

- Simulate values of r_τ from Eq. (C.42), which are used to calculate:

Table 4.8 Volatility indexes and risk-premium assumptions

	λ				
	-1.00	-0.85	-0.70	-0.55	-0.40
GB-VI$_{\text{Mc}}(t, T, S, \mathbb{T})$	7.548300	7.557374	7.566307	7.575100	7.583754
GB-VI$^*_{\text{Mc}}(t, T, S, \mathbb{T})$	7.548307	7.557381	7.566314	7.575107	7.583762

- $F^z_\tau(S, T_i)$ and $F_\tau(S, \mathbb{T})$ for $\tau \in [t, T)$, needed to calculate the first term of the R.H.S. of Eq. (C.41);
- $F_T(S, \mathbb{T})$, needed to calculate Eqs. (C.43) and, hence, the second term of the R.H.S. of Eq. (C.41).

- Calculate the two volatility indexes:

$$\text{GB-VI}_{\text{Mc}}(t, T, S, \mathbb{T}) \equiv \sqrt{\frac{1}{T-t}\mathbb{P}_{\text{Mc}}(t, T, S, \mathbb{T})} \quad \text{and}$$

$$\text{GB-VI}^*_{\text{Mc}}(t, T, S, \mathbb{T}) \equiv \sqrt{\frac{1}{T-t}\mathbb{P}^*_{\text{Mc}}(t, T, S, \mathbb{T})}, \tag{C.44}$$

where

$$\mathbb{P}^*_{\text{Mc}}(t, T, S, \mathbb{T}) \equiv 2\left(\int_0^{F_t(S,\mathbb{T})} \frac{\text{Put}_{\text{Mc},t}(K, T, S, \mathbb{T})}{P_t(T)} \frac{1}{K^2} dK \right.$$
$$\left. + \int_{F_t(S,\mathbb{T})}^{\infty} \frac{\text{Call}_{\text{Mc},t}(K, T, S, \mathbb{T})}{P_t(T)} \frac{1}{K^2} dK\right),$$

and $\mathbb{P}_{\text{Mc}}(t, T, S, \mathbb{T})$ is the Monte Carlo estimate of $\mathbb{P}(t, T, S, \mathbb{T})$ in Eq. (C.41), and $\text{Put}_{\text{Mc},t}(K, T, S, \mathbb{T})$ and $\text{Call}_{\text{Mc},t}(K, T, S, \mathbb{T})$ are the Monte Carlo estimates of $\text{Put}_t(K, T, S, \mathbb{T})$ and $\text{Call}_t(K, T, S, \mathbb{T})$ in Eq. (C.43).

Table 4.8 compares the value of the two indexes in Eqs. (C.44) based on the parameter values underlying the experiments in Figure 4.10, and for various levels of the risk-premium parameter, λ. The two indexes are substantially the same in this experiment.

Chapter 5
Credit

5.1 Introduction

The global financial crisis culminating in Lehman Brothers' collapse on 15 September 2008 and the subsequent sovereign debt turmoil have been marked by bursts of heightened volatility in the perceived credit risk of private and public debt. These events are traced, for example, by the spread of on-the-run Markit CDX North American Investment Grade indexes, which are OTC-traded credit default swap (CDS) indexes with 125 investment grade corporates as their constituents, and are perhaps the closest credit market analogue of the S&P 500 index. Figure 5.1 shows the index spread skyrocketing from below 50 bps before the crisis to well above 250 bps at the height of it, with large swings in realized volatility as shown in Fig. 5.2. Index levels have stayed far below their crisis highs after rallying through much of 2009, but have experienced a couple bouts of widening in subsequent years.

While one may reasonably expect credit volatility to be the most correlated with equity volatility among the fixed income volatilities treated in this book, Fig. 5.3 shows that they are still quite distinct. The correlation between 20-day realized volatility for SPY and CDX is a mere 64 %. Moreover, during the recent credit crisis, CDX volatility was more exaggerated compared to both SPY volatility and the VIX. While CDX spreads and realized volatility are positively correlated, they only display a modest correlation of 49 %—even less in absolute value than the correlation between SPY and its realized volatility, which is −55 %.

The previous facts provide motivation for the creation of volatility indexes for various CDS indexes that reflect the market's expectation of future volatility of broad-based credit baskets, much like the CBOE's VIX for equity indexes and the other indexes in this book. Currently, Markit publishes 20-, 60-, and 90-day realized volatility indexes (Markit VolX) for their CDS indexes, but they are backward-looking by design. In contrast, we derive indexes for implied future volatility based on options traded on CDS indexes.

We derive the model-free price of expected volatility of CDS index spreads, and designs for CDS option-based volatility indexes based on the theory of Chap. 2. Furthermore, we build upon well-established theory (Pedersen 2003; Rutkowski and

© Springer International Publishing Switzerland 2015
A. Mele, Y. Obayashi, *The Price of Fixed Income Market Volatility*, Springer Finance,
DOI 10.1007/978-3-319-26523-0_5

Fig. 5.1 On-the-run CDX
North American Investment
Grade index, in basis
points—daily data from 27
September 2004 to March 27,
2012, for a total of 1881
observations

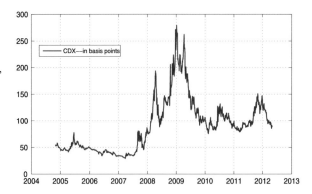

Armstrong 2009; Morini and Brigo 2011) with two key ingredients—CDS indexes
and CDS index options—and the concept of a loss-adjusted forward position in a
credit default swap index to treat the "front end protection" feature of CDS index
options.

We introduce a credit variance swap contract on a loss-adjusted forward position
in an index spread, and derive its model-free price—the fair market value of future
index spread volatility—expressed in terms of traded CDS index option prices. Finally, the contract price leads to our design for implied future volatility indexes for
credit and credit indexes. We consider both percentage and basis point indexes to
express implied volatility in both log-normal and normal terms, respectively. The
setup and derivations require the concept of survival contingent probability to account for default risk in the underlying process, which is absent in all derivations of
the previous chapters.

Since we developed the initial credit volatility pricing methodology detailed in
Sects. 5.2 through 5.4, ISDA released a modification to the standard CDS agreement, known as the "Big Bang Protocol," which requires adjustments to the index
formulae presented in Sect. 5.4. We propose two alternative methods for dealing
with these changes in market conventions; however, we note that they are both
model-dependent, much like the American–European and maturity mismatch corrections in Chap. 4 for Government bonds, which in turn makes the post-correction
index no longer model-free. Numerical estimations for the magnitudes of these corrections will be useful in weighing the upside of theoretical correctness against the
downside of added complications and model-dependency when implementing the
index in practice.

The plan of the chapter is as follows. The next section describes standard pricing methodology for relevant credit instruments. Section 5.3 introduces three credit
variance contracts evaluated in a model-free fashion that relies on the traded derivatives dealt with in Sect. 5.2, and describes marking-to-market and hedging issues
related to credit variance contracts. Section 5.4 develops credit volatility indexes
based on the fair value of the variance contracts of Sect. 5.3. Section 5.5 provides
index corrections dealing with the Big Bang Protocol as well as additional issues regarding index coupon adjustments. Appendix D provides technical details omitted
from the main text.

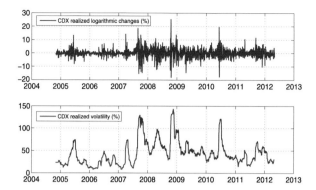

Fig. 5.2 *Top panel*: realized daily logarithmic changes in the on-the-run CDX North American Investment Grade index depicted in Fig. 5.1, in percent. *Bottom panel*: estimates of 20-day realized volatility of the changes in the CDX index, annualized, percent, $100\sqrt{12\sum_{t=1}^{21}\ln^2\frac{\text{cdx}_{t-i+1}}{\text{cdx}_{t-i}}}$, where cdx_t denotes the CDX index on day t. The sample includes daily data from 27 September 2004 to March 27, 2012, for a total of 1881 observations

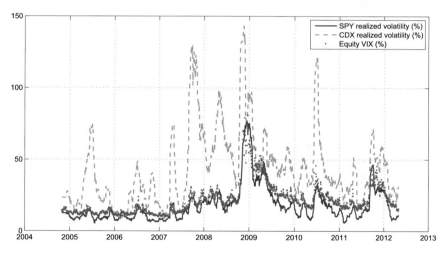

Fig. 5.3 Estimates of 20-day realized volatility of the daily logarithmic changes in the CDX North American Investment Grade index (*dashed line*), and the SPY daily returns (*solid line*), calculated as in Fig. 5.2. The *dotted line* is the VIX. The sample includes daily data from 27 September 2004 to March 27, 2012, for a total of 1881 observations

5.2 Existing Credit Trading Practices

A buyer of protection on a CDS index pays a periodic premium to the seller in exchange for insurance against losses arising from defaults by any of the index's single-name constituents during the term of the contract. Whenever a constituent defaults, the defaulted obligor is removed from the index and the index continues to be traded with a prorated notional amount. The CDX index is also known as a basket or portfolio CDS, and the premium paid is referred to as the CDS index spread. Options on CDS indexes are European-style options to buy (payers) or sell (receivers) protection on the CDS index at the strike spread upon option expiry.

The next section details the assumptions underlying the credit risk of single-name constituents over the life of an index. Section 5.2.2 describes CDS indexes, and Sect. 5.2.3 contains details regarding the pricing framework underlying CDS index options.

5.2.1 Assumptions

The single name constituents of the index can experience credit events over a referenced period of regular intervals (T_{i-1}, T_i) with length equal to $\frac{1}{b}$, for $i = 1, \ldots, bM$, where M denotes the index's time to maturity in years, and T_0 is the time of the index origination; for example, $b = 4$ corresponds to quarterly intervals. We assume that: (i) loss-given-default (LGD) is a constant; (ii) the short-term rate, r_τ, is a diffusion process; and (iii) default arrives for each name in the index as a Cox process, with the same intensity λ_τ adapted to r_τ.

5.2.2 CDS Indexes

Let n be the number of constituents in the index created at $t \equiv T_0$, each having a notional value equal to $\frac{1}{n}$ and the same LGD and default intensity λ. Denote by $\mathcal{S}_{T_i}$ the number of constituents having survived up to a generic date T_i,

$$\mathcal{S}_{T_i} \equiv \sum_{j=1}^{n} (1 - \mathbb{I}_{\{\tau_j \leq T_i\}}),$$

where τ_j is the time at which constituent j defaults, and $\mathbb{I}_{\{\mathcal{A}\}}$ is the indicator of a generic event $\mathcal{A}$, which is equal to one if $\mathcal{A}$ is true and zero otherwise. Should obligor j default, the loss in the index occurring at time τ_j, and paid off by the protection seller, is $\text{LGD}\frac{1}{n}\mathbb{I}_{\{t \leq \tau_j \leq T_{bM}\}}$. On the other hand, $\frac{1}{b}C_t\frac{1}{n}\mathcal{S}_{T_i}$ is the premium paid by the protection buyer at T_i, which equals the constant premium (the "coupon") determined at t, $\frac{1}{b}C_t$, times the outstanding notional at T_i, $\frac{1}{n}\mathcal{S}_{T_i}$.

At index origination, the index value is the value of the protection leg minus that of the premium leg over the life of the index:

$$\text{DSX}_t = \mathbb{E}_t\left[\sum_{j=1}^{n} e^{-\int_t^{\tau_j} r_\tau d\tau}\left(\text{LGD}\frac{1}{n}\mathbb{I}_{\{t \leq \tau_j \leq T_{bM}\}}\right)\right]$$

$$- \mathbb{E}_t\left[\sum_{i=1}^{bM} e^{-\int_t^{T_i} r_\tau d\tau}\left(\frac{1}{b}C_t\frac{1}{n}\mathcal{S}_{T_i}\right)\right]$$

$$= \text{LGD} \cdot v_{0t} - \frac{1}{b}C_t \cdot v_{1t}, \tag{5.1}$$

where $\mathbb{E}_t$ denotes the risk-neutral expectation given the information available up to time t, and v_{0t} and v_{1t} are the values of securities indexed to default events of a hypothetical obligor representing the average constituent with a credit spread equal to the index spread. They are defined as $v_{0t} \equiv v_{0,t,t}$, where for a given T and $\tau \in [t, T]$,

$$v_{0,\tau,T} \equiv \mathbb{E}_\tau \left(e^{-\int_\tau^{\tau_*} r_s ds} \mathbb{I}_{\{T \le \tau_* \le T_{bM}\}} \right) \quad \text{and} \quad v_{1t} \equiv \sum_{i=1}^{bM} \mathbb{E}_t \left(e^{-\int_t^{T_i} r_\tau d\tau} \cdot \mathbb{I}_{\{\text{Surv at } T_i\}} \right),$$

$$(5.2)$$

where τ_* denotes the default time of the representative constituent with default intensity λ, and $\mathbb{I}_{\{\text{Surv at } T_i\}}$ is the indicator of the event that this name survives up to time T_i.

One may interpret v_{0t} as the value at time t of one dollar paid off at the time of default of a hypothetical representative constituent with default intensity λ, provided default occurs prior to the expiry of the index, T_{bM}. Note that Eq. (5.2) introduces the more general notation $v_{0,\tau,T}$, interpreted as the present value at any $\tau \in [t, T]$ of receiving one dollar if the representative name defaults between T and T_{bM}. We will utilize this notation below while deriving the value of a forward position in an index (see Eq. (5.3)). v_{1t} is interpreted as the value at t of an annuity of one dollar paid at the dates $T_1, \ldots, T_{bM}$, until default of the representative name or expiry of the index, whichever occurs first. In other words, v_{1t} is the value at t of a basket of defaultable bonds with zero recovery value that are issued by a representative obligor. In the sequel, we shall refer to v_{1t} as the *defaultable price value of the basis point*, or defaultable-PVBP.

Appendix D.1 provides a further expression for the premium leg, which covers the situation in which if a constituent j defaults at some point in time $\tau_j \in [T_i, T_{i+1}]$, the protection seller still receives an accrued premium calculated upon S_{T_i} for the time from T_i to τ_j, and a premium calculated upon $S_{T_{i+1}}$, for the time remaining from τ_j to T_{i+1}.

5.2.3 CDS Index Options

5.2.3.1 Forward Positions

Consider a *forward position* at time t in an index that starts at time $T \equiv T_0 \ge t$, with $\frac{1}{b} C_t$ being the constant premium determined at t. This construct serves as the underlying of credit index options. By a straightforward generalization of Eq. (5.1), the value of the protection leg minus that of the premium leg over the life of the index is

$$\text{DSX}_t(t, T) \equiv \text{LGD} \cdot v_{0,t,T} - \frac{1}{b} C_t \cdot v_{1t}, \quad t \le T,$$

where $v_{0,t,T}$ and v_{1t} are defined as in Eqs. (5.2). Once again, the interpretation of $v_{0,t,T}$ is the same as that of v_{0T} (see Eq. (5.1)), but with evaluation time equal to

$t \leq T$. Note the arguments (t, T) in $\mathrm{DSX}_t(t, T)$: they emphasize that the forward position is initiated at t and the maturity is T. In Appendix D.1, we show that the value at $\tau \in [t, T]$ of the previously initiated forward position is

$$\mathrm{DSX}_\tau(t, T) = \mathcal{N}_\tau \left(\mathrm{LGD} \cdot v_{0,\tau,T} - \frac{1}{b} C_t \cdot v_{1\tau} \right), \quad \tau \in [t, T], \tag{5.3}$$

where $\mathcal{N}_\tau$ denotes the outstanding notional

$$\mathcal{N}_\tau = \frac{1}{n} \mathcal{S}_\tau, \quad \mathcal{N}_t \equiv 1. \tag{5.4}$$

The value of the forward position in the index collapses to the index level when $t = T$ (see Eq. (5.1)), viz

$$\mathrm{DSX}_t(t, t) = \mathrm{DSX}_t.$$

5.2.3.2   Front-End Protection and Credit Default Options

A CDS index option payer gives the holder the right but not obligation to enter a CDS index at some $T \geq t$ as a protection buyer with an index strike spread equal to some K. Note that a forward position in the index does not protect from losses occurring prior to T. Instead, and upon exercise, the protection buyer would receive a *front-end protection*, defined as the losses that occurred between the option origination and the exercise date.

Let t be the date of the option origination and $T = T_0$ be the option maturity date, which is the same as the date at which the index starts. The front end protection is $F_T = \mathrm{LGD} \frac{1}{n} \mathcal{D}(t, T)$, where $\mathcal{D}(t, T) \equiv \sum_{j=1}^n \mathbb{I}_{\{\tau_j \in [t, T]\}}$ is the number of defaults that occurred over the time interval $[t, T]$. It can be shown (see, e.g., Mele 2014, Chap. 13) that the value of the front-end protection is

$$v_\tau^{\mathrm{F}} = \mathbb{E}_\tau \left(e^{-\int_\tau^T r_u du} F_T \right) = \mathrm{LGD} \left(\frac{1}{n} \mathcal{D}(t, \tau) P_\tau(T) + \mathcal{N}_\tau \left(P_\tau(T) - P_{\mathrm{def},\tau}(T) \right) \right), \tag{5.5}$$

where $P_t(T)$ denotes, as usual, the time t price of a non-defaultable zero-coupon bond expiring at T, and $P_{\mathrm{def},t}(T)$ denotes its defaultable counterpart with zero recovery value and default intensity λ. The underlying of a payer equals

$$\mathrm{DSX}_T^L(t, T) \equiv \mathrm{DSX}_T(t, T) + F_T.$$

Accordingly, define the *loss-adjusted forward position* in the index as $\mathrm{DSX}_\tau^L(t, T) \equiv \mathrm{DSX}_\tau(t, T) + v_\tau^{\mathrm{F}}$, and let $\mathrm{CDX}_\tau(M)$ denote the value of C_τ that makes a forward position originated at τ worthless, viz $\mathrm{DSX}_\tau^L(\tau, T) = 0$. Evaluating Eq. (5.3) in $t = \tau$, and setting $\mathrm{DSX}_\tau^L(\tau, T)$ equal to zero, we have

$$\frac{1}{b} \mathrm{CDX}_\tau(M) = \mathrm{LGD} \frac{v_{0,\tau,T}}{v_{1\tau}} + \frac{v_\tau^{\mathrm{F}}}{\mathcal{N}_\tau v_{1\tau}}. \tag{5.6}$$

Substituting Eq. (5.6) back into the definition of $\mathrm{DSX}_\tau^L(t, T)$ leaves the following expression for the loss-adjusted forward position in the index originated at t:

$$\mathrm{DSX}_\tau^L(t, T) = \frac{1}{b}\mathcal{N}_\tau v_{1\tau}\big(\mathrm{CDX}_\tau(M) - C_t\big), \quad \tau \in [t, T]. \tag{5.7}$$

Naturally, $\mathrm{CDX}_t(M) = C_t$. We now introduce options referenced to $\mathrm{DSX}_\tau^L(t, T)$ at $\tau = T$ with strikes K replacing C_t. We shall remove this assumption in Sect. 5.5. We wish to use $\mathcal{N}_\tau v_{1\tau}$ as a numéraire so that $\mathrm{CDX}_t(M)$ is a martingale under a suitable probability.[1] Define the probability Q_{sc} through the Radon–Nikodym derivative:

$$\left.\frac{dQ_{sc}}{dQ}\right|_{\mathbb{F}_T^r} = e^{-\int_\tau^T r_u du}\frac{\mathcal{N}_T v_{1T}}{\mathcal{N}_\tau v_{1\tau}}, \tag{5.8}$$

where $\mathbb{F}_T^r$ denotes the information set at time T, which includes the path of the short-term rate only.[2]

The probability Q_{sc} is the index counterpart to the so-called *survival contingent probability*. It can be shown that $\mathrm{CDX}_t(M)$ in Eq. (5.6) is a martingale under Q_{sc} (see e.g., Schönbucher 2003, Chap. 7; Lando 2004, Chap. 5; and Mele 2014, Chap. 13). For simplicity, we shall keep referring to Q_{sc} as the "survival contingent probability." The prices of a payer and receiver options with strike K and expiry T are, for any $\tau \in [t, T]$,

$$\mathrm{SW}_\tau^p(K, T; M) \equiv \frac{1}{b}\mathcal{N}_\tau v_{1\tau}\mathbb{E}_\tau^{sc}\big[(\mathrm{CDX}_T(M) - K)^+\big], \quad \text{and}$$
$$\mathrm{SW}_\tau^r(K, T; M) \equiv \frac{1}{b}\mathcal{N}_\tau v_{1\tau}\mathbb{E}_\tau^{sc}\big[(K - \mathrm{CDX}_T(M))^+\big], \tag{5.9}$$

where $\mathbb{E}_\tau^{sc}[\cdot]$ denotes the time τ conditional expectation under the survival contingent probability Q_{sc}. We know $\mathrm{CDX}_\tau(M)$ is a martingale under Q_{sc}. We can use Black (1976) to evaluate the previous expression once we assume that, under Q_{sc}, $\mathrm{CDX}_\tau(M)$ is a geometric Brownian motion with constant volatility (see Sect. 5.4.4).

---

[1] The definition of $\mathrm{CDX}_\tau(M)$ in Eq. (5.6) is problematic as it relates to a "denominator problem"— the possibility of a total collapse of the index, $N_\tau = 0$. The occurrence of such an event has been taken into account by Rutkowski and Armstrong (2009) and Morini and Brigo (2011), but will not be incorporated into the credit variance contract design.

[2] See, e.g., Mele (2014, Chap. 13), for a proof that dQ_{sc} does indeed integrate to one.

5.3 Credit Variance Contracts

We assume that the loss-adjusted forward spread, $\text{CDX}_\tau(M)$ in Eq. (5.6), is a jump-diffusion process with stochastic volatility:

$$\frac{d\text{CDX}_\tau(M)}{\text{CDX}_\tau(M)} = -\left(\mathbb{E}_\tau^{\text{sc}}\left(e^{j_\tau(M)} - 1\right)\eta_\tau\right)d\tau + \sigma_\tau(M) \cdot dW_\tau^{\text{sc}} + \left(e^{j_\tau(M)} - 1\right)dJ_\tau^{\text{sc}},$$

$$\tau \in [t, T], \tag{5.10}$$

where W_τ^{sc} is a multidimensional Brownian motion, $\sigma_\tau(M)$ is a diffusion component adapted to W_τ^{sc}, J_τ^{sc} is a Cox process with intensity equal to η_τ, $j_\tau(M)$ is the logarithmic jump size, and both W_τ^{sc} and J_τ^{sc} are defined under the survival contingent probability. By applying Itô's lemma for jump-diffusion processes (just as in the general case of Chap. 2; see Eq. (2.40)),

$$d\ln\text{CDX}_\tau(M) = (\cdots)d\tau + \sigma_\tau(M) \cdot dW_\tau^{\text{sc}} + j_\tau(M)dJ_\tau^{\text{sc}},$$

so that the realized variance of the *logarithmic changes* in the CDX index over a time interval $[t, T]$, i.e., *percentage variance*, is[3]

$$V_M(t, T) \equiv \int_t^T \left\|\sigma_\tau(M)\right\|^2 d\tau + \int_t^T j_\tau^2(M)dJ_\tau^{\text{sc}}. \tag{5.11}$$

We shall also consider the realized variance of the *arithmetic changes* in the CDX index over $[t, T]$, i.e., *basis point variance*:

$$V_M^{\text{bp}}(t, T) \equiv \int_t^T \text{CDX}_\tau^2(M)\left\|\sigma_\tau(M)\right\|^2 d\tau + \int_t^T \text{CDX}_\tau^2(M)\left(e^{j_\tau(M)} - 1\right)^2 dJ_\tau^{\text{sc}}. \tag{5.12}$$

This section provides designs of variance contracts for both percentage and basis point variance. These contracts are priced in a model-free fashion, and lead to model-free indexes of expected volatility over the reference period $[t, T]$ (see Sect. 5.4). Note that these variance contracts are developed with respect to credit indexes, but also cover the single-name case simply by setting $n = 1$.

We consider three contracts: a forward agreement that requires an initial payment, and two variance swaps that settled at expiration. Section 5.3.1 considers the percentage versions of the contracts, and Sect. 5.3.2 develops their basis point counterparts.

5.3.1 Percentage

Consider the following agreement:

---

[3]In the statistical literature, $V_M(t, T)$ and $V_M^{\text{bp}}(t, T)$ are typically referred to as *total variations*, rather than *variances*, as explained in Chap. 2 (see Sect. 2.7, Remark 2.5).

Definition 5.1 (Credit Variance Forward Agreement)   At time t, counterparty A promises to pay counterparty B the product of the credit variance realized over the interval $[t, T]$ times the defaultable-PVBP over the outstanding notional that will prevail at time T, i.e., the value $V_M(t, T) \times \mathcal{N}_T v_{1T}$. The price counterparty B shall pay counterparty A for this agreement at time t is denoted by $\mathbb{F}_{\text{var},M}(t, T)$.

The defaultable-PVBP over the outstanding notional, $\mathcal{N}_T v_{1T}$, is unknown at time T. The contract design underlying Definition 5.1 is consistent with the general framework of Chap. 2, and admits a model-free price of credit volatility and, at the same time, pins down the loss-adjusted forward position in the index, the value of defaultable bonds, and index default swap options.

Next, we consider the following contract cast in a "swap format," whereby counterparty A promises to pay counterparty B the following difference at time T:

$$\text{Var-Swap}_M(t, T) \equiv V_M(t, T) \times \mathcal{N}_T v_{1T} - \mathbb{P}_{\text{var},M}(t, T), \qquad (5.13)$$

where $\mathbb{P}_{\text{var},M}(t, T)$ is a fixed variance swap rate such that the following zero value condition holds at time t:

Definition 5.2 (Credit Variance Swap Rate)   The credit variance swap rate is the fixed variance swap rate $\mathbb{P}_{\text{var},M}(t, T)$ that makes $\text{Var-Swap}_M(t, T)$ in Eq. (5.13) equal to zero.

In the final contract we consider, we introduce a notion of credit variance swap rate alternative to that in the previous definition. This notion is used to define credit volatility indexes (see Sect. 5.4), and parallels similar definitions given in previous chapters. Consider the following payoff:

$$\text{Var-Swap}_M^*(t, T) \equiv \left[V_M(t, T) - \mathbb{P}_{\text{var},M}^*(t, T) \right] \times \mathcal{N}_T v_{1T}, \qquad (5.14)$$

where $\mathbb{P}_{\text{var},M}^*(t, T)$, the fixed variance swap rate, is set to satisfy a zero value condition at time t:

Definition 5.3 (Standardized Credit Variance Swap Rate)   The standardized credit variance swap rate is the fixed variance swap rate $\mathbb{P}_{\text{var},M}^*(t, T)$, which makes $\text{Var-Swap}_M^*(t, T)$ in Eq. (5.14) equal to zero.

We now price these contracts. In Appendix D.2, we show that: (i) the value of the credit variance forward agreement in Definition 5.1 is approximated by:

$$\mathbb{F}_{\text{var},M}(t, T)$$

$$= 2 \left(\sum_{i:K_i < \text{CDX}_t(M)} \frac{\text{SW}_t^r(K_i, T; M)}{K_i^2} \Delta K_i + \sum_{i:K_i \geq \text{CDX}_t(M)} \frac{\text{SW}_t^p(K_i, T; M)}{K_i^2} \Delta K_i \right);$$

$$(5.15)$$

and that (ii) the credit variance swap rate of Definition 5.2 and the standardized credit variance swap rate of Definition 5.3 are approximated by

$$\mathbb{P}_{\text{var},M}(t,T) = \frac{\mathbb{F}_{\text{var},M}(t,T)}{P_t(T)}, \tag{5.16}$$

and

$$\mathbb{P}^*_{\text{var},M}(t,T) = \frac{\mathbb{F}_{\text{var},M}(t,T)}{v_{1t}}. \tag{5.17}$$

In Eq. (5.15), $\Delta K_i = \frac{1}{2}(K_{i+1} - K_{i-1})$ for $i \geq 1$, $\Delta K_0 = (K_1 - K_0)$, $\Delta K_Z = (K_Z - K_{Z-1})$, where K_0 and K_Z are the lowest and the highest traded strikes, with $Z + 1$ denoting the total number of traded options.

Note that we make two approximations. First, we use a finite number of CDS index options in Eq. (5.15), as opposed to a continuum as in the theoretical derivations (see Appendix D.2). Second, we disregard elements of the contract that span the realized variance in Eq. (5.11), which include the jump component in Eq. (5.10). This jump component results in a more general formula (see Eq. (D.6) in Appendix D.2): one with a model-dependent correction term to Eq. (5.15).

5.3.2 Basis Point

We define basis point counterparts to the credit variance contracts of the previous section as follows:

Definition 5.4 (Basis point contracts)   Consider the following basis point (BP) credit variance contracts and swap rates, replacing those in Definitions 5.1, 5.2, and 5.3:

(a) **(BP-Credit Variance Forward Agreement)**. At time t, counterparty A promises to pay counterparty B the product of the BP credit variance realized over the interval $[t, T]$ times the defaultable-PVBP over the outstanding notional that will prevail at time T, i.e., the value $V_M^{\text{bp}}(t, T) \times \mathcal{N}_T v_{1T}$. The price counterparty B shall pay counterparty A for this agreement at time t is denoted by $\mathbb{F}_{\text{var},n}^{\text{bp}}(t, T)$.

(b) **(BP-Credit Variance Swap Rate)**. The BP-credit variance swap rate is the fixed variance swap rate $\mathbb{P}_{\text{var},n}^{\text{bp}}(t, T)$, which makes the current value of $V_M^{\text{bp}}(t, T) \times \mathcal{N}_T v_{1T} - \mathbb{P}_{\text{var},M}^{\text{bp}}(t, T)$ equal to zero.

(c) **(Standardized BP-Credit Variance Swap Rate)**. The standardized BP-credit variance swap rate is the fixed variance swap rate $\mathbb{P}_{\text{var},M}^{*\text{bp}}(t, T)$, which makes the current value of $[V_M^{\text{bp}}(t, T) - \mathbb{P}_{\text{var},M}^{*\text{bp}}(t, T)] \times \mathcal{N}_T v_{1T}$ equal to zero.

In Appendix D.2, we show that: (i) the value of the BP-credit variance forward agreement in Definition 5.4-(a) is approximated by

$$
\begin{aligned}
&\mathbb{F}^{\text{bp}}_{\text{var},M}(t,T)\\
&\equiv 2\left(\sum_{i:K_i<\text{CDX}_t(M)}\text{SW}^r_t(K_i,T;M)\Delta K_i + \sum_{i:K_i\geq\text{CDX}_t(M)}\text{SW}^p_t(K_i,T;M)\Delta K_i\right),
\end{aligned}
\tag{5.18}
$$

where the strike differences, ΔK_i, are as in Eq. (5.15); and (ii) the BP-credit variance swap rate of Definition 5.4-(b) and the standardized BP-credit variance swap rate of Definition 5.4-(c) are

$$
\mathbb{P}^{\text{bp}}_{\text{var},M}(t,T)=\frac{\mathbb{F}^{\text{bp}}_{\text{var},M}(t,T)}{P_t(T)} \quad\text{and}\quad \mathbb{P}^{*\text{bp}}_{\text{var},M}(t,T)=\frac{\mathbb{F}^{\text{bp}}_{\text{var},M}(t,T)}{v_{1t}}.
\tag{5.19}
$$

Chapter 2 develops a general framework and explains how different option weights arise according to whether a variance contract references a percentage or a basis point notion of variance (see Sect. 2.3.4). These explanations apply to the fair value of credit variance forwards (percentage in Eq. (5.15) and basis point in Eq. (5.18)), in addition to the variance contracts in Chap. 3 (interest rate swaps) and Chap. 4 (government bonds and time deposits).

Note that the only approximation we make to derive Eq. (5.18) and Eqs. (5.19) relies on the use of a finite number of CDS index options. In particular, we are not disregarding the jump component in Eq. (5.10). Unlike the percentage contracts of Sect. 5.3.1, the basis point contracts of this section can be evaluated in a model-free fashion, even in the presence of jumps. These properties parallel those regarding contracts for other asset classes as one would expect from Sect. 2.7 of Chap. 2 in the context of general numéraires: basis point variance contracts are resilient to jumps while percentage contracts are not.

5.3.3 Marking to Market

Appendix D.2 shows that the value at $\tau\in[t,T]$ of the *percentage* credit variance contract, struck at time t at the swap rate $\mathbb{P}_{\text{var},M}(t,T)$ (see Eq. (5.16)), is

$$
\text{M-Var}_M(t,\tau,T)\equiv V_M(t,\tau)\times\mathcal{N}_\tau v_{1\tau} - P_\tau(T)\big[\mathbb{P}_{\text{var},M}(t,T)-\mathbb{P}_{\text{var},M}(\tau,T)\big].
\tag{5.20}
$$

The mark-to-market calculations of the contract struck at the standardized swap rate $\mathbb{P}^*_{\text{var},M}(t,T)$ in Eq. (5.17) is, instead,

$$
\text{M-Var}^*_M(t,\tau,T)\equiv\mathcal{N}_\tau v_{1\tau}\times\big[V_M(t,\tau)-\big(\mathbb{P}^*_{\text{var},M}(t,T)-\mathbb{P}^*_{\text{var},M}(\tau,T)\big)\big].
\tag{5.21}
$$

Marks to market for the corresponding *basis point* credit variance contracts are formally the same as $\text{M-Var}_M(t,\tau,T)$ and $\text{M-Var}^*_M(t,\tau,T)$, but with the basis point variance $V_M^{\text{bp}}(t,\tau)$ replacing the percentage variance $V_M(t,\tau)$, the basis point swap rate $\mathbb{P}_{\text{var},M}^{\text{bp}}(\cdot,\cdot)$ replacing $\mathbb{P}_{\text{var},M}(\cdot,\cdot)$ in Eq. (5.20), and the basis point standardized swap rate $\mathbb{P}_{\text{var},M}^{*\text{bp}}(\cdot,\cdot)$ replacing $\mathbb{P}_{\text{var},M}^{*}(\cdot,\cdot)$ in Eq. (5.21), where $\mathbb{P}_{\text{var},M}^{\text{bp}}(\cdot,\cdot)$ and $\mathbb{P}_{\text{var},M}^{*\text{bp}}(\cdot,\cdot)$ are as in Eqs. (5.19).

5.3.3.1 Hedging

We provide details regarding the hedging of credit variance contracts under a number of approximations. First, we neglect the jump component in Eq. (5.10), and set $j_\tau(M) \equiv 0$, so that the realized variance in Eq. (5.11) collapses to $V_M(t,T) \equiv \int_t^T \|\sigma_\tau(M)\|^2 d\tau$. Second, we assume that we can either (i) trade an asset with a price that is perfectly correlated with the loss-adjusted forward spread $\text{CDX}_\tau(M)$ in Eq. (5.6), or that (ii) we can hedge the variance contract with only the first term of $\text{CDX}_\tau(M)$, say

$$\widehat{\text{CDX}}_\tau(M) \equiv \text{LGD}\frac{v_{0,\tau,T}}{v_{1\tau}}, \tag{5.22}$$

and that even in this case, the hedging error arising from jumps in the value of $v_{1\tau}$ in Eq. (5.2) is small.[4,5]

Under additional conditions, $\widehat{\text{CDX}}_\tau(M)$ can indeed be replicated but only the part of it that does not include jumps in $v_{1\tau}$, although not with a self-financed strategy. More specifically, assume that λ is constant, and that default events are independent of the short-term rate. The probability density a representative obligor defaults at any time $x \in [T,T_{bM}]$ given it has survived at time $\tau < T$, is $\phi_\tau(x) \equiv \Pr\{\tau_j \in (x,x+dx)|\text{Surv at } \tau\} = \lambda e^{-\lambda(x-\tau)}dx$, so that the value of $v_{0,\tau,T}$ in the first of Eqs. (5.2) can be expressed as,

$$
\begin{aligned}
v_{0,\tau,T} &= \mathbb{E}_\tau\left[\mathbb{E}_\tau\left(e^{-\int_\tau^x r_s ds}\mathbb{I}_{\{T \le x \le T_{bM}\}}\Big|\mathbb{F}_\tau^r\right)\right] \\
&= \int_T^{T_{bM}} P_\tau(x)\phi_\tau(x)dx \\
&= \lambda \cdot \int_T^{T_{bM}} P_{\text{def},\tau}(x)dx,
\end{aligned}
\tag{5.23}
$$

where $\mathbb{F}_\tau^r$ is the information set including only the path of the short-term rate. Therefore, the value of $v_{0,\tau,T}$ is the same as either (i) the value of a portfolio of a con-

---

[4]Note that v_{1t} is the value at t of a basket of securities paying off contingent upon default of a representative name not having occurred prior to time T, with $T \le T_0$, in which case the value drops to zero. It can be shown that $Lv_1 - rv_1 + \lambda(0 - v_1) = 0$, where L is the infinitesimal generator for jump diffusions, and the value "0" in $\lambda(0 - v_1)$ captures the jump to zero due to default of the representative obligor prior to time T.

[5]As explained below, variance contracts need to be hedged through portfolios that include assets that have a value correlated with the forward positions in the CDS indexes.

Table 5.1 Replication of the credit variance contract of Definition 5.3. FS-CDX stands for forward starting position in the CDX index, and CDX options stands for CDS index options

Portfolio	Value at t	Value at T
(i) long self-financed portfolio of FS-CDX	0	$(V_M(t,T) + 2\ln\frac{\text{CDX}_T(M)}{\text{CDX}_t(M)}) \times \mathcal{N}_T v_{1T}$
(ii) short FS-CDX, long OTM CDX options	$-\mathbb{F}_{\text{var},M}(t,T)$	$-2\ln\frac{\text{CDX}_T(M)}{\text{CDX}_t(M)} \times \mathcal{N}_T v_{1T}$
(iii) borrow $\mathbb{F}_{\text{var},M}(t,T)$	$+\mathbb{F}_{\text{var},M}(t,T)$	$-\frac{\mathbb{F}_{\text{var},M}(t,T)}{P_t(T)} = -\mathbb{P}_{\text{var},M}(t,T)$
Net cash flows	0	$V_M(t,T) \times \mathcal{N}_T v_{1T} - \mathbb{P}_{\text{var},M}(t,T)$

Table 5.2 Replication of the standardized credit variance contract of Definition 5.3. FS-CDX stands for forward position in the CDX index, CDX options stands for CDS index options, and DZCB stands for defaultable zero coupon bonds

Portfolio	Value at t	Value at T
(i) long self-financed portfolio of FS-CDX	0	$(V_M(t,T) + 2\ln\frac{\text{CDX}_T(M)}{\text{CDX}_t(M)}) \times \mathcal{N}_T v_{1T}$
(ii) short FS-CDX, long OTM CDX options	$-\mathbb{F}_{\text{var},M}(t,T)$	$-2\ln\frac{\text{CDX}_T(M)}{\text{CDX}_t(M)} \times \mathcal{N}_T v_{1T}$
(iii) borrow basket of DZCB for $\mathbb{P}^*_{\text{var},M}(t,T) \times v_{1t}$	$+\mathbb{F}_{\text{var},M}(t,T)$	$-\mathbb{P}^*_{\text{var},M}(t,T) \times \mathcal{N}_T v_{1T}$
Net cash flows	0	$[V_M(t,T) - \mathbb{P}^*_{\text{var},M}(t,T)] \times \mathcal{N}_T v_{1T}$

tinuum of non-defaultable bonds maturing at $x \in [T, T_{bM}]$, with weights $\phi_\tau(x)$ as in the second line of Eq. (5.23), or (ii) the value of a portfolio of a continuum of defaultable bonds maturing at $x \in [T, T_{bM}]$, with weights λ as in the third line of Eq. (5.23). In Appendix D.3, we show how $\widehat{\text{CDX}}_\tau(M)$ in Eq. (5.22) can be replicated up to the jump component of $v_{1\tau}$. Once again, this replication strategy is not self-financed.

To hedge the forward variance contract in Definition 5.1, we use the portfolios in rows (i) and (ii) of Tables 5.1 and 5.2. To hedge against the contract underlying the credit variance swap rate of Definition 5.2, $\mathbb{P}_{\text{var},M}(t,T)$, we utilize two identical zero-cost portfolios, which include the following positions:

(i) A dynamic position in the forward starting index (FS-CDX, for brevity), aiming to replicate $\int_t^T \frac{d\text{CDX}_s(M)}{\text{CDX}_s(M)} = \frac{1}{2}V_M(t,T) + \ln\frac{\text{CDX}_T(M)}{\text{CDX}_t(M)}$, rescaled by the defaultable-PVBP over the outstanding nominal, $\mathcal{N}_T v_{1T}$ at T.
(ii) Static positions in the FS-CDX starting at T and CDS index options expiring at T, aiming to replicate the payoff of a *log-contract* (see Chap. 2) on the FS-CDX, rescaled by $\mathcal{N}_T v_{1T}$ at T. The positions are as follows:

   (ii.1) Short $1/\text{CDX}_t(M)$ units of a FS-CDX struck at $\text{CDX}_t(M)$.
   (ii.2) Long out-of-the-money (OTM) CDS index options with weights equal to $\frac{\Delta K_i}{K_i^2}$, where ΔK_i and K_i are as in Eq. (5.15).

(iii) A static borrowing position financing the portfolio in (ii.2).

Table 5.3 Replication of the BP-credit variance contract of Definition 5.4-(b). FS-CDX stands for forward position in the CDX index, CDX options stands for CDS index options, and DZCB stands for defaultable zero coupon bonds

Portfolio	Value at t	Value at T
(i) short self-financed portfolio of FS-CDX	0	$[V_M^{bp}(t,T) - (\text{CDX}_T^2(M) - \text{CDX}_t^2(M))] \times \mathcal{N}_T v_{1T}$
(ii) long FS-CDX, long OTM CDX options	$-\mathbb{F}_{var,M}^{bp}(t,T)$	$(\text{CDX}_T^2(M) - \text{CDX}_t^2(M)) \times \mathcal{N}_T v_{1T}$
(iii) borrow $\mathbb{F}_{var,M}^{bp}(t,T)$	$+\mathbb{F}_{var,M}^{bp}(t,T)$	$-\frac{\mathbb{F}_{var,M}^{bp}(t,T)}{P_t(T)} = -\mathbb{P}_{var,M}^{bp}(t,T)$
Net cash flows	0	$V_M^{bp}(t,T) \times \mathcal{N}_T v_{1T} - \mathbb{P}_{var,M}^{bp}(t,T)$

Table 5.4 Replication of the standardized BP-credit variance contract of Definition 5.4-(c). FS-CDX stands for forward position in the CDX index, CDX options stands for CDS index options, and DZCB stands for defaultable zero coupon bonds

Portfolio	Value at t	Value at T
(i) short self-financed portfolio of FS-CDX	0	$[V_M^{bp}(t,T) - (\text{CDX}_T^2(M) - \text{CDX}_t^2(M))] \times \mathcal{N}_T v_{1T}$
(ii) long swaps, long OTM CDX options	$-\mathbb{F}_{var,M}^{bp}(t,T)$	$(\text{CDX}_T^2(M) - \text{CDX}_t^2(M)) \times \mathcal{N}_T v_{1T}$
(iii) borrow basket of DZCB for $\mathbb{P}_{var,n}^{*bp}(t,T) \times v_{1t}$	$+\mathbb{F}_{var,M}^{bp}(t,T)$	$-\mathbb{P}_{var,M}^{*bp}(t,T) \times \mathcal{N}_T v_{1T}$
Net cash flows	0	$[V_M^{bp}(t,T) - \mathbb{P}_{var,M}^{*bp}(t,T)] \times \mathcal{N}_T v_{1T}$

Table 5.1 provides the value of these portfolios both at inception and at time T, which match the payoff in Eq. (5.13). Appendix D.3 provides additional details regarding the entries in Table 5.1.

Finally, to replicate the payoff in Eq. (5.14) underlying the standardized contract in Definition 5.3, we proceed as in Table 5.1, with the exception that row (iii) is now replaced with a borrowing position comprised of defaultable zero coupon bonds with value v_{1t}, notional $\mathbb{P}_{var,n}^*(t,T)$, default intensity equal to that of the representative obligor, and zero recovery value, as in Table 5.2. By Eq. (5.17), this borrowing position amounts to $\mathbb{F}_{var,M}(t,T)$, which is exactly what is needed to finance the CDS index options position in row (ii). Come time T, it will be unwound for a value of $-\mathbb{P}_{var,M}^*(t,T) \times \mathcal{N}_T v_{1T}$. The net cash flows are precisely those of the contract with the standardized credit variance swap rate.

Next, we deal with basis point contracts, which require positioning on *quadratic* contracts (see Chap. 2) on the FS-CDX and not on *log-contracts* as in the percentage contract cases of Tables 5.1 and 5.2. In the context of this chapter, quadratic contracts are those that deliver $\text{CDX}_T^2(M) - \text{CDX}_t^2(M)$. They are the counterparts to the quadratic contracts used in Chaps. 3 and 4 while dealing with interest rate swaps, government bonds and time deposits.

The BP-forward in Definition 5.4-(a) can be replicated through the positions in rows (i) and (ii) of Tables 5.3 and 5.4. To replicate the payoff relating to the BP-credit variance swap rate of Definition 5.4-(b), we use portfolios with the following positions:

(i) A dynamic, short position in the forward starting index, FS-CDX, aiming to replicate $-\int_t^T \text{CDX}_s(M)d\text{CDX}_s(M) = \frac{1}{2}[V_M^{\text{bp}}(t,T) - (\text{CDX}_T^2(M) - \text{CDX}_t^2(M))]$, rescaled by the defaultable-PVBP over the outstanding nominal, $\mathcal{N}_T v_{1T}$ at time T.
(ii) Static positions in the FS-CDX starting at T and CDS index options expiring at T, aiming to replicate the payoff of a *quadratic contract* on the FS-CDX, rescaled by $\mathcal{N}_T v_{1T}$ at T. The positions are:

(ii.1) Long $2\text{CDX}_t(M)$ units of a FS-CDX struck at $\text{CDX}_t(M)$.
(ii.2) Long OTM CDS index options with weights equal to $2\Delta K_i$, where ΔK_i and K_i are as in Eq. (5.15).

(iii) A static borrowing position aimed to finance the OTM CDS index option positions in (ii.2).

Table 5.3 provides costs and payoffs of these portfolios, resulting in a replication of the contract payoff in Definition 5.4-(b).

Finally, Table 5.4 provides details regarding the replication of the standardized BP-credit variance contract of Definition 5.4-(c). It is obtained by modifying the arguments leading to Table 5.3, similarly as we did when deriving Table 5.2.

5.4 Credit Volatility Indexes

5.4.1 Definitions

The credit variance contracts introduced in Sect. 5.3 naturally lead to two indexes of expected volatility. The first index is based on the *percentage* standardized credit variance swap rate of Definition 5.3, and is defined as

$$\text{C-VI}_M(t,T) \equiv 100 \times \sqrt{\frac{1}{T-t}\mathbb{P}^*_{\text{var},M}(t,T)} \tag{5.24}$$

where $\mathbb{P}^*_{\text{var},M}(t,T)$ is as in Eq. (5.17). The second index, based on the *basis point* standardized credit variance swap rate of Definition 5.4-(c), is defined as

$$\text{C-VI}_M^{\text{bp}}(t,T) \equiv 100^2 \times \sqrt{\frac{1}{T-t}\mathbb{P}^{*\text{bp}}_{\text{var},M}(t,T)} \tag{5.25}$$

where $\mathbb{P}^{*\text{bp}}_{\text{var},M}(t,T)$ is as in Eq. (5.19). Naturally, both indexes, by reflecting the fair value of the credit variance contracts of Sect. 5.3, also cover the single-name case when $n=1$.

5.4.2 Forward Premium Adjustments

When quotes for ATM price are unavailable, the theoretical value of the previous indexes can be approximated similarly as in the government bond and time deposit case dealt with in Chap. 4 (see, e.g., Sect. 4.2.7). Let K_0 denote the first strike below $CDX_t(M)$, where in case $CDX_t(M)$ is not observed, it is approximated by the strike price at which the absolute difference between the payer and receiver prices is smallest. (If there is an option struck at $CDX_t(M)$, then, $K_0 = CDX_t(M)$.)

In Appendix D.2, we show that the percentage index and the basis point index are approximated by the following expressions:

$$
C\text{-VI}_{M,o}(t, T) \equiv 100 \times \sqrt{\frac{\mathbb{P}^*_{o,\text{var},M}(t, T)}{T - t}},
$$

$$
C\text{-VI}^{bp}_{M,o}(t, T) \equiv 100^2 \times \sqrt{\frac{\mathbb{P}^{*bp}_{o,\text{var},M}(t, T)}{T - t}}
$$

where

$$
\mathbb{P}^*_{o,\text{var},M}(t, T) \equiv \frac{2}{v_{1t}} \left(\sum_{i:K_i < CDX_t(M)} SW^r_t(K_i, T, M) \frac{1}{K_i^2} \Delta K_i \right.
$$

$$
\left. + \sum_{i:K_i \geq CDX_t(M)} SW^p_t(K_i, T, M) \frac{1}{K_i^2} \Delta K_i \right)
$$

$$
- \left(\frac{CDX_t(M) - K_0}{K_0} \right)^2, \tag{5.26}
$$

and

$$
\mathbb{P}^{*bp}_{o,\text{var},M}(t, T) \equiv \frac{2}{v_{1t}} \left(\sum_{i:K_i < CDX_t(M)} SW^r_t(K_i, T, M) \Delta K_i \right.
$$

$$
\left. + \sum_{i:K_i \geq CDX_t(M)} SW^p_t(K_i, T, M) \Delta K_i \right)
$$

$$
- \left(CDX_t(M) - K_0 \right)^2. \tag{5.27}
$$

5.4.3 Differences with Respect to Other Fixed Income Volatility Gauges

The credit volatility indexes in this section share similarities with the CBOE SRVIX index of interest rate swap volatility (see Chap. 3, and Mele and Obayashi 2012):

they are built on derivatives (options to enter into default swaps), just as the SRVIX is built on interest rate swaptions (options to enter into interest rate swaps), both of which are only traded over-the-counter at the time of writing.

Despite such similarities, the risks inherent in the transactions underlying these indexes are fundamentally different. For credit spread volatility, the main underlying risk is default risk, and requires variance contracts specifically designed to capture these risks. Specifically, the credit event feature requires a design of variance contracts valued as martingales under the *survival contingent probability*. This probability relates to a notion more general than the *swap probability* underlying the variance contracts on interest rate swaps. Note that the survival contingent probability collapses to the swap probability only when default events have zero probability.

Finally, credit volatility relates to baskets of defaultable names, which raises additional issues not covered in the designs of Chap. 3, such as the shrinking notional of the underlying reference entities, which is fully reflected into the payoffs of the resulting variance contracts in this chapter.

5.4.4 Implementation Example

We implement one illustration of the main steps required to calculate the credit volatility indexes C-VI$_M(t, T)$ and C-VI$_M^{bp}(t, T)$ in Eqs. (5.24) and (5.25). We utilize data reflecting hypothetical market conditions on April 20, 2012, regarding percentage implied volatilities for CDS index options maturing in 2 months on an on-the-run 5-year tenor index. The first two columns of Table 5.5 report strikes K in basis points, and percentage implied volatilities for each strike, denoted by IV(K) ("Percentage Implied Vol"). The at-the-money (ATM) strike is $K = 105$.

The indexes are implemented by first plugging the "skew" IV(K) into the Black (1976) formula, and then inserting Black's formula in Eqs. (5.24) and (5.25), leaving

C-VI$_M(t, T)$

$$
= 100 \times \sqrt{\frac{2}{T-t} \left[\sum_{i:K_i < \text{CDX}_t} \frac{\hat{Z}(\text{CDX}_t, T, K_i; (T-t)\text{IV}^2(K_i))}{K_i^2} \Delta K_i + \sum_{i:K_i \geq \text{CDX}_t} \frac{Z(\text{CDX}_t, T, K_i; (T-t)\text{IV}^2(K_i))}{K_i^2} \Delta K_i \right]},
$$
(5.28)

and

C-VI$_M^{bp}(t, T)$

$$
= 100^2 \times \sqrt{\frac{2}{T-t} \left[\sum_{i:K_i < \text{CDX}_t} \hat{Z}(\text{CDX}_t, T, K_i; (T-t)\text{IV}^2(K_i)) \Delta K_i + \sum_{i:K_i \geq \text{CDX}_t} Z(\text{CDX}_t, T, K_i; (T-t)\text{IV}^2(K_i)) \Delta K_i \right]},
$$
(5.29)

where

$$
\hat{Z}(\text{CDX}, T, K_i; (T-t)\text{IV}^2(K_i))
$$
$$
= Z(\text{CDX}, T, K_i; (T-t)\text{IV}^2(K_i)) + K - \text{CDX},
$$
(5.30)

Table 5.5 Black's skew and prices

Strike (in basis points)	Percentage implied vol	Black's prices	
		Receiver options ($\hat{Z}$)	Payer options (Z)
80	48.00	$6.7430 \cdot 10^{-5}$	$2.5674 \cdot 10^{-3}$
85	47.50	$0.1279 \cdot 10^{-3}$	$2.1279 \cdot 10^{-3}$
90	49.50	$0.2512 \cdot 10^{-3}$	$1.7512 \cdot 10^{-3}$
95	50.50	$0.4151 \cdot 10^{-3}$	$1.4154 \cdot 10^{-3}$
100	54.00	$0.6714 \cdot 10^{-3}$	$1.1714 \cdot 10^{-3}$
105 (ATM)	55.00	$0.9385 \cdot 10^{-3}$	$0.9385 \cdot 10^{-3}$
110	56.00	$1.2484 \cdot 10^{-3}$	$0.7484 \cdot 10^{-3}$
115	57.50	$1.6035 \cdot 10^{-3}$	$0.6035 \cdot 10^{-3}$
120	59.50	$1.9967 \cdot 10^{-3}$	$0.4967 \cdot 10^{-3}$
125	61.50	$2.4130 \cdot 10^{-3}$	$0.4130 \cdot 10^{-3}$
130	62.50	$2.8339 \cdot 10^{-3}$	$0.3339 \cdot 10^{-3}$
135	63.50	$3.2710 \cdot 10^{-3}$	$0.2710 \cdot 10^{-3}$
140	65.50	$3.7319 \cdot 10^{-3}$	$0.2319 \cdot 10^{-3}$
145	66.50	$4.1910 \cdot 10^{-3}$	$0.1910 \cdot 10^{-3}$

$$Z(\text{CDX}, T, K; V) = \text{CDX} \cdot \Phi(d) - K\Phi(d - \sqrt{V}), \quad d = \frac{\ln \frac{\text{CDX}}{K} + \frac{1}{2}V}{\sqrt{V}}, \quad (5.31)$$

Φ denotes the cumulative standard normal distribution, and CDX and K are expressed in percentage terms.

Equations (5.28) and (5.29) follow, because the value of each CDS index option payer is

$$\mathbb{E}_t\left[e^{-\int_t^T r_\tau d\tau} \mathcal{N}_T v_{1T}\left(\text{CDX}_T(M) - K\right)^+\right]$$
$$= v_{1t} \cdot \mathbb{E}_t^{\text{sc}}\left(\text{CDX}_T(M) - K\right)^+$$
$$= v_{1t} \cdot Z\left(\text{CDX}_t(M), T, K; (T-t)\text{IV}^2(K)\right),$$

so that the contribution of each OTM CDS index options payer to the index is simply $Z(\cdot)$ in both Eq. (5.24) and Eq. (5.25). The same reasoning applies to the out-of-the-money CDS index option receivers.

Percentage implied volatilities, $\text{IV}(K)$, are plugged into Eqs. (5.30)–(5.31), to obtain values for $\hat{Z}$ and Z. The third and fourth column of Table 5.5 provide CDS index option prices divided by the defaultable-PVBP; that is, the values of $\hat{Z}$ and Z ("Black's prices"), for each strike.

Table 5.6 contains details of the index calculation: the second column displays the type of out-of-the-money CDS index option entering into the calculation; the third column has the Black's price corresponding to the used CDS index option; the fourth and fifth columns report the weights each Black's price bears towards the final

Table 5.6 Calculation of volatility indexes

Strike (in basis points)	Option type	Price	Weights Basis point ΔK_i	Percentage $\Delta K_i / K_i^2$	Contributions to strikes Basis point contribution	Percentage contribution
80	Receiver	$6.7430 \cdot 10^{-5}$	0.0005	7.8125	$3.3715 \cdot 10^{-8}$	$0.5268 \cdot 10^{-3}$
85	Receiver	$0.1279 \cdot 10^{-3}$	0.0005	6.9204	$6.3951 \cdot 10^{-8}$	$0.8851 \cdot 10^{-3}$
90	Receiver	$0.2512 \cdot 10^{-3}$	0.0005	6.1728	$1.2563 \cdot 10^{-7}$	$1.5510 \cdot 10^{-3}$
95	Receiver	$0.4151 \cdot 10^{-3}$	0.0005	5.5401	$2.0774 \cdot 10^{-7}$	$2.3018 \cdot 10^{-3}$
10	Receiver	$0.6714 \cdot 10^{-3}$	0.0005	5.0000	$3.3574 \cdot 10^{-7}$	$3.3574 \cdot 10^{-3}$
105	ATM	$0.9385 \cdot 10^{-3}$	0.0005	4.5351	$4.6929 \cdot 10^{-7}$	$4.2566 \cdot 10^{-3}$
110	Payer	$0.7484 \cdot 10^{-3}$	0.0005	4.1322	$3.7421 \cdot 10^{-7}$	$3.0926 \cdot 10^{-3}$
115	Payer	$0.6035 \cdot 10^{-3}$	0.0005	3.7807	$3.0179 \cdot 10^{-7}$	$2.2820 \cdot 10^{-3}$
120	Payer	$0.4967 \cdot 10^{-3}$	0.0005	3.4722	$2.4837 \cdot 10^{-7}$	$1.7248 \cdot 10^{-3}$
125	Payer	$0.4130 \cdot 10^{-3}$	0.0005	3.2000	$2.0651 \cdot 10^{-7}$	$1.3216 \cdot 10^{-3}$
130	Payer	$0.3339 \cdot 10^{-3}$	0.0005	2.9585	$1.6695 \cdot 10^{-7}$	$0.9879 \cdot 10^{-3}$
135	Payer	$0.2710 \cdot 10^{-3}$	0.0005	2.7434	$1.3551 \cdot 10^{-7}$	$0.7435 \cdot 10^{-3}$
140	Payer	$0.2319 \cdot 10^{-3}$	0.0005	2.5510	$1.1597 \cdot 10^{-7}$	$0.5917 \cdot 10^{-3}$
145	Payer	$0.1910 \cdot 10^{-3}$	0.0005	2.3781	$9.5530 \cdot 10^{-8}$	$0.4543 \cdot 10^{-3}$
SUMS					$2.8809 \cdot 10^{-6}$	$2.4077 \cdot 10^{-2}$

index calculation, before the final rescaling of $\frac{2}{T-t}$; finally, the sixth and seventh columns report each out-of-the-money CDS index option price corrected by the appropriate weight: each price in the third column multiplied by the corresponding weight in the fourth column ("Basis Point Contribution") and each price in the third column multiplied by the corresponding weight in the fifth column ("Percentage Contribution").

Finally, we evaluate Eq. (5.28) and Eq. (5.29), obtaining:

$$C\text{-VI} \equiv 100 \times \sqrt{\frac{2}{(2/12)} \times 2.4077 \cdot 10^{-2}} = 53.7524,$$

$$C\text{-VI}_M^{bp} \equiv 100^2 \times \sqrt{\frac{2}{(2/12)} \times 2.8809 \cdot 10^{-6}} = 58.7975.$$

In comparison, ATM implied volatilities are 55.00 % and $55 \times 1.05 = 57.75$ basis points.

5.4.5 Index Design Through Option Cycles

In practice, CDS and CDS index options may trade in cycles based on standardized roll dates. For example, most standardized CDS and CDS indexes have quarterly and semi-annual roll dates, respectively, in over-the-counter trading. In these cases, the time to expiry of traded options changes everyday, leading to the same issues encountered while dealing with the government bond and time-deposit volatility indexes in Chap. 4 (see Sect. 4.5).

To calculate a credit volatility index reflecting a constant maturity, we can rely on the two estimators introduced in Sect. 4.5: (i) a "sandwich combination," which weighs options with different maturities such that the resulting weighted volatility index has a constant maturity at each point in time, e.g. three months; and (ii) a "rolling index," which uses options in the same cycle to estimate the expected volatility left to the exhaustion of this cycle.

5.5 Post "Big-Bang" Conventions and Index Adjustments

In 2009, the ISDA released a protocol aiming to revamp the credit market through modifications to standardized CDS contracts.[6] One of the ensuing (post-"Big-Bang") conventions relies on the assumption of a constant hazard rate for each

---

[6]See the 2009 ISDA Credit Derivatives Determinations Committees and Auction Settlement CDS Protocol ("Big Bang" Protocol).

maturity. We describe how this assumption affects some of the evaluation formulae in Sect. 5.2. Moreover, the pricing protocol in Sect. 5.2 relies on the assumption that index swaptions are struck as if they were cash-settled at any strike K or that swaptions exist for strikes equal to any arbitrary coupon C_t. We review this assumption and ways to deal with alternatives that lead to corrections to the index formulae in Sect. 5.4.

5.5.1 Index Values Under Constant Hazard Rates

As in Sect. 5.2, we consider indexes with fixed coupons C_t, such that at origination, the index value, DSX_t, is as in Eq. (5.1). The *index spread* is defined as the value of the coupon C_t that makes $\mathrm{DSX}_t = 0$, i.e.,

$$S_t \equiv \mathrm{LGD} \frac{v_{0t}}{\frac{1}{b} v_{1t}}, \tag{5.32}$$

so that the index value can be written as

$$\mathrm{DSX}_t = \frac{1}{b}(S_t - C_t) v_{1t}. \tag{5.33}$$

We now evaluate v_{0t} and v_{1t} based on some assumptions. First, we assume a constant hazard rate λ. By Eqs. (5.2) and derivations in Eq. (5.23), we know that

$$v_{0,\tau,T}(\lambda) = \lambda \int_T^{T_{bM}} e^{-\lambda(x-\tau)} P_\tau(x) dx \quad \text{and} \quad v_{1\tau}(\lambda) = \sum_{i=1}^{bM} \mathbb{E}_t \left(e^{-\int_t^{T_i}(r_u+\lambda)du} \right),$$

where we are emphasizing that both $v_{0,\tau,T}$ and $v_{1\tau}$ are functions of λ.

Assuming also that the short-term rate is constant and equal to r leaves

$$
\begin{aligned}
v_{0,\tau,T}(\lambda) &= e^{-(r+\lambda)(T-\tau)} \frac{\lambda}{\lambda+r} \left(1 - e^{-(\lambda+r)M}\right), \\
v_{1\tau}(\lambda) &= e^{-(r+\lambda)(T-\tau)} \frac{e^{-\frac{1}{b}(\lambda+r)}}{1 - e^{-\frac{1}{b}(\lambda+r)}} \left(1 - e^{-(\lambda+r)M}\right),
\end{aligned}
\tag{5.34}
$$

where we recall that $v_{0t} \equiv v_{0,t,t}$.

Expressed as a function of λ, the index value in Eq. (5.1) is

$$\mathrm{DSX}_t(\lambda) = \mathrm{LGD} \cdot v_{0t}(\lambda) - C_t \frac{1}{b} v_{1t}(\lambda).$$

Note that $v_{0t}(\lambda)$ is increasing in λ and $v_{1t}(\lambda)$ is decreasing in λ. Therefore, the mapping $\lambda \longmapsto \mathrm{DSX}_t(\lambda)$ is increasing so that for LGD large enough, there exists a positive $\bar{\lambda}$ that makes $\mathrm{DSX}_t(\bar{\lambda}) = \mathrm{DSX}_t^\$$—the index market value. In terms of this $\bar{\lambda}$, the index spread in Eq. (5.32) is

$$\bar{S}_t \equiv s_t(\bar{\lambda}) \equiv \mathrm{LGD}\frac{v_{0t}(\bar{\lambda})}{\frac{1}{b}v_{1t}(\bar{\lambda})}, \qquad (5.35)$$

which is referred to as the *quoted spread* or *flat spread*. Alternatively, note that one can define $\bar{\lambda}$ as the number that yields the flat spread $\bar{S}_t$ through Eq. (5.35). That is, define the mapping $S_t \longmapsto \lambda = s_t^{-1}[S_t]$, where $s_t^{-1}[\cdot]$ denotes the inverse function of $s_t(\lambda)$ in Eq. (5.35). Then, clearly, $\bar{\lambda} = s_t^{-1}[\bar{S}_t]$. In terms of $\bar{S}_t$, then, Eq. (5.33) is

$$\mathrm{DSX}_t(\bar{S}_t) = \frac{1}{b}(\bar{S}_t - C_t)v_{1t}(\bar{S}_t),$$

where, with a slight abuse of notation, we set

$$v_{1t}(\bar{S}_t) \equiv v_{1t}\big(s_t^{-1}[\bar{S}_t]\big), \qquad (5.36)$$

and the R.H.S. of this identity is obtained through the second of Eqs. (5.34). We refer to $v_{1t}(x)$ as the *flat defaultable annuity for a spread level equal to* x.

5.5.2 Forward Positions

Consider a forward starting position (at T) in an index initiated at t and with fixed coupons C_t. The value at any time τ of this *forward starting index* is given by $\mathrm{DSX}_\tau(\tau, T)$ in Eq. (5.3). Define Fw_τ as the coupon C_τ at τ such that $\mathrm{DSX}_\tau(\tau, T) = 0$. It is easy to verify that

$$\mathrm{Fw}_\tau = \mathrm{LGD}\frac{v_{0,\tau,T}}{\frac{1}{b}v_{1\tau}}, \qquad \mathrm{Fw}_T \equiv \bar{S}_T, \qquad (5.37)$$

where $\bar{S}_T$ satisfies Eq. (5.35). We have

$$\mathrm{DSX}_\tau(t, T) = \frac{1}{b}\mathcal{N}_\tau v_{1\tau}(\mathrm{Fw}_\tau - C_t), \quad \tau \in [t, T]. \qquad (5.38)$$

We refer to Fw_τ as the *forward spread*. As explained in Sect. 5.2.3, the holder of a payer option on an index is entitled to a front-end protection, leading to the definition of the *loss-adjusted forward spread*, $\mathrm{CDX}_\tau(M)$ in Eq. (5.6) and the expression for the *loss-adjusted forward starting index value*, $\mathrm{DSX}_\tau^L(t, T)$ in Eq. (5.7), which at the option expiry (T say) is

$$\mathrm{DSX}_T^L(t, T) = \frac{1}{b}\mathcal{N}_T v_{1T}\big(\mathrm{CDX}_T(M) - C_t\big). \qquad (5.39)$$

Equation (5.39) forms the basis for option evaluation as elaborated below.

5.5.3 Option Payoffs and Evaluation

Options to enter into an on-the-run index may be struck at spreads differing from the initial contractual coupon, C_t. The contract then requires a strike adjustment proportional to the difference between some value K and the coupon C_t. Only when $K = C_t$ would the option payoffs collapse to those considered in Sect. 5.2.3. Precisely, the strike adjustment is contractually equal to $\frac{1}{b} H_T(C_t, K)$, where

$$H_T(C_t, K) \equiv (K - C_t) v_{1T}(K), \tag{5.40}$$

and $v_{1T}(K)$ denotes the flat defaultable annuity for spread level equal to K (see Eq. (5.36)). The final payoff of, say, a payer option is $(\mathrm{DSX}_T^L(t, T) - \frac{1}{b} H_T(C_t, K))^+$, where $\mathrm{DSX}_T^L(t, T)$ is as in Eq. (5.39).[7]

The rationale behind the strike adjustment in Eq. (5.40) is that once the holder of a payer swaption exercises, he will be long the on-the-run index, and be compensated for the difference between the index value and the value of the non-traded index with spread K. Accordingly, the payer and receiver prices that are the counterparts to Eqs. (5.9) are:

$$\mathrm{SW}_\tau^p \big(\mathcal{K}_T(C_t, K), T; M \big) \equiv \frac{1}{b} \mathcal{N}_\tau v_{1\tau} \mathbb{E}_\tau^{\mathrm{sc}} \big[(\mathrm{CDX}_T(M) - \mathcal{K}_T(C_t, K))^+ \big], \quad \text{and}$$

$$\mathrm{SW}_\tau^r \big(\mathcal{K}_T(C_t, K), T; M \big) \equiv \frac{1}{b} \mathcal{N}_\tau v_{1\tau} \mathbb{E}_\tau^{\mathrm{sc}} \big[(\mathcal{K}_T(C_t, K) - \mathrm{CDX}_T(M))^+ \big], \tag{5.41}$$

where

$$\mathcal{K}_T(C_t, K) \equiv C_t + \frac{K - C_t}{\mathcal{N}_T v_{1T}} v_{1T}(K). \tag{5.42}$$

Naturally, when $K = C_t$ (as in Sect. 5.2), $\mathcal{K}_T(C_t, K)|_{K=C_t} = K$ so that Eqs. (5.41) collapse to Eqs. (5.9). However, when $K \neq C_t$, we cannot rely on $\mathcal{N}_\tau v_{1\tau}$ as a numéraire, similarly as in Sect. 5.2 and in previous chapters. The pricing formulae in Eqs. (5.41) are indeed "exact," in that they rely on no other assumptions than those in Sect. 5.2. At the same time, they suggest that swaptions can be thought of as fictitious options with a random strike, $\mathcal{K}_T(C_t, K)$: even assuming that interest rates are deterministic, the notional-adjusted risky annuity in (5.42), $\mathcal{N}_T v_{1T}$, is unknown at t. This randomness complicates evaluation. We now explain two non-limitative approaches to dealing with these complications: a first, based on approximations that leads to a standard Black formula with modified strikes; and a second, based on a model that aims to explicitly take into account the intricacies regarding the previous "random" strikes.

---

[7]Alternatively, one can consider $v_{1T}(K)$ in Eq. (5.40) as being a flat annuity on a reduced notional, i.e., $v_{1T}^{\mathcal{N}}(K) \equiv v_{1T}(K) \mathcal{N}_T$. An approximation similar to that in Eqs. (5.43) would follow while approximating v_{1T} with $\frac{v_{1t}}{P_t(T)}$.

5.5.3.1 Modified Market Formula

We approximate the option prices in (5.41) with ones struck at deterministic strikes. Namely, replace $\mathcal{N}_T v_{1T}$ in Eq. (5.42) with its conditional expectation under Q_{sc}, $\mathbb{E}_t^{sc}(\mathcal{N}_T v_{1T})$. Under the assumption that r is constant, one has that $\mathcal{N}_T v_{1T} \approx \mathbb{E}_t^{sc}(\mathcal{N}_T v_{1T}) = \frac{\mathcal{N}_\tau v_{1\tau}}{P_\tau(T)}$, which leaves the following approximation to $\mathrm{SW}_t^p(\cdot)$ and $\mathrm{SW}_t^r(\cdot)$:

$$\widehat{\mathrm{SW}}_\tau^p\big(\hat{\mathcal{K}}_T(C_t, K), T; M\big) \equiv \frac{1}{b}\mathcal{N}_\tau v_{1\tau}\mathbb{E}_\tau^{sc}\big[\big(\mathrm{CDX}_T(M) - \hat{\mathcal{K}}_T(C_t, K)\big)^+\big], \quad \text{and}$$

$$\widehat{\mathrm{SW}}_\tau^r\big(\hat{\mathcal{K}}_T(C_t, K), T; M\big) \equiv \frac{1}{b}\mathcal{N}_\tau v_{1\tau}\mathbb{E}_\tau^{sc}\big[\big(\hat{\mathcal{K}}_T(C_t, K) - \mathrm{CDX}_T(M)\big)^+\big],$$

$$(5.43)$$

where

$$\hat{\mathcal{K}}_T(C_t, K) \equiv C_t + (K - C_t)\frac{P_\tau(T)v_{1T}(K)}{\mathcal{N}_\tau v_{1\tau}}.$$

Note that Eqs. (5.43) imply that as soon as $K = C_t$, $\hat{\mathcal{K}}_T(C_t, K)|_{K=C_t} = K$, so that swaptions can be evaluated through an exact Black pricer (as in Sect. 5.2). For other strikes, Black pricers can still be used to approximate swaption values relying on strikes set equal to $\hat{\mathcal{K}}_T(C_t, K)$.

5.5.3.2 Pedersen's Model

Pedersen (2003) considers a model that has become a standard at the time of writing. Define X_T as the random variable satisfying the equation X_T : $\frac{1}{b}(X_T - C_t)v_{1T}(X_T) = \mathrm{DSX}_T^L(t, T)$, where $v_{1T}(X_T)$ denotes the flat defaultable annuity for spread X_T as defined in Eq. (5.36), and $\mathrm{DSX}_T^L(t, T)$ is as in Eq. (5.39). In terms of X_T, the price of a payer swaption can be written as:

$$\overline{\mathrm{SW}}_\tau^p\big(H_T(C_t, K), T; M\big) \equiv \frac{1}{b}P_\tau(T)\mathbb{E}_\tau^{Q_{FT}}\big[\big((X_T - C_t)v_{1T}(X_T) - H_T(C_t, K)\big)^+\big],$$

$$(5.44)$$

where $\mathbb{E}_\tau^{Q_{FT}}$ denotes the expectation under the forward probability Q_{FT} (see Chap. 4, Eq. (4.13)).

Next, define the conditional expectation $\mathcal{F}_\tau(T) \equiv b \cdot \mathbb{E}_\tau^{Q_{FT}}(\mathrm{DSX}_T^L(t, T))$, and the ATM strike as the strike K_{atm} that equalizes payer and receiver swaptions. By the put-call parity, K_{atm} satisfies

$$H_T(C_t, K_{\mathrm{atm}}) = \mathcal{F}_\tau(T) = \mathbb{E}_\tau^{Q_{FT}}\big[(X_T - C_t)v_{1T}(X_T)\big], \qquad (5.45)$$

where the second equality follows by the definition of X_T. We refer to $\mathcal{F}_\tau(T)$ as the (annualized) *ATM forward price*. It is the time τ price for "delivery" of the forward starting index value (at T) and equals

$$\mathcal{F}_\tau(T) = b \cdot \mathbb{E}_\tau^{Q_{FT}} \left(\mathrm{DSX}_T^L(t, T) \right)$$

$$= \mathbb{E}_\tau^{Q_{FT}} \left[\mathcal{N}_T v_{1T} \left(\mathrm{CDX}_T(M) - C_t \right) \right]$$

$$= \frac{1}{P_\tau(T)} \mathcal{N}_\tau v_{1\tau} \mathbb{E}_\tau^{sc} \left(\mathrm{CDX}_T(M) - C_t \right)$$

$$= \frac{1}{P_\tau(T)} \mathcal{N}_\tau v_{1\tau} \left(\mathrm{CDX}_\tau(M) - C_t \right)$$

$$= \frac{1}{P_\tau(T)} \left[\mathcal{N}_\tau v_{1\tau} (\mathrm{Fw}_\tau - C_t) + b v_\tau^F \right],$$

where the second line follows by Eq. (5.39), the third by a change of probability, the fourth by the martingale property of $\mathrm{CDX}_\tau(M)$ under Q_{sc}, the fifth by Eq. (5.6), the definition of Fw_τ in Eq. (5.37) and by Eq. (5.38), and v_τ^F is as in Eq. (5.5).

Note that Eq. (5.44) holds for any distribution of X_T that satisfies Eq. (5.45). The key assumption underlying Pedersen's model is that under the forward probability,

$$X_T^{x^o} \equiv X_T = X_t \exp \left(-\frac{1}{2} s^2 (T - t) + s \sqrt{T - t} \omega \right), \quad X_t \equiv x^o, \qquad (5.46)$$

where ω is a standard Gaussian variable and s is a constant volatility parameter.

Given a value for s, the initial condition x^o can be calibrated by solving the following nonlinear equation:

$$\mathcal{F}_\tau(T) = \mathbb{E}_\tau^{Q_{FT}} \left[\left(X_T^{x^o} - C_t \right) v_{1T} \left(X_T^{x^o} \right) \right]. \qquad (5.47)$$

Swaption payers and receivers can now be evaluated based on the value of x^o in Eq. (5.47) which matches the ATM forward price, $\hat{x}^o$. Their prices are:

$$\overline{\mathrm{SW}}_\tau^p \left(H_T(C_t, K), T; M \right)$$

$$\equiv \frac{1}{b} P_\tau(T) \mathbb{E}_\tau^{Q_{FT}} \left[\left(\left(X_T^{\hat{x}^o} - C_t \right) v_{1T} \left(X_T^{\hat{x}^o} \right) - H_T(C_t, K) \right)^+ \right], \quad \text{and}$$

$$\overline{\mathrm{SW}}_\tau^r \left(H_T(C_t, K), T; M \right) \tag{5.48}$$

$$\equiv \frac{1}{b} P_\tau(T) \mathbb{E}_\tau^{Q_{FT}} \left[\left(H_T(C_t, K) - \left(X_T^{\hat{x}^o} - C_t \right) v_{1T} \left(X_T^{\hat{x}^o} \right) \right)^+ \right].$$

These prices can be determined once we are given an estimate of the volatility parameter, i.e. s in Eq. (5.46). For example, one can estimate s through the historical volatility of the spread of the underlying index. Alternatively, one could solve for the values of s and x^o that match market prices, as proposed below.

5.5.4 Index Corrections

We develop two modifications of the indexes formulae in Sect. 5.4 based on Eqs. (5.43) and (5.48) in Sect. 5.5.3.

5.5.4.1 Correction Based on the Modified Black Formula

Consider the modified market formulae in Sect. 5.5.3.1. Let $\sigma_{\hat{\mathcal{K}}(C_t, K)}$ denote the Black implied volatility for strike $\hat{\mathcal{K}}(C_t, K)$ obtained by inverting Eqs. (5.43) with respect to market price. Based on $\sigma_{\hat{\mathcal{K}}(C_t, K)}$, define the mappings $k \longmapsto \widehat{SW}_t^p(k, T; M)$ and $k \longmapsto \widehat{SW}_t^r(k, T; M)$, where $\widehat{SW}_t(\cdot)$ are defined in Eqs. (5.43). These mappings can be approximated through interpolations in correspondence of missing values for $\hat{\mathcal{K}}(\cdot)$. Our first index correction consists of calculating the indexes in Sect. 5.4 with these values of $\widehat{SW}_t^p(\cdot, T; M)$ and $\widehat{SW}_t^r(\cdot, T; M)$ plugged into the index formulae of Sect. 5.4.

5.5.4.2 Correction Based on Pedersen's Model

Next, we consider index corrections based on Pedersen's model in Sect. 5.5.3.2. Define:

$$\bar{H}_T(C_t, K) \equiv (K - C_t) v_{1T}\left(X_T^{x^o}\right) \quad \text{and} \quad h_T(C_t, K) \equiv (K - C_t) v_{1T}\left(x^o\right).$$

In words, $\bar{H}_T(C_t, K)$ is the hypothetical strike adjustment such that the swaption prices in this section collapse to the swaption prices in Sect. 5.2; the function $h_T(C_t, K)$ is an approximation to $\bar{H}_T(C_t, K)$ where the unknown spread at T, $X_T^{x^o}$, is replaced with its time t expectation, x^o. We have:

$$SW_t^p(K, T; M) = \frac{1}{b} P_\tau(T) \mathbb{E}_\tau^{Q_{FT}}\left[\left(\left(X_T^{\hat{x}^o} - C_t\right) v_{1T}\left(X_T^{\hat{x}^o}\right) - \bar{H}_T(C_t, K)\right)^+\right]$$

$$\approx \frac{1}{b} P_\tau(T) \mathbb{E}_\tau^{Q_{FT}}\left[\left(\left(X_T^{\hat{x}^o} - C_t\right) v_{1T}\left(X_T^{\hat{x}^o}\right) - h_T(C_t, K)\right)^+\right]$$

$$= \overline{SW}_t^p\left(h_T(C_t, K), T; M\right),$$

where $SW_t^p(K, T; M)$ is the payer price defined in the first of Eqs. (5.9), and $\overline{SW}_t^p(h_T(C_t, K), T; M)$ is Pedersen's model price when the strike is equal to $h_T(\cdot)$ (see Eq. (5.48)).

Next, for each option strike K, define the implied volatility parameter and the initial condition, i.e. the pair (s_K, x_K^o), such that Eqs. (5.47) and (5.48) are matched to their market counterparts. Based on (s_K, x_K^o), and similarly as in the previous Black-based index correction, define the mapping $h \longmapsto \overline{SW}_t^p(h, T; M)$, which can be approximated through interpolations in correspondence of missing values for $h_T(\cdot)$. One reconstructs the mapping $h \longmapsto \overline{SW}_t^r(h, T; M)$ similarly. Our second index correction is obtained by calculating the index formulae in Sect. 5.4 using these values for $\overline{SW}_t^p(\cdot, T; M)$ and $\overline{SW}_t^r(\cdot, T; M)$.

Appendix D: Appendix on Credit Markets

D.1 Preliminary Facts Concerning CDS Indexes

ACCRUALS OF PREMIUMS. We consider an extension of Eq. (5.1), which takes into account accruals of premiums occurring over the reset times of the index. To develop intuition, assume, initially, that the short-term rate $r = 0$, so that the premium made available to the protection seller over a generic time interval $[T_{i-1}, T_i]$ is

$$\frac{1}{b} C_T \cdot Z_{T_i},$$

where

$$Z_{T_i} \equiv \frac{1}{n} \mathcal{S}_{T_{i-1}} \mathbb{I}_{\text{no-defaults in } [T_{i-1}, T_i]}$$

$$+ \sum_{j=1}^{n} \left(\frac{\tau_j - T_{i-1}}{T_i - T_{i-1}} \frac{1}{n} \mathcal{S}_{T_{i-1}} + \frac{T_i - \tau_j}{T_i - T_{i-1}} \frac{1}{n} (\mathcal{S}_{T_{i-1}} - \mathbb{I}_{\tau_j \in [T_{i-1}, T_i]}) \right) \mathbb{I}_{\tau_j \in [T_{i-1}, T_i]},$$

$$(D.1)$$

and $\mathbb{I}_{\text{no-defaults in } [T_{i-1}, T_i]}$ is the indicator of the event that no defaults have occurred over the time interval $[T_{i-1}, T_i]$. Over this interval, either no defaults occur at all or at least one name defaults. If there are no defaults, the protection seller receives $\frac{1}{b} C_T$ times the first term on the R.H.S. of Eq. (D.1). If there is at least one default, the protection seller receives $\frac{1}{b} C_T$ times the second term on the R.H.S. of Eq. (D.1). The latter is the sum of two components: (i) the premium accrued from T_{i-1} to the time of default τ_j of the j-th name, calculated upon the notional value at time T_{i-1}, $\frac{1}{n} \mathcal{S}_{T_{i-1}}$, and (ii) the premium calculated upon the outstanding nominal at time T_i, for the time remaining from τ_j to T_i.

In the case of stochastic interest rates, define the random present value of Z_{T_i}

$$\hat{Z}_t(T_i) \equiv e^{-\int_t^{T_i} r_\tau d\tau} \frac{1}{n} \mathcal{S}_{T_i} \mathbb{I}_{\text{no-defaults in } [T_{i-1}, T_i]}$$

$$+ \sum_{j=1}^{n} \left(e^{-\int_t^{\tau_j} r_\tau d\tau} \frac{\tau_j - T_{i-1}}{T_i - T_{i-1}} \frac{1}{n} \mathcal{S}_{T_{i-1}} \right.$$

$$\left. + e^{-\int_t^{T_i} r_\tau d\tau} \frac{T_i - \tau_j}{T_i - T_{i-1}} \frac{1}{n} (\mathcal{S}_{T_{i-1}} - \mathbb{I}_{\tau_j \in [T_{i-1}, T_i]}) \right) \mathbb{I}_{\tau_j \in [T_{i-1}, T_i]},$$

so that the value of the premium leg is

$$\frac{1}{b} C_t \cdot v_{1t}, \quad v_{1t} = \sum_{i=1}^{bM} \mathbb{E}_t \left(\hat{Z}_t(T_i) \right).$$

FORWARD POSITION IN CDX INDEXES. We prove Eq. (5.3). Conditional upon the information set at time $\tau \leq T$, we have that for $i = 1, \ldots, bM$,

$$\text{LGD} \frac{1}{n} \mathbb{E}_\tau \left(\sum_{j=1}^{n} e^{-\int_\tau^{\tau_j} r_s ds} \mathbb{I}_{\{\text{Surv}_j \text{ at } \tau\}} \mathbb{I}_{\{T \leq \tau_j \leq T_{bM}\}} \right)$$

$$= \text{LGD} \frac{1}{n} \sum_{j=1}^{n} \mathbb{I}_{\{\text{Surv}_j \text{ at } \tau\}} \mathbb{E}_\tau \left(e^{-\int_\tau^{\tau_j} r_s ds} \mathbb{I}_{\{T \leq \tau_j \leq T_{bM}\}} \right)$$

$$= \text{LGD} \cdot \mathcal{N}_\tau \mathbb{E}_\tau \left(e^{-\int_\tau^{\tau_*} r_s ds} \mathbb{I}_{\{T \leq \tau_* \leq T_{bM}\}} \right),$$

where the last equality follows by (i) the definition of the outstanding notional value in Eq. (5.4), and (ii) the assumption that all the names in the index have the same credit quality. The previous expression is the first term in Eq. (5.3).

The second term in Eq. (5.3) follows by the following equalities:

$$\mathbb{E}_\tau \left(e^{-\int_\tau^{T_i} r_s ds} \cdot \mathbb{I}_{\{\text{Surv}_j \text{ at } T_i\}} \right)$$

$$= \mathbb{E}_\tau \left(e^{-\int_\tau^{T_i} r_s ds} \cdot \mathbb{I}_{\{\text{Surv}_j \text{ at } \tau\}} \mathbb{I}_{\{\text{Surv}_j \text{ at } T_i | \text{Surv}_j \text{ at } \tau\}} \right)$$

$$= \mathbb{I}_{\{\text{Surv}_j \text{ at } \tau\}} \mathbb{E}_\tau \left(e^{-\int_\tau^{T_i} r_s ds} \mathbb{I}_{\{\text{Surv}_j \text{ at } T_i | \text{Surv}_j \text{ at } \tau\}} \right),$$

and summing over the reset dates and all names j, using the definition of the outstanding notional value in Eq. (5.4), and noting, again, that credit risk is the same for all names.

D.2 Spanning Credit Variance Contracts

PRICING. We derive pricing results by relying on the general framework of Chap. 2. The derivations in this appendix are made to help the reader become acquainted with the specific details arising while pricing credit variance and make the chapter self-contained. To alleviate the notation, we set $\text{CDX}_\tau \equiv \text{CDX}_\tau(M)$, although we shall keep on emphasizing the dependence on M of other objects of interest. First, we deal with percentage contracts, and then with basis point contracts.

As for the percentage contracts of Sect. 5.3.1, we have, by the usual Taylor expansion with remainder,

$$\ln \frac{\mathrm{CDX}_T}{\mathrm{CDX}_t} = \frac{1}{\mathrm{CDX}_t}(\mathrm{CDX}_T - \mathrm{CDX}_t) - \int_0^{\mathrm{CDX}_t} (K - \mathrm{CDX}_T)^+ \frac{1}{K^2} dK$$

$$- \int_{\mathrm{CDX}_t}^\infty (\mathrm{CDX}_T - K)^+ \frac{1}{K^2} dK. \tag{D.2}$$

Multiplying both sides of the previous equation by $\frac{1}{b} v_{1t}$, taking expectations under Q_{sc}, and using Eq. (5.9),

$$v_{1t} \mathbb{E}_t^{\mathrm{sc}} \left(\ln \frac{\mathrm{CDX}_T}{\mathrm{CDX}_t} \right) = - \int_0^{\mathrm{CDX}_t} \frac{\mathrm{SW}_t^r(K, T; M)}{K^2} dK - \int_{\mathrm{CDX}_t}^\infty \frac{\mathrm{SW}_t^p(K, T; M)}{K^2} dK. \tag{D.3}$$

Moreover, applying Itô's lemma to Eq. (5.10) leaves

$$d \ln \mathrm{CDX}_\tau = -\left(\mathbb{E}_\tau^{\mathrm{sc}} \left(e^{j_\tau(M)} - 1 \right) \eta_\tau \right) d\tau - \frac{1}{2} \left\| \sigma_\tau(M) \right\|^2 d\tau$$

$$+ \sigma_\tau(M) \cdot dW_\tau^{\mathrm{sc}} + j_\tau(M) dJ_\tau^{\mathrm{sc}}$$

$$= -\frac{1}{2} \left(\left\| \sigma_\tau(M) \right\|^2 d\tau + j_\tau^2(M) dJ_\tau^{\mathrm{sc}} \right) + \sigma_\tau(M) \cdot dW_\tau^{\mathrm{sc}}$$

$$- \left(\mathbb{E}_\tau^{\mathrm{sc}} \left(e^{j_\tau(M)} - 1 \right) \eta_\tau \right) d\tau + j_\tau(M) dJ_\tau^{\mathrm{sc}} + \frac{1}{2} j_\tau^2(M) dJ_\tau^{\mathrm{sc}}.$$

We have

$$\mathbb{E}_\tau^{\mathrm{sc}} \left(e^{j_\tau(M)} - 1 \right) dJ_\tau^{\mathrm{sc}} = \left(\mathbb{E}_\tau^{\mathrm{sc}} \left(e^{j_\tau(M)} - 1 \right) \eta_\tau \right) d\tau. \tag{D.4}$$

Therefore, using the definition of $V_M(t, T)$ in Eq. (5.11), and Eq. (D.4), leaves

$$-2 \mathbb{E}_t^{\mathrm{sc}} \left(\ln \frac{\mathrm{CDX}_T}{\mathrm{CDX}_t} \right) - 2 \mathbb{E}_t^{\mathrm{sc}} \left[\int_t^T \left(e^{j_\tau(M)} - 1 - j_\tau(M) - \frac{1}{2} j_\tau^2(M) \right) dJ_\tau^{\mathrm{sc}} \right]$$

$$= \mathbb{E}_t^{\mathrm{sc}} \left[V_M(t, T) \right]$$

$$= \frac{1}{v_{1t}} \mathbb{E}_t \left(e^{-\int_t^T r_s ds} \mathcal{N}_T v_{1T} V_M(t, T) \right)$$

$$= \frac{\mathbb{F}_{\mathrm{var}, M}(t, T)}{v_{1t}}, \tag{D.5}$$

with $\mathbb{E}_t$ denoting the risk-neutral expectation conditional on information available at t, and where the second equality follows by a change in probability, and the third follows by the definition of the credit variance forward agreement (Definition 5.1). The approximation of $\mathbb{F}_{\mathrm{var}, M}(t, T)$ in Eq. (5.15) follows by comparing Eq. (D.3) and Eq. (D.5), and disregarding the second term on the L.H.S. of Eq. (D.5), which is of order $O((\frac{d\mathrm{CDX}_t}{\mathrm{CDX}_t})^3)$, as first pointed out by Carr and Wu (2008) in the context of the derivation of an equity VIX. Note that Jiang and Tian (2005) also derive the equity VIX in a jump-diffusion model, proposing approximations similar to ours.

Note that the exact formula prevailing in the presence of jumps is

$$
\mathbb{F}_{\text{var},M}(t,T) = 2\left(\int_0^{\text{CDX}_t} \frac{\text{SW}_t^r(K,T;M)}{K^2} dK + \int_{\text{CDX}_t}^\infty \frac{\text{SW}_t^P(K,T;M)}{K^2} dK \right)
$$
$$
- 2v_{1t}\mathbb{E}_t^{\text{sc}}\left[\int_t^T \left(e^{j_\tau(M)} - 1 - j_\tau(M) - \frac{1}{2}j_\tau^2(M) \right) dJ_\tau^{\text{sc}} \right]. \quad (\text{D.6})
$$

Next, we determine the credit variance swap rate of Definition 5.2. We have

$$
0 = \mathbb{E}_t\left[e^{-\int_t^T r_s ds}\left(V_M(t,T) \times (\mathcal{N}_T v_{1T}) - \mathbb{P}_{\text{var},M}(t,T) \right) \right],
$$

or

$$
P_t(T)\mathbb{P}_{\text{var},M}(t,T) = \mathbb{E}_t\left(e^{-\int_t^T r_s ds} V_M(t,T) \times \mathcal{N}_T v_{1T} \right) = \mathbb{F}_{\text{var},M}(t,T), \quad (\text{D.7})
$$

where the last equality holds by the definition of the credit variance forward agreement (Definition 5.1). Equation (5.16) immediately follows. Finally, we derive the standardized credit variance swap rate of Definition 5.3. We have that

$$
0 = \mathbb{E}_t\left[e^{-\int_t^T r_s ds}\left(V_M(t,T) - \mathbb{P}_{\text{var},M}^*(t,T) \right) \times \mathcal{N}_T v_{1T} \right],
$$

or, using the definition of the Radon–Nikodym derivative in Eq. (5.8),

$$
\mathbb{P}_{\text{var},M}^*(t,T)v_{1t} = \mathbb{E}_t\left(e^{-\int_t^T r_s ds} V_M(t,T) \times \mathcal{N}(T)v_{1T} \right) = \mathbb{F}_{\text{var},M}(t,T), \quad (\text{D.8})
$$

where the second equality follows, again, by Definition 5.1. The expression of $\mathbb{P}_{\text{var},M}^*(t,T)$ in Eq. (5.17) follows immediately.

Next, we proceed with the pricing of basis point contracts introduced in Sect. 5.3.2. By a Taylor's expansion with remainder, we have

$$
\text{CDX}_T^2 = \text{CDX}_t^2 + 2\text{CDX}_t(\text{CDX}_T - \text{CDX}_t)
$$
$$
+ 2\left(\int_0^{\text{CDX}_t} (K - \text{CDX}_T)^+ dK + \int_{\text{CDX}_t}^\infty (\text{CDX}_T - K)^+ dK \right). \quad (\text{D.9})
$$

Multiplying both sides of the previous equation by v_{1t}, and taking expectations under Q_{sc}, leaves

$$
v_{1t}\mathbb{E}_t^{\text{sc}}\left(\text{CDX}_T^2 - \text{CDX}_t^2 \right)
$$
$$
= 2\left(\int_0^{\text{CDX}_t} \text{SW}_t^r(K,T;M)dK + \int_{\text{CDX}_t}^\infty \text{SW}_t^P(K,T;M)dK \right). \quad (\text{D.10})
$$

Next, note that by Itô's lemma for jump-diffusion processes,

$$\frac{d\text{CDX}_\tau^2}{\text{CDX}_\tau^2} = -2\big(\mathbb{E}_\tau^{\text{sc}}\big(e^{j_\tau(M)} - 1\big)\eta_\tau\big)d\tau + 2\sigma_\tau(M) \cdot dW_\tau^{\text{sc}}$$

$$+ \|\sigma_\tau(M)\|^2 d\tau + \big(e^{2j_\tau(M)} - 1\big)dJ_\tau^{\text{sc}}$$

$$= -2\big(\mathbb{E}_\tau^{\text{sc}}\big(e^{j_\tau(M)} - 1\big)\eta_\tau\big)d\tau + 2\big(e^{j_\tau(M)} - 1\big)dJ_\tau^{\text{sc}} + 2\sigma_\tau(M) \cdot dW_\tau^{\text{sc}}$$

$$+ \|\sigma_\tau(M)\|^2 d\tau + \big(e^{j_\tau(M)} - 1\big)^2 dJ_\tau^{\text{sc}}, \tag{D.11}$$

where the second equality follows by rearranging terms. Note that the first three terms of the second equality in Eq. (D.11) form a martingale under Q_{sc}, due to Eq. (D.4) and the fact that W_τ^{sc} is obviously a martingale under Q_{sc}. Then, by integrating, taking expectations under Q_{sc}, and using the definition of basis point variance, $V_M(t, T)$ in Eq. (5.12), leaves

$$\mathbb{E}_t^{\text{sc}}\big(\text{CDX}_T^2 - \text{CDX}_t^2\big) = \mathbb{E}_t^{\text{sc}}\big[V_M^{\text{bp}}(t, T)\big]. \tag{D.12}$$

Substituting Eq. (D.12) into Eq. (D.10) yields

$$2\bigg(\int_0^{\text{CDX}_t} \text{SW}_t^r(K, T; M)dK + \int_{\text{CDX}_t}^\infty \text{SW}_t^p(K, T; M)dK\bigg)$$

$$= v_{1t}\mathbb{E}_t^{\text{sc}}\big(\text{CDX}_T^2 - \text{CDX}_t^2\big)$$

$$= v_{1t}\mathbb{E}_t^{\text{sc}}\big[V_M^{\text{bp}}(t, T)\big]$$

$$= \mathbb{E}_t\big[e^{-\int_t^T r_s ds}(\mathcal{N}_T V_{1T}) \times V_M^{\text{bp}}(t, T)\big]$$

$$= \mathbb{F}_{\text{var},M}^{\text{bp}}(t, T),$$

where the third equality follows by a change of probability and the last by the definition of the credit variance forward agreement (Definition 5.1).

Finally, the expression for the basis point variance swap rates in Eqs. (5.19) follows by arguments nearly identical to those leading to Eq. (D.7) and Eq. (D.8).

MARKING TO MARKET. To derive the update in Eq. (5.20), we evaluate the risk-neutral expectation of the discounted payoff Var-Swap$_M(t, T)$ in Eq. (5.13), for any given $\tau \in [t, T]$,

$$\mathbb{E}_\tau\big(e^{-\int_\tau^T r_u du}\text{Var-Swap}_M(t, T)\big)$$

$$= \mathbb{E}_\tau\big[e^{-\int_\tau^T r_u du}\big(V_M(t, \tau) + V_M(\tau, T)\big) \times \mathcal{N}_T v_{1T}\big] - P_\tau(T)\mathbb{P}_{\text{var},M}(t, T)$$

$$= V_M(t, \tau)\mathcal{N}_\tau v_{1\tau} + \mathbb{F}_{\text{var},M}(\tau, T) - P_\tau(T)\mathbb{P}_{\text{var},M}(t, T), \tag{D.13}$$

where $\mathbb{E}_\tau$ is the expectation under Q, conditional on the information up to time τ. The second equality follows by the definition of the Radon–Nikodym derivative in

Eq. (5.8) and the definition of the credit variance forward agreement. Substituting
the expression for $\mathbb{P}_{\text{var},n}(t,T)$ in Eq. (5.16) into Eq. (D.13) gives Eq. (5.20).

Next, we derive Eq. (5.21) by taking the risk-neutral expectation of the discounted payoff, Var-Swap$^*_M(t,T)$ in Eq. (5.14),

$$\mathbb{E}_\tau\left(e^{-\int_\tau^T r_u du}\text{Var-Swap}^*_M(t,T)\right)$$

$$= \mathbb{E}_\tau\left[e^{-\int_\tau^T r_u du}\left(V_M(t,\tau)+V_M(\tau,T)-\mathbb{P}^*_{\text{var},M}(t,T)\right)\times\mathcal{N}_T v_{1T}\right]$$

$$= \mathcal{N}_\tau v_{1\tau}\left(V_M(t,\tau)+\mathbb{P}^*_{\text{var},M}(\tau,T)-\mathbb{P}^*_{\text{var},M}(t,T)\right),$$

where the second equality follows by Eq. (D.8) and the definition of the Radon–
Nikodym derivative in Eq. (5.8).

D.3 Hedging

We first derive the portfolio positions in Table 5.1 regarding the replication of the
contracts in Definitions 5.1 and 5.2. The additional arguments leading to Table 5.2
are in the main text. If CDX$_\tau(M)$ in Eq. (5.6) were actually traded and not subject
to jumps, we could perfectly hedge these contracts by the following arguments. Itô's
lemma gives us

$$\mathcal{N}_T v_{1T} V_M(t,T) = 2\mathcal{N}_T v_{1T}\int_t^T \frac{d\text{CDX}_s}{\text{CDX}_s}-2\mathcal{N}_T v_{1T}\left(\ln\frac{\text{CDX}_T}{\text{CDX}_t}\right). \qquad (D.14)$$

By Eq. (D.2), the second term of Eq. (D.14) is the time T payoff of a static portfolio
created at t, and corresponding to row (ii) of Table 5.1, which is two times: (a) short
$1/\text{CDX}_t$ units a forward of a FS-CDX, struck at CDX$_t$, and (b) long a continuum
of out-of-the-money CDS index options with weights $K^{-2}dK$. The value of this
portfolio is $\mathbb{F}_{\text{var},M}(t,T)$, due to Eqs. (D.2), (D.3), and (5.15), which we finance by
borrowing $\mathbb{F}_{\text{var},M}(t,T)$ at time t, in turn repaid at time T, as in row (iii) of Table 5.1.

As for the self-financing portfolio of row (i) in Table 5.1, we want it to be
worthless at time t, and to replicate the first term of Eq. (D.14). Because we are
assuming that CDX$_t$ is traded, we can consider a self-financed strategy that is
long (a) the FS-CDX, and (b) a money market account, with value M_τ satisfying
$dM_\tau = r_\tau M_\tau d\tau$. The strategy at time τ is worth: $v_\tau = \theta_\tau\text{CDX}_\tau + \psi_\tau M_\tau$, where θ_τ
are the units invested in the FS-CDX, and ψ_τ are the units invested in the money
market account. Let

$$\hat{\theta}_\tau\text{CDX}_\tau = \mathcal{N}_\tau v_{1\tau}, \qquad \hat{\psi}_\tau M_\tau = \mathcal{N}_\tau v_{1\tau}\left(\int_t^\tau \frac{d\text{CDX}_s}{\text{CDX}_s}-1\right), \qquad (D.15)$$

so that $\hat{v}_\tau = \hat{\theta}_\tau\text{CDX}_\tau + \hat{\psi}_\tau M_\tau$ satisfies

$$\hat{v}_\tau = \mathcal{N}_\tau v_{1\tau}\int_t^\tau \frac{d\text{CDX}_s}{\text{CDX}_s}, \qquad (D.16)$$

and

$$\hat{v}_t = 0, \quad \text{and} \quad \hat{v}_T = \mathcal{N}_T v_{1T} \int_t^T \frac{d\text{CDX}_s}{\text{CDX}_s}.$$

Therefore, we can replicate the first term on the R.H.S. of Eq. (D.14) (and, then, by the previous results, the R.H.S. of Eq. (D.14)) through a long position in the two portfolios $(\hat{\theta}_\tau, \hat{\psi}_\tau)$, provided $(\hat{\theta}_\tau, \hat{\psi}_\tau)$ is self-financed. To check that $(\hat{\theta}_\tau, \hat{\psi}_\tau)$ is self-financed, note that

$$\begin{aligned}
d\hat{v}_\tau &= \hat{\theta}_\tau \text{CDX}_\tau \frac{d\text{CDX}_\tau}{\text{CDX}_\tau} + \hat{\psi}_\tau M_\tau \frac{dM_\tau}{M_\tau} \\
&= \mathcal{N}_\tau v_{1\tau} \left(\frac{d\text{CDX}_\tau}{\text{CDX}_\tau} - r_\tau d\tau \right) + \left(\mathcal{N}_\tau v_{1\tau} \int_t^\tau \frac{d\text{CDX}_s}{\text{CDX}_s} \right) r_\tau d\tau \\
&= \mathcal{N}_\tau v_{1\tau} \left(\frac{d\text{CDX}_\tau}{\text{CDX}_\tau} - r_\tau d\tau \right) + r_\tau \hat{v}_\tau d\tau \\
&= \hat{\theta}_\tau \text{CDX}_\tau \left(\frac{d\text{CDX}_\tau}{\text{CDX}_\tau} - r_\tau d\tau \right) + r_\tau \hat{v}_\tau d\tau,
\end{aligned} \tag{D.17}$$

where the second line follows by Eqs. (D.15), the third by Eq. (D.16), and the fourth, again, by Eq. (D.15). The dynamics of $\hat{v}_\tau$ in Eq. (D.17) are indeed those of a self-financed strategy.

Next, we derive the portfolio positions in Table 5.3, which leads to the replication of the contracts in Definitions 5.4-(a) and 5.4-(b), with the additional arguments concerning Table 5.4 found in the main text.

Itô's lemma gives us

$$\mathcal{N}_T v_{1T} V_M^{\text{bp}}(t, T) = -2\mathcal{N}_T v_{1T} \int_t^T \text{CDX}_s d\text{CDX}_s + \mathcal{N}_T v_{1T} (\text{CDX}_T^2 - \text{CDX}_t^2).$$
$$\tag{D.18}$$

By Eq. (D.9), the second term on the R.H.S. of Eq. (D.18) is the payoff at T of a portfolio set up at t, which is: (a) long 2CDX_t units a FS-CDX struck at CDX_t, and (b) long a continuum of out-of-the-money CDS index options with weights $2dK$. It is the static position (ii) in Table 5.3 of the main text. By Eq. (5.18), its cost is $\mathbb{F}_{\text{var}, M}^{\text{bp}}(t, T)$, which we borrow at t, to repay it back at T, as in row (iii) of Table 5.3. The self-financed portfolio to be shorted, as indicated by row (i) of Table 5.3, is obtained similarly as the portfolio in row (i) of Table 5.1, but with the portfolio

$$\hat{\psi}_\tau M_\tau = \mathcal{N}_\tau v_{1\tau} \left(\int_t^\tau 2\text{CDX}_s d\text{CDX}_s - 1 \right),$$

replacing that in Eq. (D.15).

When $\text{CDX}_\tau(M)$ is not traded, we may replicate its approximation based on $\widehat{\text{CDX}}_\tau(M)$ in Eq. (5.22) of the main text, based on the assumption in the main text, and at least up to the jump components possibly affecting $v_{1\tau}$, as follows.

First, relying on the third equality in (5.23), consider a self-financed strategy into (i) a continuum of defaultable zero coupon bonds, say $\theta_\tau(x)$ for a continuum of maturities $x \in [T, T_{bM}]$, (ii) in the defaultable annuity $v_{1\tau}$, say $\theta_{1\tau}$, and (iii) in a money market account, say $\theta_{2\tau}$. The value of this strategy satisfies

$$dV_\tau = \int_T^{T_{bM}} \theta_\tau(x) d_\tau P_{\text{def},\tau}(x) dx + \theta_{1\tau} dv_{1\tau} + \theta_{2\tau} dM_\tau.$$

On the other hand, $C_\tau \equiv \widehat{\text{CDX}}_\tau(M)$ satisfies

$$dC_\tau = \text{LGD} \cdot \lambda \frac{\int_T^{T_{bM}} d_\tau P_{\text{def},\tau}(x) dx}{v_{1\tau}} - \frac{C_\tau}{v_{1\tau}} dv_{1\tau} + \text{VC}_\tau^{cr} d\tau,$$

for some τ-measurable process VC_τ^{cr}. Therefore, by neglecting jumps that $v_{1\tau}$ possibly experiences, we have that with

$$\theta_\tau(x) = \frac{\text{LGD} \cdot \lambda}{v_{1\tau}}, \quad \text{for each } x \in [T, T_{bM}], \qquad \theta_{1\tau} = -\frac{\widehat{\text{CDX}}_\tau(M)}{v_{1\tau}},$$

one has that

$$dV_\tau - dC_\tau = \left(r_\tau \theta_{2\tau} M_\tau - \text{VC}_\tau^{cr} \right) d\tau,$$

where $\theta_{2\tau}$ is chosen to replicate $\widehat{\text{CDX}}_\tau(M)$; that is,

$$\theta_{2\tau} M_\tau = \widehat{\text{CDX}}_\tau(M) - \int_T^{T_{bM}} \theta_\tau(x) d_\tau P_{\text{def},\tau}(x) dx - \theta_{\tau 1} v_{1\tau} = \widehat{\text{CDX}}_\tau(M).$$

While this portfolio replicates $\widehat{\text{CDX}}_\tau(M)$, it leads to a hedging cost equal to $\varepsilon_\tau^{cr} \equiv \widehat{\text{CDX}}_\tau(M) - V_\tau$, which satisfies $d\varepsilon_\tau^{cr} = (\text{VC}_\tau^{cr} - r_\tau \widehat{\text{CDX}}_\tau(M)) d\tau$.

ESTIMATES BASED ON FORWARD PREMIUM APPROXIMATIONS. Let K_0 be the first strike below CDX_t, as defined in the main text. We expand $\ln \text{CDX}_T$ around K_0, as follows:

$$\ln \frac{\text{CDX}_T}{\text{CDX}_t} = \ln \frac{K_0}{\text{CDX}_t} + \frac{\text{CDX}_T - K_0}{K_0}$$
$$- \left(\int_0^{K_0} (K - \text{CDX}_T)^+ \frac{1}{K^2} dK + \int_{K_0}^\infty (\text{CDX}_T - K)^+ \frac{1}{K^2} dK \right),$$

$$(\text{D.19})$$

so that the fair value of the standardized credit variance swap rate of Definition 5.3 satisfies:

$$\frac{\mathbb{F}_{\text{var},M}(t,T)}{v_{1t}} + 2\mathbb{E}_t^{\text{sc}} \left[\int_t^T \left(e^{j_\tau(M)} - 1 - j_\tau(M) - \frac{1}{2} j_\tau^2(M) \right) dJ_\tau^{\text{sc}} \right]$$
$$= -2\mathbb{E}_t^{\text{sc}} \left(\ln \frac{\text{CDX}_T}{\text{CDX}_t} \right)$$

$$= -2\left(\ln \frac{K_0}{\text{CDX}_t} + \frac{\text{CDX}_t - K_0}{K_0}\right)$$

$$+ \frac{2}{v_{1t}}\left(\int_0^{K_0} \text{SW}_t^r(K, T, M)\frac{1}{K^2}dK + \int_{K_0}^\infty \text{SW}_t^p(K, T, M)\frac{1}{K^2}dK\right).$$

$$(D.20)$$

Next, consider a second order expansion of the function $-\ln \frac{\text{CDX}_t}{K_0}$ about K_0:

$$-\ln \frac{\text{CDX}_t}{K_0} \approx -\frac{1}{K_0}(\text{CDX}_t - K_0) + \frac{1}{2}\frac{1}{K_0^2}(\text{CDX}_t - K_0)^2.$$

By substituting this approximation into the first term in the R.H.S. of Eq. (D.20), disregarding the jump component terms, and proceeding as usual with a discretization, leaves the expression of $\mathbb{P}^*_{o,\text{var},M}(t, T)$ in Eq. (5.26).

Next, consider the correction applying to the basis point index. Similarly as for Eq. (D.19), expand CDX_T^2 around K_0,

$$\text{CDX}_T^2 - \text{CDX}_t^2 = K_0^2 - \text{CDX}_t^2 + 2K_0(\text{CDX}_T - K_0)$$

$$+ 2\left(\int_0^{K_0}(K - \text{CDX}_T)^+dK + \int_{K_0}^\infty(\text{CDX}_T - K)^+dK\right),$$

so that the standardized BP-variance swap rate of Definition 5.4-(c) is

$$\frac{\mathbb{F}^{\text{bp}}_{\text{var},M}(t, T)}{v_{1t}} = \mathbb{E}_t^{\text{sc}}\left(\text{CDX}_T^2 - \text{CDX}_t^2\right)$$

$$= K_0^2 - \text{CDX}_t^2 + 2K_0(\text{CDX}_t - K_0)$$

$$+ \frac{2}{v_{1t}}\left(\int_0^{K_0} \text{SW}_t^r(K, T, M)dK + \int_{K_0}^\infty \text{SW}_t^p(K, T, M)dK\right).$$

$$(D.21)$$

A second order expansion of the function CDX_t^2 about K_0 yields

$$\text{CDX}_t^2 - K_0^2 \approx 2K_0(\text{CDX}_t - K_0) + (\text{CDX}_t - K_0)^2.$$

Substituting this approximation into Eq. (D.21) and discretizing the integral as usual yields the approximation in Eq. (5.27).

References

Aït-Sahalia, Y., & Jacod, J. (2014). *High frequency financial econometrics*. Princeton: Princeton University Press.

Andersen, T., Bollerslev, T., & Diebold, F. (2010). Parametric and nonparametric measurements of volatility. In Y. Aït-Sahalia & L. P. Hansen (Eds.), *Tools and techniques: Vol. 1. Handbook of financial econometrics* (pp. 67–137). Amsterdam: North-Holland.

Bakshi, G., & Madan, D. (2000). Spanning and derivative security evaluation. *Journal of Financial Economics, 55*, 205–238.

Bakshi, G., Kapadia, N., & Madan, D. (2003). Stock return characteristics, skew laws, and differential pricing of individual equity options. *The Review of Financial Studies, 16*, 101–143.

Bank of England (2008). *Minutes of the Monetary Policy Committee Meeting*, 6 and 7 August.

Barone-Adesi, G., & Whaley, R. E. (1987). Efficient analytical approximation of American option values. *The Journal of Finance, 42*, 301–320.

Bikbov, R., & Chernov, M. (2011). Yield curve and volatility: lessons from Eurodollar futures and options. *Journal of Financial Econometrics, 9*, 66–105.

Bibkov, R., & Misra, P. (2012). SRVX volatility index. Bank of America Merrill Lynch US Rates Weekly, 15.

Black, F. (1976). The pricing of commodity contracts. *Journal of Financial Economics, 3*, 167–179.

Black, F., & Scholes, M. (1973). The pricing of options and corporate liabilities. *Journal of Political Economy, 81*, 637–659.

Bollerslev, T., Tauchen, G., & Zhou, H. (2009). Expected stock returns and variance risk premia. *The Review of Financial Studies, 22*, 4463–4492.

Bossu, S., Strasser, E., & Guichard, R. (2005). Just what you need to know about variance swaps. JPMorgan Equity Derivatives.

Brace, A., Gatarek, D., & Musiela, M. (1997). The market model of interest rate dynamics. *Mathematical Finance, 7*, 127–155.

Brenner, M., & Subrahmanyam, M. G. (1988). A simple formula to compute implied standard deviation. *Financial Analysts Journal, 5*, 80–83.

Brigo, D., & Mercurio, F. (2006). *Interest rate models—theory and practice, with smile, inflation and credit* (2nd ed.). Berlin: Springer.

Britten-Jones, M., & Neuberger, A. (2000). Option prices, implied price processes and stochastic volatility. *The Journal of Finance, 55*, 839–866.

Broadie, M., Chernov, M., & Johannes, M. (2007). Model specification and risk premia: evidence from futures options. *The Journal of Finance, 62*, 1453–1490.

Carr, P., & Corso, A. (2001). Commodity covariance contracting. *Energy and Power Risk Management, 4*, 42–45.

© Springer International Publishing Switzerland 2015

A. Mele, Y. Obayashi, *The Price of Fixed Income Market Volatility*, Springer Finance, DOI 10.1007/978-3-319-26523-0

Carr, P., & Madan, D. (2001). Optimal positioning in derivative securities. *Quantitative Finance*, *1*, 19–37.

Carr, P., & Wu, L. (2006). A tale of two indices. *The Journal of Derivatives*, *13*, 13–29.

Carr, P., & Wu, L. (2008). Variance risk premiums. *The Review of Financial Studies*, *22*, 1311–1341.

Carr, P., & Lee, R. (2009). Volatility derivatives. *Annual Review of Financial Economics*, *1*, 1–21.

Chicago Board Options Exchange (2009). The CBOE volatility index®—VIX®. White paper underlying the CBOE-VIX volatility index. Available from: http://www.cboe.com/micro/VIX/vixwhite.pdf.

Chicago Board Options Exchange (2012). CBOE interest rate swap volatility index^{SM}: http://www.cboe.com/micro/srvix/default.aspx.

Chicago Board Options Exchange (2013). CBOE/CBOT 10-year U.S. treasury note volatility index^{SM}: http://www.cboe.com/micro/volatility/tyvix/default.aspx.

Collin-Dusfresne, P., & Goldstein, R. S. (2002). Do bonds span the fixed-income markets? Theory and evidence for unspanned stochastic volatility. *The Journal of Finance*, *57*, 1685–1730.

Corradi, V., Distaso, W., & Mele, A. (2013). Macroeconomic determinants of stock volatility and volatility premiums. *Journal of Monetary Economics*, *60*, 203–220.

Cox, J. C., Ingersoll, J. E. Jr., & Ross, S. A. (1985). A theory of the term structure of interest rates. *Econometrica*, *53*, 385–407.

Demeterfi, K., Derman, E., Kamal, M., & Zou, J. (1999a). A guide to variance swaps. *Risk*, *12*(6), 54–59.

Demeterfi, K., Derman, E., Kamal, M., & Zou, J. (1999b). A guide to volatility and variance swaps. *The Journal of Derivatives*, *6*(4), 9–32.

Dumas, B. (1995). The meaning of the implicit volatility function in case of stochastic volatility. Unpublished appendix of: Dumas, B., Fleming, J., & Whaley, R. E. (1998). Implied Volatility Functions: Empirical Tests. *The Journal of Finance*, *53*, 2059–2106.

El Karoui, N., Jeanblanc-Picqué, M., & Shreve, S. (1998). Robustness of the Black and Scholes formula. *Mathematical Finance*, *8*, 93–126.

Engle, R. F. (2004). Nobel lecture: risk and volatility: econometric models and financial practice. *The American Economic Review*, *94*, 405–420.

Flesaker, B. (1993). Testing the Heath-Jarrow-Morton/Ho-Lee model of interest rate contingent claims pricing. *Journal of Financial and Quantitative Analysis*, *28*, 483–495.

Geman, H. (1989). The importance of the forward neutral probability in a stochastic approach to interest rates. Unpublished working paper, ESSEC.

Geman, H., El Karoui, N., & Rochet, J.-C. (1995). Changes of numéraire, changes of probability measures and pricing of options. *Journal of Applied Probability*, *32*, 443–458.

Fornari, F. (2010). Assessing the compensation for volatility risk implicit in interest rate derivatives. *Journal of Empirical Finance*, *17*, 722–743.

Hagan, P. S. (2003). Convexity conundrums: pricing CMS swaps, caps, and floors. *Wilmott Magazine*, 38–44.

Hagan, P. S., Kumar, D., Lesniewski, A. S., & Woodward, D. E. (2002). Managing smile risk. *Wilmott Magazine*, 84–108.

Heston, S. L. (1993). A closed form solution for options with stochastic volatility with applications to bond and currency options. *The Review of Financial Studies*, *6*, 327–344.

Ho, T. S. Y., & Lee, S. B. (1986). Term structure movements and the pricing of interest rate contingent claims. *The Journal of Finance*, *41*, 1011–1029.

Hull, J., & White, A. (1987). The pricing of options on assets with stochastic volatilities. *The Journal of Finance*, *42*, 281–300.

Jacod, J., & Shiryaev, A. N. (1987). *Limit theorems for stochastic processes*. Berlin: Springer.

Jamshidian, F. (1989). An exact bond option pricing formula. *The Journal of Finance*, *44*, 205–209.

Jamshidian, F. (1997). Libor and swap market models and measures. *Finance and Stochastics*, *1*, 293–330.

Jiang, G. J., & Tian, Y. S. (2005). The model-free implied volatility and its information content. *The Review of Financial Studies, 18*, 1305–1342.

Jiang, R. (2011). An analysis of gamma trading strategy in swaption market. Working paper, London School of Economics.

Lando, D. (2004). *Credit risk modeling—theory and applications*. Princeton: Princeton University Press.

Lee, R. (2010). Weighted variance swaps. In *Encyclopedia of quantitative finance*. New York: Wiley.

Longstaff, F. A., & Schwartz, E. S. (2001). Valuing American options by simulation: a simple least-squares approach. *The Review of Financial Studies, 14*, 113–147.

Martin, I. (2013). Simple variance swaps. Working paper, Stanford University.

Mele, A. (2003). Fundamental properties of bond prices in models of the short-term rate. *The Review of Financial Studies, 16*, 679–716.

Mele, A. (2014). Lectures on financial economics. Book manuscript. Available from: http://www.antoniomele.org.

Mele, A., & Obayashi, Y. (2012). An interest rate swap volatility index and contract. Technical white paper underlying the CBOE interest rate swap volatility index. Available from: http://www.cboe.com/micro/srvix/default.aspx.

Mele, A., & Obayashi, Y. (2013a). The price of government bond volatility. Swiss Finance Institute Research Paper No. 13–27.

Mele, A., & Obayashi, Y. (2013b). Volatility indexes and contracts for Eurodollar and related deposits. Swiss Finance Institute Research Paper No. 13–25.

Mele, A., & Obayashi, Y. (2013c). Credit variance swaps and volatility indexes. Swiss Finance Institute Research Paper No. 13–24.

Mele, A., & Obayashi, Y. (2014). Interest rate variance swaps and the pricing of fixed income volatility. *GARP Risk Professional: Quant Perspectives*, 1–8.

Mele, A., & Obayashi, Y. (2015). Interest rate derivatives and volatility. In P. Veronesi (Ed.), *Handbook series in financial engineering and econometrics. The handbook of fixed income securities*, New York: Wiley. Forthcoming.

Mele, A., Obayashi, Y., & Shalen, C. (2015a). Rate fears gauges and the dynamics of interest rate swap and equity volatility. *Journal of Banking & Finance, 52*, 256–265.

Mele, A., Obayashi, Y., & Yang, S. (2015b). Pricing options and futures on a government bond volatility index. Working paper, Swiss Finance Institute.

Mercurio, F. (2009). Interest rates and the credit crunch: new formulas and market models. Available from: http://papers.ssrn.com/sol3/papers.cfm?abstract_id=1332205.

Mercurio, F. (2012). The widening of the basis: new market formulas for swaps, caps and swaptions. In C. S. Wehn, C. Hoppe, & G. N. Gregoriou (Eds.), *Rethinking valuation and pricing models: lessons learned from the crisis and future challenges* (pp. 137–145). Amsterdam: Elsevier.

Mercurio, F., & Pallavicini, A. (2006). Swaption skews and convexity adjustments. Banca IMI.

Merener, N. (2012). Swap rate variance swaps. *Quantitative Finance, 12*, 249–261.

Miltersen, K. R., Sandmann, K., & Sondermann, D. (1997). Closed form solutions for term structure derivatives with lognormal interest rate. *The Journal of Finance, 52*, 409–430.

Morini, M., & Brigo, D. (2011). No-Armageddon arbitrage-free equivalent measure for index options in a credit crisis. *Mathematical Finance, 21*, 573–593.

Mueller, P., Vedolin, A., & Yen, Y.-M. (2012). Bond variance risk premia. Working paper, Institute of Economics, Academia Sinica.

Neuberger, A. (1994). Hedging volatility: the case for a new contract. *The Journal of Portfolio Management, 20*, 74–80.

Øksendal, B. (1998). *Stochastic differential equations—an introduction with applications*. Berlin: Springer.

Pedersen, C. M. (2003). Valuation of portfolio credit default swaptions. Lehman Brothers Quantitative Credit Research.

Renault, E. (1997). Econometric models of option pricing errors. In D. Kreps & K. Wallis (Eds.), *Advances in economics and econometrics—theory and applications*, Cambridge: Cambridge University Press.

Rhoads, R. (2011). *Trading VIX derivatives*. New York: Wiley.

Rutkowski, M., & Armstrong, A. (2009). Valuation of credit default swaptions and credit default index swaptions. *International Journal of Theoretical and Applied Finance, 12*, 1027–1053.

Schönbucher, P. J. (2003). *Credit risk pricing models—models, pricing and implementation*. Chichester: Wiley Finance.

Trolle, A. B., & Schwartz, E. S. (2013). The swaption cube. Working paper, EPFL and Swiss Finance Institute.

Vasicek, O. A. (1977). An equilibrium characterization of the term structure. *Journal of Financial Economics, 5*, 177–188.

Veronesi, P. (2010). *Fixed income securities*. New York: Wiley.

Walras, L. (1874). *Éléments d'économie politique pure, ou théorie de la richesse sociale*. Lausanne: Corbaz & Cie. English translation: *Elements of Pure Economics*, Homewood, Il.: R.D. Irwin (1954).

Whaley, R. E. (1993). Derivatives on market volatility: hedging tools long overdue. *The Journal of Derivatives, 1*, 71–84.